Marine Tephrochronology

Geological Society books refereeing procedures

The Society makes every effort to ensure that the scientific and production quality of its books matches that of its journals. Since 1997, all book proposals have been refereed by specialist reviewers as well as by the Society's Books Editorial Committee. If the referees identify weaknesses in the proposal, these must be addressed before the proposal is accepted.

Once the book is accepted, the Society Book Editors ensure that the volume editors follow strict guidelines on refereeing and quality control. We insist that individual papers can only be accepted after satisfactory review by two independent referees. The questions on the review forms are similar to those for *Journal of the Geological Society*. The referees' forms and comments must be available to the Society's Book Editors on request.

Although many of the books result from meetings, the editors are expected to commission papers that were not presented at the meeting to ensure that the book provides a balanced coverage of the subject. Being accepted for presentation at the meeting does not guarantee inclusion in the book.

More information about submitting a proposal and producing a book for the Society can be found on its website: www.geolsoc.org.uk.

It is recommended that reference to all or part of this book should be made in one of the following ways:

AUSTIN, W. E. N., ABBOTT, P. M., DAVIES, S. M., PEARCE, N. J. G. & WASTEGÅRD, S. (eds) 2014. *Marine Tephrochronology*. Geological Society, London, Special Publications, **398**.

PEARCE, N. J. G., ABBOTT, P. M. & MARTIN-JONES, C. 2014. Microbeam methods for the analysis of glass in fine-grained tephra deposits: a SMART perspective on current and future trends. *In:* AUSTIN, W. E. N., ABBOTT, P. M., DAVIES, S. M., PEARCE, N. J. G. & WASTEGÅRD, S. (eds) 2014. *Marine Tephrochronology*. Geological Society, London, Special Publications, **398**, 29–46. First published online February 27, 2014, http://dx.doi.org/10.1144/SP398.1

GEOLOGICAL SOCIETY SPECIAL PUBLICATION NO. 398

Marine Tephrochronology

EDITED BY

W. E. N. AUSTIN
University of St Andrews, UK
The Scottish Association for Marine Science, UK

P. M. ABBOTT
Swansea University, UK

S. M. DAVIES
Swansea University, UK

N. J. G. PEARCE
Aberystwyth University, UK

and

S. WASTEGÅRD
Stockholm University, Sweden

2014
Published by
The Geological Society
London

THE GEOLOGICAL SOCIETY

The Geological Society of London (GSL) was founded in 1807. It is the oldest national geological society in the world and the largest in Europe. It was incorporated under Royal Charter in 1825 and is Registered Charity 210161.

The Society is the UK national learned and professional society for geology with a worldwide Fellowship (FGS) of over 10 000. The Society has the power to confer Chartered status on suitably qualified Fellows, and about 2000 of the Fellowship carry the title (CGeol). Chartered Geologists may also obtain the equivalent European title, European Geologist (EurGeol). One fifth of the Society's fellowship resides outside the UK. To find out more about the Society, log on to www.geolsoc.org.uk.

The Geological Society Publishing House (Bath, UK) produces the Society's international journals and books, and acts as European distributor for selected publications of the American Association of Petroleum Geologists (AAPG), the Indonesian Petroleum Association (IPA), the Geological Society of America (GSA), the Society for Sedimentary Geology (SEPM) and the Geologists' Association (GA). Joint marketing agreements ensure that GSL Fellows may purchase these societies' publications at a discount. The Society's online bookshop (accessible from www.geolsoc.org.uk) offers secure book purchasing with your credit or debit card.

To find out about joining the Society and benefiting from substantial discounts on publications of GSL and other societies worldwide, consult www.geolsoc.org.uk, or contact the Fellowship Department at: The Geological Society, Burlington House, Piccadilly, London W1J 0BG: Tel. +44 (0)20 7434 9944; Fax +44 (0)20 7439 8975; E-mail: enquiries@geolsoc.org.uk.

For information about the Society's meetings, consult *Events* on www.geolsoc.org.uk. To find out more about the Society's Corporate Affiliates Scheme, write to enquiries@geolsoc.org.uk.

Published by The Geological Society from:
The Geological Society Publishing House, Unit 7, Brassmill Enterprise Centre, Brassmill Lane, Bath BA1 3JN, UK

The Lyell Collection: www.lyellcollection.org
Online bookshop: www.geolsoc.org.uk/bookshop
Orders: Tel. +44 (0)1225 445046, Fax +44 (0)1225 442836

British Library Cataloguing in Publication Data

A catalogue record for this book is available from the British Library.
ISBN 978-1-86239-641-8
ISSN 0305-8719

Distributors
For details of international agents and distributors see:
www.geolsoc.org.uk/agentsdistributors

Typeset by Techset Composition India (P) Ltd., Bangalore and Chennai, India.
Printed by Berforts Information Press Ltd, Oxford, UK

Contents

Foreword

Tephra layers provide unique isochrons for the geological record since the explosive eruptions, particularly those of large magnitude, distribute volcanic ash over huge areas. Tephra can usually be precisely dated and in most cases the major and trace element geochemistry of the glass and minerals are unique for a particular volcano and quite often for a particular eruption of that volcano. This means that once a tephra layer is identified in sediments it can also provide a precise age. However, the study of tephra layer markers is broader than tephrochronology because tephra can record information about many diverse processes. In volcanology the study of tephra thickness and grain size variations enables estimates to be made about the magnitude and intensity of past eruptions and also atmospheric conditions of wind directions and speeds. Tephra is commonly re-distributed by wind, ocean currents, mass wasting and biological processes on a wide variety of spatial scales, and this has the potential to help unravel surface and near surface processes.

The diverse applications of tephra studies are well illustrated in this significant special publication volume on Marine Tephrochronology, which includes some of the major advances in methodologies. Application of microbeam techniques such as ICP-MS enables very tiny glassy ash particles to be uniquely characterized from trace element patterns. Ages of tephras can help define carbon 14 residences times in the ocean, ice sheet melting and aspects of ocean circulation, the magnitudes of the largest explosive eruptions and insights into biotubation. Cryptotephra studies provide an expanding frontier for major improvement of age models for marine sediments. This is an excellent collection of papers that shows how tephra studies continue to innovate and develop.

Steve Sparks, FRS
University of Bristol
August 2013

Marine tephrochronology: an introduction to tracing time in the ocean

WILLIAM E. N. AUSTIN[1,2*], PETER M. ABBOTT[3], SIWAN DAVIES[3], NICHOLAS J. G. PEARCE[4] & STEFAN WASTEGÅRD[5]

[1]*School of Geography and Geosciences, University of St Andrews, St Andrews, Fife KY16 9AL, UK*

[2]*Scottish Association of Marine Science, Scottish Marine Institute, Oban, Argyll PA37 1QA, UK*

[3]*Department of Geography, College of Science, Swansea University, Singleton Park, Swansea SA2 8PP, UK*

[4]*Department of Geography and Earth Science, Aberystwyth University, Aberystwyth SY23 3DB, UK*

[5]*Department of Physical Geography and Quaternary Geology, Stockholm University, SE-10691 Stockholm, Sweden*

**Corresponding author (e-mail: bill.austin@st-andrews.ac.uk)*

This volume represents a selection of papers presented at the Geological Society of London meeting 'Marine Tephrochronology', held at Burlington House, London, UK on 26 October 2011. The meeting itself was sponsored by the Geological Society of London's Marine Studies Group, with additional co-sponsorship from the Quaternary Research Association. The aim of this meeting was to bring together a wide range of speakers who would contribute evidence and data to illustrate the diverse range of tephrochronology applications, helping to spotlight recent advances in the field of marine tephrochronology. In particular, the motivation for the conference organizers resulted from their involvement in the SMART project (Synchronising Marine And ice-core Records using Tephrochronology) – one of the first systematic initiatives to investigate the cryptotephra record preserved within North Atlantic marine sequences (Abbott *et al.* 2011, 2013, 2014; Davies *et al.* in press).

This volume consists of 10 papers (plus preface, personal perspective and introduction), which have been arranged thematically in terms of methods, stratigraphy and processes. The three thematic sections concern: (a) geochemical fingerprinting (preparation and analytical methods); (b) stratigraphy; and (c) depositional processes and the broader issues that concern the integrity and interpretation of tephra isochrons.

The volume begins with a foreword by **Steve Sparks**, highlighting the significance and wide range of applications of tephra in the Earth, Ocean and Atmospheric sciences. Professor Sparks has made a significant career contribution to this field and was a keynote speaker at the original conference in October 2011; his support of the current project is gratefully acknowledged.

David Lowe's (2014) personal perspective offers a fascinating insight and, significantly, sets out three phases in the broad development of tephra (and associated Quaternary) studies spanning the last 40 years. This paper offers a unique perspective from the Southern Hemisphere, while also noting major advances, such as the widespread recognition of cryptotephra in distal marine deposits. It also introduces some of the recent internationally coordinated projects, such as the EU COST action INTIMATE (Integration of Ice-core, Marine and Terrestrial Records of the North Atlantic), highlighting the significant role (correlation, synchronization and dating) that tephrochronology now plays in our understanding of many palaeoclimate records.

In this introductory paper we highlight the main themes of the papers within the volume and offer a perspective on the significant future potential of marine tephrochronology studies. Two papers deal specifically with sample preparation methods and the advanced analytical techniques that currently underpin geochemical fingerprinting of marine tephra. **Hall & Hayward (2014)** describe the emerging challenges that have arisen in recent years as the analytical capability for quantitative microbeam

From: AUSTIN, W. E. N., ABBOTT, P. M., DAVIES, S. M., PEARCE, N. J. G. & WASTEGÅRD, S. (eds) 2014. *Marine Tephrochronology*. Geological Society, London, Special Publications, **398**, 1–5.
First published online July 15, 2014, updated August 29, 2014, http://dx.doi.org/10.1144/SP398.12

chemical analysis by electron microprobe and laser ablation ICP-MS (inductively coupled plasma-mass spectrometry) has allowed very small tephra grains to be studied. These authors outline an innovative and inexpensive method for the preparation of tephra grains as small as 10–50 μm and highlight the importance of obtaining high-quality polished surfaces as an essential pre-requisite for the collection of high-quality geochemical data. **Pearce *et al.* (2014)** expand on these analytical challenges, pointing out that, under appropriate analytical conditions, reliable electron microprobe data are now being obtained with beam diameters as small as 3 μm on basaltic or moderately hydrated rhyolitic glass. These authors make a recommendation that analytical data are normalized to an anhydrous basis in order to facilitate robust correlations from different environments. Reviewing recent advances in trace element analysis by laser ablation-ICP-MS, they highlight the potential to analyse nearly 30 elements at <1 ppm from ablation craters 10 μm in diameter. Such analytical advances enable smaller and increasingly more distal material to be analysed, thus expanding the geographical range over which reliable tephra correlations can be achieved. Trace element analysis holds considerable promise for scrutinizing tephra correlations based on major element results alone and represents an additional and valuable discrimination tool (e.g. Abbott *et al.* 2014; Davies *et al.* in press).

A series of papers deal with stratigraphic applications of marine tephrochronology, including the use of marine horizons to deduce dispersal patterns of tephra, the identification of new tephra horizons in the marine realm and the use of tephrochronology in the study of the marine radiocarbon reservoir effects. The first of these, **Engwell *et al.* (2014)**, discusses the problems of subaerial reworking and the often limited exposure of tephra deposits as the rationale for an important new study on the preservation of tephra within deep sea sequences. They focus on the distribution and thickness of marine tephra deposits from the large (magnitude ~7.7) Campanian Ignimbrite eruption (*c.* 39 kyr BP). They show an exponential decrease in tephra thickness with distance from source, and highlight the value of marine tephrochronology in resolving such trends, pointing out that tephra grain size distributions are much less affected than tephra thicknesses by post-depositional processes and are therefore very useful in understanding eruption, transport and deposition processes.

A short paper by **Hibbert *et al.* (2014)** points to the discovery of a potentially new Marine Isotope Stage (MIS) 6 age Icelandic tephra within the North East Atlantic, which the authors tentatively correlate to the Kerlingarfjöll volcanic system. Based on the size sorting and distinctive geochemical characteristics, these authors argue that the deposit resulted from primary airfall and highlight its potential as a new regional chronostratigraphic marker with an age of 181 ± 6 kyr BP. **Wastegård & Rasmussen (2014)** also present evidence to support the discovery of a new basaltic tephra layer, up to 20 cm thick, in marine cores from the Faroe Islands margin, which they name the Faroe Marine Ash Zone (FMAZ) IV. While mainly deposited and redistributed by strong bottom currents, the authors show from the rather uniform geochemistry that the tephra is relatively unmixed and assign an age of 46.8 ± 1 kyr BP; they suggest that FMAZ IV is correlative to ash zone V5, previously described from the Reykjanes Ridge by Lackschewitz & Wallrabe-Adams (1997). A follow-up investigation of this deposit is also outlined in Griggs *et al.* (in press). The tephra discoveries in both of these papers highlight how marine sequences may contain evidence for volcanic eruptions not preserved in ice-core and terrestrial sequences owing to the continuous and long-term nature of the records.

Olsen *et al.* (2014) highlight an innovative approach to the estimation of marine radiocarbon reservoir ages during the last glacial period, focussing their attention on the abrupt reorganization of North Atlantic circulation associated with Heinrich Event (HE) 4. Using a tie-point approach to link a marine event stratigraphy to the NGRIP Greenland ice core, anchored (and tested) by tephra isochrons, these authors show that age models can be refined and age uncertainties calculated. The authors show that marine radiocarbon reservoir ages rose to nearly 2000 years during HE4, adding significant uncertainty to age control that depends on radiocarbon ages that currently use the 'global' ocean average of 400 years (Reimer *et al.* 2013).

A final group of papers consider a wide range of transport and depositional processes and their wider implications for understanding tephra and their potential in constraining marine stratigraphy. **Larsen *et al.* (2014)** highlight a range of interesting and sometimes overlooked mechanisms that control the dispersal of airborne tephra once they have reached the marine environment on the North Iceland shelf. The rafting of pumice, for example, is known to be recorded in the stratigraphic record of sediments from the North Iceland shelf and eastern Norwegian Sea. **Kuhs *et al.* (2014)** continue the theme of rafting in a detailed study of an ice-rafted deposit of MIS 6 age from Integrated Ocean Drilling Project Site U1304 on the Gardar Drift, North Atlantic. They show that the heterogeneous geochemical composition, large shard sizes and co-variance with IRD all point to the material being iceberg-rafted from Iceland to Site U1304.

This iceberg-rafting, as opposed to sea-ice-rafting, mechanism is an important distinction because many primary air-fall tephra deposits are likely to have been transported by sea-ice during glacial periods. The authors argue that the widespread use of the descriptive term 'Icelandic glass' in the characterization of the tephra components within North Atlantic IRD can be improved through quantitative geochemical characterization and the attribution of a source to material, opening up the potential to improve understanding of surface ocean circulation and Icelandic Ice Sheet history.

Jennings *et al.* (2014) continue the theme of tephra dispersal from Icelandic sources, contrasting the geochemistry of tephra from cores around South East Greenland and the North Icelandic shelf. The results are potentially an important step in linking the Icelandic tephrochronological framework with material sourced from Alaska, so that long-range climate correlations may become possible in the future. Finally, in a novel experimental study, **Todd *et al.* (2014)** undertake a 'tracer' study of bioturbation in an intertidal environment by simulating an ash fall event. The authors show that bioturbation, which acts to blur the stratigraphic record of tephra isochrons, is rapid, but occurs to a limited depth (18 mm) in these fine-grained, low-energy, high-oxygen-demand sediments; the results imply a strong ecological control on bioturbation. This study provides a methodology for the future study of contrasting depositional environments.

Future perspectives

This volume and a number of recent studies (e.g. Brendryen *et al.* 2010; Bourne *et al.* 2010; Cage *et al.* 2011; Gudmundsdóttir *et al.* 2011; Griggs *et al.* in press) demonstrate the diversity of studies currently being undertaken within the realm of marine tephrochronology and the significant advances that are currently being made in the field.

The papers presented in this volume represent key concepts and aspects that pave the way for the future direction of marine tephrochronology. In particular, the advances in analytical methods (e.g. Hall & Hayward 2014; Pearce *et al.* 2014) together with the development of cryptotephra extraction techniques for marine sediments (e.g. Bourne *et al.* 2010; Abbott *et al.* 2013, 2014) highlight exciting possibilities for this technique. It is now possible to obtain a precise geochemical fingerprint from a low concentration of fine-grained tephra particles isolated from marine sediments and several studies including Jennings *et al.* (2014) have demonstrated that the geographical extent of cryptotephra deposits is far wider than previously realized. As such, tephrochronology has far-reaching applications within the fields of palaeoclimatology, archaeology, volcanic hazard assessment and environmental change. The use of this technique within the marine environment, however, is not without challenges for the future.

A key focus for future investigations is an assessment of the role of tephra transport, sedimentation and post-depositional processes in the marine environment. In particular, iceberg rafting can impart a temporal delay in the deposition of tephra and, thus, disentangling the primary and secondary depositional processes is crucial to determine the value and potential application of a particular deposit. Increased understanding of the physical and geochemical characteristics of tephra layers, for example, Engwell *et al.* (2014) and Abbott *et al.* (2014), will help in interpreting the complex transport and depositional processes that impact tephra in the marine environment (e.g. Kuhs *et al.* 2014; Wastegård & Rasmussen 2014). In addition, inspired by results presented by Matthews *et al.* (2011) at the conference in October 2011, Griggs *et al.* (in press) outlines the value of combining the physical and geochemical tephra characteristics with thin-section micromorphology to assess the stratigraphic integrity of marine tephra deposits. Griggs *et al.* (in press) also put forward a valuable protocol that holds considerable future promise for assessing the integrity of marine tephra deposits. A further challenge is the correct placement of the isochron in marine sequences – an issue that can be addressed by integration of approaches outlined in this volume. For example, combining an understanding of how tephra material can be affected by bioturbation (e.g. Todd *et al.* 2014) with studies of palaeohorizons (e.g. Jennings *et al.* 2014; Larsen *et al.* 2014) has the potential to improve the placement of the marine isochron position. Pinpointing the precise position of the isochron that reflects the timing of the eruption, rather than any post-depositional processes, is crucial because tephrochronology is increasingly being advocated as a key technique for the high-precision integration of marine records with other proxy archives (e.g. Davies *et al.* 2012; Austin & Hibbert 2012).

If isochrons can be traced between the ice and marine records, then future work can adopt the approach outlined by Olsen *et al.* (2014) to constrain the estimates of the marine radiocarbon reservoir effect. Initial results indicate that several new and previously unknown tephras are preserved within the Greenland ice-core records (Abbott *et al.* 2012; Bourne *et al.* 2013) and thus their identification within the North Atlantic marine realm will allow improved estimation of temporal and spatial variability in this effect and the past rate of mixing between surface and subsurface waters. This approach has been demonstrated in some studies (e.g. Sikes *et al.* 2000; Eiríksson *et al.* 2004; Siani *et al.* 2013), highlighting the potential application to other geographical areas. Identifying linkages between the marine and ice-core records also holds much promise for obtaining an insight into the phase relationships between the atmospheric and marine realms during periods of abrupt climatic change (e.g. Davies *et al.* in press). Large initiatives such as the EU-COST Action INTIMATE project and the TRACE project (Tephra constraints on Rapid Climatic Events) are likely to advance this work, providing independent constraints for the integration of disparate proxy records.

Tephrochronology is by no means a new technique, but the search for cryptotephra deposits has changed the scope and reach of this correlative tool. Marine tephrochronology and crypto-tephrochronlogy promise to facilitate the synchronization of records and yield new information on volcanic history, the distribution of their products and past oceanic variability.

We would like to thank all the participants at the Geological Society of London meeting 'Marine Tephrochronology', held at Burlington House, London, UK on 26 October 2011. We are particularly grateful to L. Dyer and colleagues at the Geological Society of London for their help in organizing the event, supported by E. Stahly and H. Olaez at the University of St Andrews. A. Hills and colleagues at the Geological Society Publishing House are gratefully thanked for their support and patience throughout the preparation of the current volume. We also acknowledge funding received from Natural Environment Research Council for the SMART project (NE/F020600/1; NE/F02116X/1; NE/F021445/1), the European Research Council (TRACE project 259253) and the EU COST INTIMATE action (ES0907).

References

Abbott, P. M., Davies, S. M., Austin, W. E. N., Pearce, N. J. G. & Hibbert, F. D. 2011. Identification of cryptotephra horizons in a North East Atlantic marine record spanning marine isotope stages 4 and 5a (~60,000–82,000 a b2k). *Quaternary International*, **246**, 177–189.

Abbott, P. M., Davies, S. M. *et al.* 2012. A detailed framework of marine isotope stage 4 and 5 volcanic events recorded in two Greenland ice-cores. *Quaternary Science Reviews*, **36**, 59–77.

Abbott, P. M., Austin, W. E. N., Davies, S. M., Pearce, N. J. G. & Hibbert, F. D. 2013. Cryptotephrochronology of the Eemian and the last interglacial-glacial transition in the North East Atlantic. *Journal of Quaternary Science*, **28**, 501–514.

Abbott, P. M., Austin, W. E. N., Davies, S. M., Pearce, N. J. G., Rasmussen, T. L., Wastegård, S. & Brendryen, J. 2014. Re-evaluation and extension of the Marine Isotope Stage 5 tephrostratigraphy of the Faroe Islands Region: the cryptotephra record. *Palaeogeography, Palaeoclimatology, Palaeoecology*, **409**, 153–168.

Austin, W. E. N. & Hibbert, F. 2012. Tracing time in the ocean: a brief review of chronological constraints (60–8 kyr) on North Atlantic marine event-based stratigraphies. *Quaternary Science Reviews*, **36**, 28–37.

Bourne, A. J., Lowe, J. J. *et al.* 2010. Distal tephra record for the last 105,000 years from PRAD 1-2 in the Central Adriatic sea: implications for marine tephrostratigraphy. *Quaternary Science Reviews*, **29**, 3079–3094.

Bourne, A. J., Davies, S. M., Abbott, P. M., Rasmussen, S. O., Steffensen, J. P. & Svensson, A. 2013. Revisiting the Faroe Marine Ash Zone III in two Greenland ice cores: implications for marine-ice-core correlations. *Journal of Quaternary Science*, **28**, 641–646.

Brendryen, J., Haflidason, H. & Sejrup, H. P. 2010. Norwegian Sea tephrostratigraphy of marine isotope stages 4 and 5: prospects and problems for tephrochronology in the North Atlantic region. *Quaternary Science Reviews*, **29**, 847–864.

Cage, A. G., Davies, S. M., Wastegard, S. & Austin, W. E. N. 2011. Identification of the Icelandic Landnam tephra (AD 871 ± 2) in Scottish fjordic sediment. *Quaternary International*, **246**, 168–176, http://dx.doi.org/10.1016/j.quaint.2011.08.016

Davies, S. M., Abbott, P. M., Pearce, N. J. G., Wastegård, S. & Blockley, S. P. E. 2012. Integrating the INTIMATE records using tephrochronology: rising to the challenge. *Quaternary Science Reviews*, **36**, 11–27.

Davies, S. M., Abbott, P. M. *et al.* in press. A North Atlantic tephra framework for 130–60 ka b2k: new tephra discoveries, marine based-correlations and future challenges. *Quaternary Science Reviews*.

Eiríksson, J., Larsen, G., Knudsen, K. L., Heinemeier, J. & Símonarson, L. A. 2004. Marine reservoir age variability and water mass distribution in the Iceland Sea. *Quaternary Science Reviews*, **23**, 2247–2268.

Engwell, S. L., Sparks, R. S. J. & Carey, S.2014. Physical characteristics of tephra layers in the deep sea realm: the Campanian Ignimbrite eruption. *In*: Austin, W. E. N., Abbott, P. M., Davies, S. M., Pearce, N. J. G. & Wastegård, S. (eds) *Marine Tephrochronology*. Geological Society, London, Special Publications, **398**. First published online February 6, 2014, http://dx.doi.org/10.1144/SP398.7

Griggs, A. J., Davies, S. M., Abbott, P. M., Rasmussen, T. L. & Palmer, A. P. in press. Optimising the use of marine tephrochronology in the North Atlantic: a detailed investigation of the Faroe Marine Ash Zones II, III and IV. *Quaternary Science Reviews*, http://dx.doi.org/10.1016/j.quascirev.2014.04.031

Gudmundsdóttir, E. R., Eiríksson, J. & Larsen, G. 2011. Identification and definition of primary and reworked tephra in Late Glacial and Holocene marine shelf sediments off North Iceland. *Journal of Quaternary Science*, **26**, 589–602.

Hall, M. & Hayward, C. 2014. Preparation of micro- and crypto-tephras for quantitative microbeam analysis. *In*: Austin, W. E. N., Abbott, P. M., Davies, S. M., Pearce, N. J. G. & Wastegård, S. (eds) *Marine Tephrochronology*. Geological Society, London,

Special Publications, **398**. First published online February 27, 2014, http://dx.doi.org/10.1144/SP398.5

Hibbert, F. D., Wastegård, S., Gwynn, R. & Austin, W. E. N. 2014. Identification of a MIS 6 age (c. 180 ka) Icelandic tephra within NE Atlantic sediments: a new potential chronostratigraphic marker. *In*: Austin, W. E. N., Abbott, P. M., Davies, S. M., Pearce, N. J. G. & Wastegård, S. (eds) *Marine Tephrochronology*. Geological Society, London, Special Publications, **398**. First published online February 20, 2014, http://dx.doi.org/10.1144/SP398.10

Jennings, A., Thordarson, T., Zalzal, K., Stoner, J., Hayward, C., Geirsdóttir, Á. & Miller, G. 2014. Holocene tephra from Iceland and Alaska in SE Greenland Shelf Sediments. *In*: Austin, W. E. N., Abbott, P. M., Davies, S. M., Pearce, N. J. G. & Wastegård, S. (eds) *Marine Tephrochronology*. Geological Society, London, Special Publications, **398**. First published online February 27, 2014, http://dx.doi.org/10.1144/SP398.6

Kuhs, M., Austin, W. E. N., Abbott, P. M. & Hodell, D. A. 2014. Iceberg-rafted tephra as a potential tool for the reconstruction of ice-sheet processes and ocean surface circulation in the glacial North Atlantic. *In*: Austin, W. E. N., Abbott, P. M., Davies, S. M., Pearce, N. J. G. & Wastegård, S. (eds) *Marine Tephrochronology*. Geological Society, London, Special Publications, **398**. First published online February 6, 2014, http://dx.doi.org/10.1144/SP398.8

Lackschewitz, K. S. & Wallrabe-Adams, H. J. 1997. Composition and origin of volcanic ash zones in Late Quaternary sediments from the Reykjanes Ridge: evidence for ash fallout and ice-rafting. *Marine Geology*, **136**, 209–224.

Larsen, G., Eiríksson, J. & Gudmundsdóttir, E. R. 2014. Last millennium dispersal of air-fall tephra and ocean-rafted pumice towards the north Icelandic shelf and the Nordic seas. *In*: Austin, W. E. N., Abbott, P. M., Davies, S. M., Pearce, N. J. G. & Wastegård, S. (eds) *Marine Tephrochronology*. Geological Society, London, Special Publications, **398**. First published online March 3, 2014, http://dx.doi.org/10.1144/SP398.4

Lowe, D. J. 2014. Marine tephrochronology: a personal perspective. *In*: Austin, W. E. N., Abbott, P. M., Davies, S. M., Pearce, N. J. G. & Wastegård, S. (eds) *Marine Tephrochronology*. Geological Society, London, Special Publications, **398**. First published online April 16, 2014, http://dx.doi.org/10.1144/SP398.11

Matthews, I. P., Palmer, A. P., Folkes, H. & Trincardi, F. 2011. Understanding tephra layers in marine sediments, the potential of thin section micromorphology to investigate depositional and taphonomic processes. Abstract. Presented at the Marine Tephrochronology Meeting. Geological Society London, UK, 26 October, 2011.

Olsen, J., Rasmussen, T. L. & Reimer, P. J. 2014. North Atlantic marine radiocarbon reservoir ages through Heinrich event H4: a new method for marine age model construction. *In*: Austin, W. E. N., Abbott, P. M., Davies, S. M., Pearce, N. J. G. & Wastegård, S. (eds) *Marine Tephrochronology*. Geological Society, London, Special Publications, **398**. First published online February 6, 2014, http://dx.doi.org/10.1144/SP398.2

Pearce, N. J. G., Abbott, P. M. & Martin-Jones, C. 2014. Microbeam methods for the analysis of glass in fine-grained tephra deposits: a SMART perspective on current and future trends. *In*: Austin, W. E. N., Abbott, P. M., Davies, S. M., Pearce, N. J. G. & Wastegård, S. (eds) *Marine Tephrochronology*. Geological Society, London, Special Publications, **398**. First published online February 27, 2014, http://dx.doi.org/10.1144/SP398.1

Reimer, P. J., Bard, E. *et al.* 2013. INTCAL13 and MARINE13 radiocarbon age calibration curves 0–50,000 years cal BP. *Radiocarbon*, **55**, 1869–1887.

Siani, G., Michel, E. *et al.* 2013. Carbon isotope records reveal precise timing of enhanced Southern Ocean upwelling during the last deglaciation. *Nature Communications*, **4**, 2758.

Sikes, E. L., Samson, C. R., Guilderson, T. P. & Howard, W. R. 2000. Old radiocarbon ages in the southwest Pacific Ocean during the last deglaciation. *Nature*, **405**, 555–559.

Todd, J. A., Austin, W. E. N. & Abbott, P. M. 2014. Quantifying bioturbation of a simulated ash fall event. *In*: Austin, W. E. N., Abbott, P. M., Davies, S. M., Pearce, N. J. G. & Wastegård, S. (eds) *Marine Tephrochronology*. Geological Society, London, Special Publications, **398**. First published online February 5, 2014, http://dx.doi.org/10.1144/SP398.9

Wastegård, S. & Rasmussen, T. L. 2014. Faroe Marine Ash Zone IV: a new MIS 3 ash zone on the Faroe Islands margin. *In*: Austin, W. E. N., Abbott, P. M., Davies, S. M., Pearce, N. J. G. & Wastegård, S. (eds) *Marine Tephrochronology*. Geological Society, London, Special Publications, **398**. First published online March 4, 2014, http://dx.doi.org/10.1144/SP398.3

Marine tephrochronology: a personal perspective

DAVID J. LOWE[1,2]

[1]*School of Science, Faculty of Science and Engineering, University of Waikato, Private Bag 3105, Hamilton 3240, New Zealand (e-mail: d.lowe@waikato.ac.nz)*

[2]*President, International Focus Group on Tephrochronology and Volcanism (INTAV), Hamilton, New Zealand*

This special volume on marine tephrochronology is remarkable, and timely, because it marks a concerted step towards what might be informally termed 'phase 3' of a revolution in Quaternary geosciences that began around 40 years ago. The 10 articles collectively represent a re-focussed examination of tephras and cryptotephras preserved in ocean sediments at various locations and the authors describe their significance for a range of subdisciplines. Eight articles provide a new understanding of the origin, distribution and ages of various tephra and cryptotephra deposits and their stratigraphic inter-relationships; how the terrestrial ages of the tephra/crypotephra deposits relate to those of enclosing sediments and inform the ongoing development of the marine radiocarbon time-scale; mechanisms for the emplacement, remobilization or bioturbation of the tephras or cryptotephras; and volcanic eruption history. Two further articles document the characterization of tephra-derived glass shards using microbeam techniques to analyse 30–40 elements from individual shards as small as 10 μm in diameter. The collection thus provides snapshots of many aspects of the latest developments and directions in tephra studies – volcanology, primary and secondary dispersal, stratigraphy, single-grain characterization, chronology – through the medium of marine sediments. My personal perspective reflects briefly on how this point was reached and identifies a few of the important milestones on the way from 'phase 1' to 'phase 3'. I am privileged to write it.

Marine science revolution

As an undergraduate in the early-mid 1970s, I recall my first real 'awakening' regarding the *dynamic* nature of science, and of Quaternary geoscience in particular, when told about deep-sea core V28-238 from the equatorial Pacific Ocean (Shackleton & Opdyke 1973; >2650 citations, Google Scholar). Analogous to the opening notes of Beethoven's 5th Symphony, perhaps the most famous quartet of notes in history, the alpha-numerical assemblage 'V28-238' is possibly the most instantly recognizable abbreviation in Quaternary science ('V' stands for *Vema*, the three-masted coring schooner of the Lamont Geological Observatory, now the Lamont–Doherty Earth Observatory of Columbia University). Lecturers Michael Selby and Cam Nelson (both now retired) at the University of Waikato, New Zealand, strongly emphasized the importance of V28-238, and of the deep-sea marine record in general, in developing a much more comprehensive understanding of changes in the Quaternary than previously attained. Although the modern phase of deep-sea drilling began in the 1950s, the formative findings emerging in the 1970s, including those of Hays *et al.* (1976; >2800 citations, Google Scholar) have, rightly, been regarded as revolutionary (Walker & Lowe 2007). At this time, remember, the substantive theory of 'plate tectonics' was barely a decade old, and articles were still being published regularly on the correlation of on-land terrace sequences with the so-called 'four main Pleistocene glacial episodes'. In their discussion on the chronology of V28-238, Shackleton & Opdyke (1973, p. 49) recorded the potential of tephrochronology by commenting on the usefulness of an Italian tephra (a 'Tuff with Black Pumices'): 'If this eruption could be confidently placed in a climatic sequence, it is possible that it could be used to render our chronology even more accurate'.

The sense of discovery and the dramatic new world of possibilities that had opened up with the marine-based research continued in the 1980s when Cam Nelson took part as a ship-board scientist on the *Glomar Challenger* on Leg 90 (December 1982–January 1983) of the Deep-Sea Drilling Project (DSDP) that included coring in both the Tasman Sea and the Pacific Ocean immediately east of New Zealand (Fig. 1; the final voyage of the *Glomar Challenger*, Leg 96, took place in October, 1983). Although not directly involved, it was exhilarating to be rubbing shoulders with Cam while he was working on material from the Leg 90 cores (Fig. 2). Cam and others including Chris Hendy, Paul Froggatt and Kerry Black soon published several critically important papers, one showing for the first time synchroneity between

From: Austin, W. E. N., Abbott, P. M., Davies, S. M., Pearce, N. J. G. & Wastegård, S. (eds) 2014. *Marine Tephrochronology*. Geological Society, London, Special Publications, **398**, 7–19.
First published online April 16, 2014, http://dx.doi.org/10.1144/SP398.11

Fig. 1. The *Glomar Challenger* in Noumea in December, 1982, before embarking on DSDP Leg 90. Photograph courtesy of Cam Nelson.

long-term glacial and interglacial signals of both hemispheres and a 'genetic link' between alpine glaciation and the marine isotope record (from analyses of core sediments from the DSDP site 594; Nelson *et al.* 1985*a*, 1986), another documenting a new spectral analysis procedure for dating Quaternary deep-sea sediments (Black *et al.* 1988), and several articles on the visible ('megascopic') tephra record preserved in the marine sediments (Nelson *et al.* 1985*b*; Froggatt *et al.* 1986; Fig. 3). A startling find was that tephra deposits could be markedly thicker 'upwind' of source: ash ascending above 20 km altitude was carried directly westward towards the Tasman Sea by easterly stratospheric winds, while coeval lower-level ash was blown eastward ('downwind') of New Zealand into the southern Pacific Ocean (Nelson *et al.* 1985*b*). A second interesting result was the recognition that the visible tephras were only a fragmentary record of Quaternary volcanism in the SW Pacific because New Zealand-derived silicic shards were a ubiquitous but disseminated minor component in many of the sediments in the cores. The disseminated glass shard concentrations, not visible as layers, today would be called cryptotephras (Lowe & Hunt 2001; Lowe 2011). In addition, pale-green laminae <2 mm thick, common and conspicuous features in many of the cores examined (Fig. 4a), were inferred to represent diagenetically altered (mainly smectitic) remnant volcanic ash layers (Gardner *et al.* 1985).

This DSDP-based work thus revealed a compelling story about tephras but it was recognized clearly as being incomplete. Subsequent work on marine sediments in the SW Pacific, which I have loosely denoted 'phase 2', expanded the Quaternary marine tephra record through analysis of cores obtained by various ships including the *JOIDES Resolution* (Ocean Drilling Program (ODP) followed by the Integrated ODP (IODP)), the *Roger Revelle* (Scripps Institute of Oceanography), the *Marion Dufresne* (French Polar Institute Paul-Emile Victor), and the New Zealand research vessel *Tangaroa* (e.g. Pillans & Wright 1992; Carter *et al.* 1995, 2002, 2004; Alloway *et al.* 2005; Shane *et al.* 2006; Allan *et al.* 2008). The orbitally tuned chronology for the ODP/IODP cores, 'calibrated' or tested by numerical ages on tephras and magnetostratigraphy, provides an important framework for associated palaeoenvironmental reconstructions (e.g. Carter 2005; Naish 2005; Holt *et al.* 2010; Alloway *et al.* 2013; Fig. 4b). Tephras preserved in marine sedimentary sequences on land were also examined (e.g. Alloway *et al.* 1993; Naish *et al.* 1996; Shane *et al.* 1998; Pillans *et al.* 2005).

Fig. 2. Cam Nelson (right) inspecting fresh cores with a colleague on *Glomar Challenger* in January, 1983, during the Leg 90 cruise. Photograph courtesy of Cam Nelson.

Phase 3 – delving into marine cryptotephras in the SW Pacific region – has barely begun, as noted by Holt *et al.* (2011), although marine records and palaeoceanographic changes in the Australian and New Zealand regions, especially since 30 000 years ago, have now been documented and interpreted in some detail (e.g. Carter *et al.* 2008; Hayward *et al.* 2011; Newnham *et al.* 2012; Bostock *et al.* 2013; Marr *et al.* 2013; Nelson *et al.* 2013).

Hence the work reported in this volume by Austin *et al.* (2014) on sediments in oceans of the Northern Hemisphere is at the forefront of marine tephra studies, and the authors and editors deserve credit and acclaim.

Tephrochronological advances

While the new tephra and palaeoenvironmental work in the marine realm was buzzing in the 1970s and 1980s, complementary advances were taking place on land in the discipline of physical

Fig. 3. Micrograph of glass shards of visible tephra no. 13 (after Nelson *et al.* 1985*b*, p. 1173) from DSDP site 594, subsequently identified by Nelson (1988) as a correlative of the very widespread and voluminous Whakamaru-caldera derived 340 000-year-old Rangitawa tephra (formerly known also as Mt Curl tephra, Ohinewai ash, and Hamilton ash 'H1' bed), which was deposited from a super-eruption near the end of Marine Oxygen Isotope Stage 10 (Kohn *et al.* 1992; Pillans *et al.* 1996; Lowe *et al.* 2001; Holt *et al.* 2010; Matthews *et al.* 2012; Alloway *et al.* 2013). Micrograph courtesy of Cam Nelson (his handwritten notes reflect the pre-digital era).

volcanology, especially relating to explosive volcanism and its products (e.g. Walker 1973, 1983; Self & Sparks 1981; Fisher & Schminke 1984; Wilson & Walker 1985; Wilson 1986, 1993), and in the thriving discipline of tephrochronology (e.g. Topping & Kohn 1973; Westgate & Fulton 1975; Machida & Arai 1978; Smalley 1980; Lowe 1990). As well as studies on distal tephra deposits preserved in sediments, and in soils or palaeosols, new laboratory methods for characterizing the glass-shard and crystal (mineral) components of tephra deposits were being developed in a period of substantial technical advancement (e.g. Kohn 1970; Sarna-Wojcicki 1976; Westgate & Gorton 1981; Beaudoin & King 1986). These methods included the innovative use of the electron microprobe, developed mainly through the 1970s and 1980s, to obtain major element compositions of individual glass shards rather than bulk samples (Smith & Westgate 1968; Froggatt 1983, 1992; Hunt & Hill 1996). Its advent in 1968 a stroke of brilliance, the microprobe method for glass analysis, initially criticized by some 'traditional' volcanic petrologists and geoanalysts, has been widely adopted today as the cornerstone analytical technique in tephra studies used alongside stratigraphic, chronological and mineralogical data to help enable correlations to be made (Froese *et al.* 2008*a*; Lowe 2011). More recent refinements of the electron microprobe method for the analysis of glass shards or melt inclusions, or for crystals, include those described by Turney *et al.* (2004), Kuehn *et al.* (2011), Matsu'ura *et al.* (2011, 2012), Hayward (2012), Hall & Hayward (2014), Marcaida *et al.* (2014) and Pearce *et al.* (2014). Subsequently, from the mid-1990s, the development of the laser-ablation inductively coupled plasma–mass spectrometric method for the analysis of trace elements in individual glass shards provided another dramatic leap forward in tephra analysis, especially that involving thin distal deposits or cryptotephras (e.g. Westgate *et al.* 1994, 2013*a*; Pearce *et al.* 2004, 2007, 2011; Tomlinson *et al.* 2010; Preece *et al.* 2011).

Alongside these methods of characterization, new or improved methods for dating tephras or cryptotephras have emerged, such as the isothermal plateau fission-track method for dating glass shards (Westgate 1989; Sandhu & Westgate 1995; Westgate *et al.* 2013*b*). New ways of developing age models for tephra layers or cryptotephra deposits in sedimentary sequences using wiggle-matching or Bayesian statistical procedures (Buck *et al.* 2003) alongside radiocarbon, optical, $^{40}Ar/^{39}Ar$ or other dating methods have become very important as well in the past decade (e.g. Hogg *et al.* 2003, 2012; Lowe *et al.* 2007; Lowe *et al.* 2013; Blockley *et al.* 2008; Blaauw & Christen 2011; Biswas *et al.* 2013; Smith *et al.* 2013). Methods by which tephras or cryptotephras could be correlated on a more quantitative basis using numerical or statistical analysis of their compositional data, along with other information such as stratigraphic position and relationship to other tephras or deposits, palaeoecological or archaeological associations, and numerical age, are currently being reviewed and evaluated.

Perhaps the most unexpected development during my career has been the flourishing of tephra studies in areas remote from volcanic sources through cryptotephra deposits (Lowe 2008). The rise of a new generation of cryptotephra specialists all around the world has been astounding. Given that my own early New Zealand work was on distal tephras, including studies on admixed 'non-visible' tephras in buried soil sequences (e.g. Lowe 1986), I should perhaps have been more prescient! The advent of *systematic* studies of cryptotephras – the identification, correlation, mapping and dating of sparse, fine-grained glass-shard and/or crystal concentrations 'hidden' within sediments or soil – over the past 25 years since Andrew Dugmore's seminal (1989) paper has undoubtedly been revolutionary. Although new and increasingly efficacious techniques for cryptotephra detection, extraction and identification were developed primarily in northwestern Europe and Scandinavia, and more lately North America (e.g. Dugmore *et al.* 1992, 1995; Turney 1998; Wastegård *et al.* 2000; Hall & Pilcher 2002; Blockley *et al.* 2005; Gehrels *et al.* 2008; Payne *et al.* 2008; Kuehn & Froese 2010; Pyne-O'Donnell *et al.* 2012), earlier work in New Zealand had demonstrated the occurrence of glass shards or crystals in concentrations not visible as layers in sedimentary deposits or soils. Robertson & Mew (1982) were the first to count glass shards in soils developed on loess and glacial deposits in western South Island, writing (p. 506): 'Although no distinct ash layers were observed in the West Coast soil profiles, it is probable that, as the amounts of glass present are low, such a layer would be difficult to detect in the field'. The shards were subsequently identified as correlatives of the *c.* 25 400 cal. year-old Kawakawa tephra; 'cryptic' occurrences in loess elsewhere were also identified through glass-shard counting and major element analysis (Mew *et al.* 1986; Eden & Froggatt 1988; Almond 1996; Vandergoes *et al.* 2013). Lowe *et al.* (1981) and Lowe (1988*a*) were among the first to use X-radiography to detect non-visible glass shard concentrations in peat and lake sediments and to characterize them using a rapid X-ray fluorescence technique, and Hogg & McCraw (1983, p. 182) used the sparse occurrence of distinct (optically dark green) aegirine crystals in ash-derived soils and palaeosols to map *c.* 7000 cal. year-old Tuhua tephra at least 80 km beyond its visible limits in the field (see also Table 2 in Lowe *et al.* 2008, p. 101).

The burgeoning cryptotephra methods, likened to 'forensic tephrochronology', have now been applied to terrestrial, ice-core and marine sedimentary sequences in all parts of the world (e.g. see Table 1 in Lowe 2008, p. 315). Cryptotephras have been discovered in peat, lake, marine and aeolian sediments (including frozen sediment, i.e. permafrost), in ice, in soils and palaeosols, and in deposits in caves. The tenacious cryptotephra studies, although not without serious limitations and problems such as taphonomic issues and quantitation uncertainties (especially when dealing with limited numbers of shards or crystals; e.g. Payne *et al.* 2005; Davies *et al.* 2007; Lowe 2011; Pyne-O'Donnell 2011; Lane *et al.* 2014), have documented tephra-fall occurrences at sub-millimetre scale at distances spanning hundreds to thousands of kilometres (now described as 'ultra-distal'), greatly extending known geographical limits to >7000 km from source (e.g. Gehrels *et al.* 2006; Davies *et al.* 2008; Brendryen *et al.* 2010; Lawson *et al.* 2012; Pyne-O'Donnell *et al.* 2012; Lane *et al.* 2013).

Marine cryptotephrochronology: the next revelatory revolution

Returning to the main theme of this volume, the value of marine sediments in providing potentially very detailed records of tephra layers or cryptotephras, and their applications tephrochronologically, are now well recognized and exploited (e.g. Chun *et al.* 2007; Hillenbrand *et al.* 2008; Abbott *et al.* 2011, 2013; Cage *et al.* 2011; Albert *et al.* 2012; Gudmundsdóttir *et al.* 2012). As recognized by Nelson *et al.* (1985*b*), cores of marine sediment have the potential to provide a detailed stratigraphic record of tephra fallout because large proportions of volcanic eruptives are deposited in the sea (Cassidy *et al.* 2013). Together with terrestrial records such as those from lakes, bogs and ice cores, the marine tephra or cryptotephra deposits can therefore help document patterns of explosive volcanism in time and space, often more comprehensively than those obtainable near volcanic eruption centres, and integrate the stratigraphic interfingering of eruptives from multiple volcanic sources (Lowe 1988*b*; Alloway *et al.* 2005; Shane 2005; Shane *et al.* 2006; Bourne *et al.* 2010, 2013; Davies *et al.* 2010; Lim *et al.* 2013; Ponomareva *et al.* 2013; Insinga *et al.* 2014).

Although distal deposits tend to be more restricted compositionally than their proximal counterparts generated from the same eruptive episode (e.g. Smith *et al.* 2005), recent work on magmatic heterogeneity at some volcanoes has shown that multiple fingerprints may arise according to tephra dispersal direction during a 'single' eruption episode, adding complexity and the need for a careful approach in making long-range correlations (Shane *et al.* 2008). Potential complications and, paradoxically, insights, thus abound especially in studying tephras and cryptotephras in the marine realm where reworking and bioturbation may be commonplace, and where 'anomalous' grain-size patterns may tell an unexpected story (e.g. Hunt *et al.* 1995; Lackschewitz & Wallrabe-Adams 1997; Lacasse *et al.* 1998; Manville & Wilson 2004; Ascough *et al.* 2005; Kristjánsdóttir *et al.* 2007; Todd *et al.* 2014).

Tephra studies globally are in an incredibly expansive and productive phase right now (e.g. Froese *et al.* 2008*b*; Lowe *et al.* 2011; Alloway *et al.* 2013; Gorbarenko *et al.* 2013; Smith 2013), with a number of specialist laboratory groups around the world leading the way in exemplary fashion. Tephra specialists (tephrochronologists) provide essential expertise in numerous projects where tephras or cryptotephras, commonly both, are key stratigraphic correlational, synchronizational or dating tools to help meet the objectives of high-resolution palaeoclimatic projects such as INTIMATE (integration of ice-core, marine and terrestrial records to reconstruct past abrupt and extreme climate changes from 60 000 to 8000 years ago), SMART (synchronizing marine and ice-core records using tephrochronology), CELL50K (calibrating environmental leads and lags over the last 50 000 years), TRACE (tephra constraints on rapid climatic events), the Lake Suigetsu 2006 varved sediment-core project (Japan) and SHAPE (Southern Hemisphere assessment of palaeoenvironments). These and other projects (such as RESET, response of humans to abrupt environmental transitions), all involving tephrochronology, are meeting the challenge laid down for Quaternary scientists by Walker & Lowe (2007, p. 1087) 'to generate integrated datasets of the highest possible quality and temporal resolution to provide meaningful baselines for predictive models of the global future'. Tephras and crypotephras are also being identified and utilized in growing archaeological and palaeoanthropological studies (e.g. Wastegård & Davies 2009; Housley *et al.* 2012; Lowe *et al.* 2012; Streeter *et al.* 2012; Riede & Thastrup 2013; Lane *et al.* 2014).

Conclusion

This benchmark volume sets the scene for the development of the next phase of integrating marine science studies with tephrochronology. Tephra or cryptotephra deposits, usually in the form of glass shard concentrations at distal or ultra-distal sites, provide a chronostratigraphic link to equivalent deposits in other environments, including the superlative ice-core records (e.g. Davies *et al.* 2008, 2010, 2014). Such work is already proving to be as stimulating as the breathtaking revelations 'hauled up' from the deep oceans *c.* 40 years ago that helped to inspire several generations of Quaternary geoscientists, myself included.

Fig. 4. (**a**) Around 12 pale-green laminae (thin dark layers, some composites) in part of a core from DSDP site 599 from central Lord Howe Rise, Tasman Sea (from Gardner *et al.* 1985, p. 1147). Core shown is about 20 cm long (scale marks on right at 5 cm intervals). The laminae comprise mainly authigenic smectite and tend to occur in association with iron sulphide. There is a general correspondence between the occurrence of laminae and the presence of volcanic glass in smear slides, hence their identification as diagentically altered remnant tephra layers (Gardner *et al.* 1985). Indistinct circular feature near the top between the first and third laminae (cutting though the second lamina) is an infilled burrow. Photograph courtesy of Cam Nelson. (**b**) Tephra beds (dark coloured) in nanofossil ooze (pale coloured) in cores from ODP site 1124 (*c.* 4000 m water depth) due east of central North Island, New Zealand. Scale at left in centimetres. Photograph courtesy of Lionel Carter (from Alloway *et al.* 2013, p. 298, with permission from Elsevier).

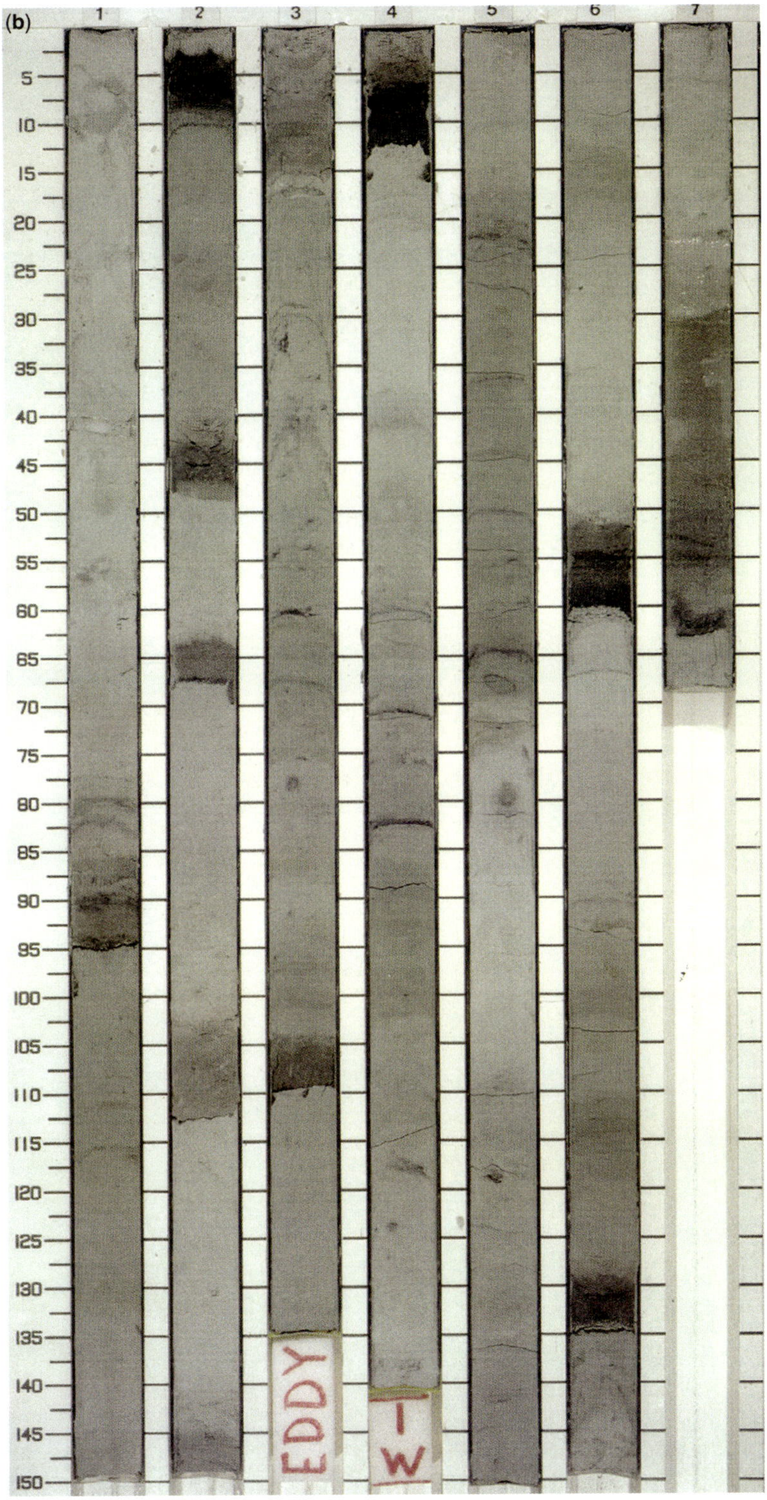

Fig. 4. *Continued.*

It is perhaps instructive here to quote from a clever, insightful short essay by Schwartz (2008) on thc need to seek answers by asking the right questions through what he termed 'productive stupidity' (p. 1171): 'Productive stupidity means being ignorant by choice. Focusing on important questions puts us in the awkward position of being ignorant' and hence 'The more comfortable we become with being stupid, the deeper we will wade into the unknown and the more likely we are to make big discoveries'.

Always accompanied by the excitement and collegiality of discovery, and lifelong respect and friendship, much is promised and surely will be delivered as we wade deeper into the unknown through marine – and complementary terrestrial – tephrochronology and cryptotephrochronology.

I thank N. Pearce for his invitation to contribute this article, and H. Bostock, C. Nelson, S. Davies and J. Hunt for reviewing it. Helpful comments from editors W. Austin and N. Pearce were also appreciated. C. Nelson and L. Carter provided the figures, Elsevier gave me permission to use Figure 4b, and S. Davies kindly furnished a pre-publication article. G. Xu and T. Ingle helped with the figures. This paper is an output of INTREPID Tephra-II (INQUA project 1307s) 'Enhancing tephrochronology as a global research tool through improved fingerprinting and correlation techniques and uncertainty modelling', an initiative of the International Focus Group on Tephrochronology and Volcanism (INTAV) supported by the Stratigraphy and Chronology Commission (SACCOM) of the International Union for Quaternary Research (INQUA).

References

Abbott, P. M., Davies, S. M., Austin, W. E. N., Pearce, N. J. G. & Hibbert, F. D. 2011. Identification of cryptotephra horizons in a north east Atlantic marine record spanning marine isotope stages 4 and 5a (~60 000–82 000 a b2k). *Quaternary International*, **246**, 177–189.

Abbott, P. M., Austin, W. E. N., Davies, S. M., Pearce, N. J. G. & Hibbert, F. D. 2013. Cryptotephrochronology of the Eemian and the last interglacial–glacial transition in the north-east Atlantic. *Journal of Quaternary Science*, **28**, 501–514.

Albert, P. G., Tomlinson, E. L. *et al.* 2012. Marine-continental tephra correlations: volcanic glass geochemistry from the Marsili Basin and the Aeolian Islands, southern Tyrrhenian Sea, Italy. *Journal of Volcanology and Geothermal Research*, **229–230**, 74–94.

Allan, A. S. R., Baker, J. A., Carter, L. & Wysoczanksi, R. J. 2008. Reconstructing the Quaternary evolution of the world's most active silicic volcanic system: insights from a ~1.65 Ma deep ocean tephra record sourced from the Taupo Volcanic Zone, New Zealand. *Quaternary Science Reviews*, **27**, 2341–2360.

Alloway, B. V., Pillans, B. J., Sandhu, A. S. & Westgate, J. A. 1993. Revision of the marine chronology in Wanganui Basin, New Zealand, based on the isothermal plateau fission-track dating of tephra horizons. *Sedimentary Geology*, **82**, 299–310.

Alloway, B. V., Pillans, B. J., Carter, L., Naish, T. & Westgate, J. A. 2005. Onshore-offshore correlation of Pleistocene rhyolitic eruptions from New Zealand: implications for TVZ eruptive history and paleoenvironmental construction. *Quaternary Science Reviews*, **24**, 1601–1622.

Alloway, B. V., Lowe, D. J., Larsen, G., Shane, P. A. R. & Westgate, J. A. 2013. Tephrochronology. *In*: Elias, S. A. & Mock, C. J. (eds) *The Encyclopaedia of Quaternary Science*, 2nd edn. Elsevier, Amsterdam, **4**, 277–304.

Almond, P. C. 1996. Loess, soil stratigraphy and Aokautere Ash on Late Pleistocene surfaces in south Westland, New Zealand: interpretation and correlation with the glacial stratigraphy. *Quaternary International*, **34–36**, 163–176.

Ascough, P., Cook, G. & Dugmore, A. 2005. Methodological approaches to determining the marine radiocarbon reservoir effect. *Progress in Physical Geography*, **29**, 532–547.

Austin, W. E. N., Abbott, P. M., Davies, S. M., Pearce, N. J. G. & Wastegård, S. (eds) 2014. *Marine Tephrochronology*. Geological Society, London, Special Publications, **398**.

Beaudoin, A. B. & King, R. H. 1986. Using discriminant function analysis to identify Holocene tephras based on magnetite composition: a case study from the Sunwapta Pass area, Jasper National Park. *Canadian Journal of Earth Science*, **2**, 804–812.

Biswas, R. H., Williams, M. A. J., Raj, R., Juyal, N. & Singhvi, A. K. 2013. Methodological studies on luminescence dating of volcanic ashes. *Quaternary Geochronology*, **17**, 14–25.

Blaauw, M. & Christen, J. A. 2011. Flexible paleoclimate age–depth models using an autoregressive gamma process. *Bayesian Analysis*, **6**, 457–474.

Black, K. P., Nelson, C. S. & Hendy, C. H. 1988. A spectral analysis procedure for dating Quaternary deep-sea cores and its application to a high-resolution Bruhnes record from the souhwest Pacific. *Marine Geology*, **83**, 21–30.

Blockley, S. P. E., Pyne-O'Donnell, S. D. F. *et al.* 2005. A new and less destructive laboratory procedure for the physical separation of distal glass tephra shards from sediments. *Quaternary Science Reviews*, **24**, 1952–1960.

Blockley, S. P. E., Bronk Ramsey, C., Lane, C. S. & Lotter, A. F. 2008. Improved age modelling approaches as exemplified by the revised chronology for the central European varved lake Soppensee. *Quaternary Science Reviews*, **27**, 61–71.

Bostock, H. C., Barrows, T. T. *et al.* 2013. A review of the Australian–New Zealand sector of the Southern Ocean over the last 30 ka (Aus-INTIMATE project). *Quaternary Science Reviews*, **74**, 35–57.

Bourne, A. J., Lowe, J. J. *et al.* 2010. Distal tephra record of the last *c.* 105,000 years from core PRAD 1–2 in the central Adriatic Sea: implications for marine tephrostratigraphy. *Quaternary Science Reviews*, **29**, 3079–3094.

Bourne, A. J., Davies, S. M., Abbott, P. M., Rasmussen, S. O., Steffensen, J. P. & Svensson, A. 2013.

Revisiting the Faroe Marine Ash Zone III in two Greenland ice cores: implications for marine-ice correlations. *Journal of Quaternary Science*, **28**, 641–646.

BRENDRYEN, J., HAFLIDASON, H. & SEJRUP, H. P. 2010. Norwegian Sea tephrostratigraphy of marine isotope stages 4 and 5: prospects and problems for tephrochronology in the North Atlantic region. *Quaternary Science Reviews*, **29**, 847–864.

BUCK, C. E., HIGHAM, T. F. G. & LOWE, D. J. 2003. Bayesian tools for tephrochronology. *The Holocene*, **13**, 639–647.

CAGE, A. G., DAVIES, S. M., WASTEGÅRD, S. & AUSTIN, W. E. N. 2011. Identification of the Icelandic Landnám tephra (AD 871 $\pm$ 2) in Scottish fjordic sediment. *Quaternary International*, **246**, 168–176.

CARTER, R. M. 2005. A New Zealand climate template back to *c*. 3.9 Ma: ODP Site 1119, Canterbury Bight, south-west Pacific Ocean, and its relationship to onland successions. *Journal of the Royal Society of New Zealand*, **35**, 9–42.

CARTER, L., NELSON, C. S., NEIL, H. L. & FROGGATT, P. C. 1995. Correlation, dispersal, and preservation of the Kawakawa Tephra and other late Quaternary tephra layers in the southwest Pacific Ocean. *New Zealand Journal of Geology and Geophysics*, **38**, 29–46.

CARTER, L., MANIGHETTI, B., ELLIOT, M., TRUSTRUM, N. & GOMEZ, B. 2002. Source, sea level and circulation effects on the sediment flux to the deep ocean over the past 15 ka off eastern New Zealand. *Global and Planetary Change*, **33**, 339–355.

CARTER, L., ALLOWAY, B. V., SHANE, P. & WESTGATE, J. A. 2004. Deep-ocean record of major late Cenozoic rhyolitic eruptions from New Zealand. *New Zealand Journal of Geology and Geophysics*, **47**, 481–500.

CARTER, L., MANIGHETTI, B., GANSSEN, G. & NORTHCOTE, L. 2008. Southwest Pacific modulation of abrupt climate change during the Antarctic Cold Reversal–Younger Dryas. *Palaeogeography, Palaeoclimatology, Palaeoecology*, **260**, 284–298.

CASSIDY, M., TROFIMOVS, J., PALMER, M. R., WATT, S. F. L. & SYMONS, W. 2013. *Cores for Concern: the Difficulty of Recording Tephra Fallout Deposits in Marine Sediments*. Abstracts, IAVCEI 2013 Scientific Assembly (20–24 July), Kagoshima, Japan, 912.

CHUN, J.-H., CHEONG, D., IKEHARA, K. & HAN, S.-J. 2007. Age of the SKP-I and SKP-II tephras from the southern East Sea/Japan Sea: implications for interstadial events recorded in sediment from marine isotope stages 3 and 4. *Palaeogeography, Palaeoclimatology, Palaeoecology*, **247**, 100–114.

DAVIES, S. M., ELMQUIST, M., BERGMAN, J., WOHLFARTH, B. & HAMMARLUND, D. 2007. Cryptotephra sedimentation processes within two lacustrine sequences from west central Sweden. *The Holocene*, **17**, 319–330.

DAVIES, S. M., WASTEGÅRD, S., RASMUSSEN, T. L., SVENSSON, A., JOHNSEN, S. J., STEFFENSEN, J. P. & ANDERSEN, K. K. 2008. Identification of the Fugloyarbanki tephra in the NGRIP ice core: a key tie-point for marine and ice-core sequences during the last glacial period. *Journal of Quaternary Science*, **23**, 409–414.

DAVIES, S. M., WASTEGÅRD, S. *ET AL*. 2010. Tracing volcanic events in the NGRIP ice-core and synchronising North Atlantic marine records during the last glacial period. *Earth and Planetary Science Letters*, **294**, 69–79.

DAVIES, S. M., ABBOTT, P. M. *ET AL*. 2014. A North Atlantic tephrostratigraphic framework for 130 000 to 60 000 years b2k: new tephra discoveries, marine based-correlations, and future challenges. *Quaternary Science Reviews* (in press).

DUGMORE, A. J. 1989. Icelandic volcanic ash in Scotland. *Scottish Geographical Magazine*, **105**, 168–172.

DUGMORE, A. J., LARSEN, G., NEWTON, A. J. & SUGDEN, D. E. 1992. Geochemical stability of fine-grained silicic Holocene tephra in Iceland and Scotland. *Journal of Quaternary Science*, **7**, 173–183.

DUGMORE, A. J., LARSEN, G. & NEWTON, A. J. 1995. Seven tephra isochrones [sic] in Scotland. *The Holocene*, **5**, 257–266.

EDEN, D. N. & FROGGATT, P. C. 1988. Identification and stratigraphic significance of distal Aokautere Ash in three cores from eastern South Island, New Zealand. *In*: EDEN, D. N. & FURKERT, R. J. (eds) *Loess: Its Distribution, Geology and Soils*. A. A. Balkema, Rotterdam, 47–58.

FISHER, R. V. & SCHMINKE, H.-U. 1984. *Pyroclastic Rocks*. Springer, Berlin.

FROESE, D. G., ALLOWAY, B. V. & LOWE, D. J. 2008*a*. John A. Westgate – global tephrochronologist, stratigrapher, mentor. *Quaternary International*, **178**, 4–9.

FROESE, D. G., LOWE, D. J., KNOTT, J. & SLATE, J. L. 2008*b*. Global tephra studies: John Westgate and Andrei Sarna-Wojcicki commemorative volume. [Preface.] *Quaternary International*, **178**, 1–3.

FROGGATT, P. C. 1983. Toward a comprehensive Upper Quaternary tephra and ignimbrite stratigraphy in New Zealand using electron microprobe analysis of glass shards. *Quaternary Research*, **19**, 188–200.

FROGGATT, P. C. 1992. Standardization of the chemical analysis of tephra deposits. Report of the ICCT working group. *Quaternary International*, **13–14**, 93–96.

FROGGATT, P. C., NELSON, C. S., CARTER, L., GRIGGS, G. & BLACK, K. P. 1986. An exceptionally large late Quaternary eruption from New Zealand. *Nature*, **319**, 578–582.

GARDNER, J. V., NELSON, C. S. & BAKER, P. A. 1985. Distribution and character of pale green laminae in sediment from Lord Howe Rise: a probable late Neogene and Quaternary tephrostratigraphic record. *In*: KENNETT, J. P. & VON DER BORCH, C. C. (eds) *Proceedings of the Ocean Drilling Program, Initial Reports*, **90**. Ocean Drilling Program, Texas A & M University, College Station, TX, 1145–1159.

GEHRELS, M. J., LOWE, D. J., HAZELL, Z. J. & NEWNHAM, R. M. 2006. A continuous 5300-yr Holocene cryptotephrostratigraphic record from northern New Zealand and implications for tephrochronology and volcanic hazard assessment. *The Holocene*, **16**, 173–187.

GEHRELS, M. J., NEWNHAM, R. M., LOWE, D. J., WYNNE, S., HAZELL, Z. J. & CASELDINE, C. 2008. Towards rapid assay of cryptotephra in peat cores: review and evaluation of various methods. *Quaternary International*, **178**, 68–84.

GORBARENKO, S., CHEBYKIN, E. P., GOLDBERG, E. L., STEPANOVA, O. G. & LU, H. 2013. Chronicle of regional volcanic eruptions recorded in Okhotsk Sea sediments over the last 350 ka. *Quaternary Geochronology*, **20**, 29–38.

Gudmundsdóttir, E. R., Eiríksson, J. & Larsen, G. 2012. Holocene marine tephrochronology on the Iceland shelf: an overview. *Jökull*, **62**, 53–72.

Hall, V. A. & Pilcher, J. R. 2002. Late Quaternary Icelandic tephras in Ireland and Great Britain: detection, characterization and usefulness. *The Holocene*, **12**, 223–230.

Hall, M. & Hayward, C. 2014. Preparation of micro- and crypto-tephras for quantitative microbeam analysis. *In*: Austin, W. E. N., Abbott, P. M., Davies, S. M., Pearce, N. J. G. & Wastegård, S. (eds) *Marine Tephrochronology*. Geological Society, London, Special Publications, **398**. First published online February 27, 2014. http://dx.doi.org/10.1144/SP398.5

Hays, J. D., Imbrie, J. & Shackleton, N. J. 1976. Variations in the Earth's orbit: pacemaker of the ice ages. *Science*, **194**, 1121–1132.

Hayward, C. 2012. High spatial resolution electron probe microanalysis of tephras and melt inclusions without beam-induced chemical modification. *The Holocene*, **22**, 119–125.

Hayward, B. W., Sabaa, A. T. *et al.* 2011. Planktic foraminifera-based sea-surface temperature record in the Tasman Sea and history of the Subtropical Front around New Zealand, over the last one million years. *Marine Micropaleontology*, **82–83**, 13–27.

Hillenbrand, C.- D., Moreton, S. G. *et al.* 2008. Volcanic time markers for Marine Isotopic Stages 6 and 5 in Southern Ocean sediments and Antarctic ice cores: implications for tephra correlations between palaeoclimatic records. *Quaternary Science Reviews*, **27**, 518–540.

Hogg, A. G. & McCraw, J. D. 1983. Late Quaternary tephras of Coromandel Peninsula, North Island, New Zealand: a mixed peralkaline and calkalkaline tephra sequence. *New Zealand Journal of Geology and Geophysics*, **26**, 163–187.

Hogg, A. G., Higham, T. F. G., Lowe, D. J., Palmer, J., Reimer, P. & Newnham, R. M. 2003. A wiggle-match date for Polynesian settlement of New Zealand. *Antiquity*, **77**, 116–125.

Hogg, A. G., Lowe, D. J., Palmer, J. G., Boswijk, G. & Bronk Ramsey, C. J. 2012. Revised calendar date for the Taupo eruption derived by ^{14}C wiggle-matching using a New Zealand kauri ^{14}C calibration data set. *The Holocene*, **22**, 439–449.

Holt, K., Wallace, R. C., Neall, V. E., Kohn, B. P. & Lowe, D. J. 2010. Quaternary tephra marker beds and their potential for palaeoenvironmental reconstruction on Chatham Islands east of New Zealand, southwest Pacific Ocean. *Journal of Quaternary Science*, **25**, 1169–1178.

Holt, K. A., Lowe, D. J., Hogg, A. G. & Wallace, R. C. 2011. Distal occurrence of mid-Holocene Whakatane Tephra on the Chatham Islands, New Zealand, and potential for cryptotephra studies. *Quaternary International*, **246**, 344–351.

Housley, R. A., Lane, C. S., Cullen, V. L., Weber, M.-J., Riede, F., Gamble, C. S. & Brock, F. 2012. Icelandic volcanic ash from the Late-glacial open-air archaeological site of Ahrenshöft LA 58 D, north Germany. *Journal of Archaeological Science*, **39**, 708–716.

Hunt, J. B. & Hill, P. G. 1996. An inter-laboratory comparison of the electron probe microanalysis of glass geochemistry. *Quaternary International*, **34–36**, 229–241.

Hunt, J. B., Fannin, N. G. T., Hill, P. G. & Peacock, J. D. 1995. The tephrochronology and radiocarbon dating of North Atlantic, Late-Quaternary sediments: an example from the St. Kilda Basin. *In*: Scrutton, R. A., Stoker, M. S., Shimmield, G. B. & Tudhope, A. W. (eds) *The Tectonics, Sedimentation and Palaeoceanography of the North Atlantic Region*. Geological Society, London, Special Publications, **90**, 227–248.

Insinga, D. D., Tamburrino, S. *et al.* 2014. Tephrochronology of the astronomically-tuned KC01B deep-sea core, Ionian Sea: insights into the explosive activity of the central Mediterranean area during the last 200 ka. *Quaternary Science Reviews*, **85**, 63–84.

Kohn, B. P. 1970. Identification of New Zealand tephra layers by emission spectrographic analysis of their titanomagnetites. *Lithos*, **3**, 361–368.

Kohn, B. P., Pillans, B. J. & McGlone, M. S. 1992. Zircon fission track age for Middle Pleistocene Rangitawa Tephra, New Zealand: stratigraphic and paleoclimatic significance. *Palaeogeography, Palaeoclimatology, Palaeoecology*, **95**, 73–94.

Kristjánsdóttir, G. B., Stoner, J. S., Jennings, A. E., Andrews, J. T. & Grönvold, K. 2007. Geochemistry of Holocene cryptotephras from the North Iceland Shelf (MD99–2269): intercalibration with radiocarbon and paleomagnetic chronostratigraphies. *The Holocene*, **17**, 155–176.

Kuehn, S. C. & Froese, D. G. 2010. Tephra from ice – a simple method to routinely mount, polish, and quantitatively analyze sparse fine particles. *Microscopy and Microanalysis*, **16**, 218–225.

Kuehn, S. C., Froese, D. G., Shane, P. A. R., INTAV INTERCOMPARISON PARTICIPANTS 2011. The INTAV intercomparison of electron-beam microanalysis of glass by tephrochronology laboratories: results and recommendations. *Quaternary International*, **246**, 19–47.

Lacasse, C., Werner, R., Paterne, M., Sigurdson, H., Carey, S. & Pinte, G. 1998. Long range transport of Icelandic tephra to the Irminger Basin, Site 919. *In*: Saunders, A. D., Larsen, H. C. & Wise, S. W. Jr (eds) *Proceedings of the Ocean Drilling Program, Scientific Results*, **152**, Ocean Drilling Program, Texas A & M University, College Station, TX, 51–65.

Lackschewitz, K. S. & Wallrabe-Adams, H.-J. 1997. Composition and origin of volcanic ash zones in Late Quaternary sediments from the Reykjanes Ridge: evidence for ash fallout and ice-rafting. *Marine Geology*, **136**, 209–224.

Lane, C. S., Chorn, B. T. & Johnson, T. C. 2013. Ash from the Toba supereruption in Lake Malawi shows no volcanic winter in East Africa at 75 ka. *Proceedings of the National Academy of Sciences of the United States of America*, **110**, 8025–8029.

Lane, C. S., Cullen, V. L., White, D., Bramham-Law, C. W. F. & Smith, V. C. 2014. Cryptotephra as a dating and correlation tool in archaeology. *Journal of Archaeological Science*, **42**, 42–50.

Lawson, I. T., Swindles, G. T., Plunkett, G. & Greenberg, D. 2012. The spatial distribution of Holocene cryptotephras in north-west Europe since 7 ka: implications for understanding ash fall events from Icelandic eruptions. *Quaternary Science Reviews*, **41**, 57–66.

Lim, C., Toyoda, K., Ikehara, K. & Peate, D. W. 2013. Late Quaternary tephrostratigraphy of Baegdusan and Ulleung volcanoes using marine sediments in the Japan Sea/East Sea. *Quaternary Research*, **80**, 76–87.

Lowe, D. J. 1986. Controls on the rates of weathering and clay mineral genesis in airfall tephras: a review and New Zealand case study. *In*: Colman, S. M. & Dethier, D. P. (eds) *Rates of Chemical Weathering of Rocks and Minerals*. Academic Press, Orlando, FL, 265–330.

Lowe, D. J. 1988*a*. Stratigraphy, age, composition, and correlation of late Quaternary tephras interbedded with organic sediments in Waikato lakes, North Island, New Zealand. *New Zealand Journal of Geology and Geophysics*, **31**, 125–165.

Lowe, D. J. 1988*b*. Late Quaternary volcanism in New Zealand: towards an integrated record using distal airfall tephras in lakes and bogs. *Journal of Quaternary Science*, **3**, 111–120.

Lowe, D. J. 1990. Tephra studies in New Zealand: an historical review. *Journal of the Royal Society of New Zealand*, **20**, 119–150.

Lowe, D. J. 2008. Globalization of tephrochronology: new views from Australasia. *Progress in Physical Geography*, **32**, 311–335.

Lowe, D. J. 2011. Tephrochronology and its application: a review. *Quaternary Geochronology*, **6**, 107–153.

Lowe, D. J. & Hunt, J. B. 2001. A summary of terminology used in tephra-related studies. *Les Dossiers de l'Archeo-Logis*, **1**, 17–22.

Lowe, D. J., Hogg, A. G. & Hendy, C. H. 1981. *Detection of Thin Tephra Deposits in Peat and Organic Lake Sediments by Rapid X-radiography and X-ray Fluorescence Techniques*. Geology Department, Victoria University of Wellington Publications, **20**, 74–77.

Lowe, D. J., Tippett, J. M., Kamp, P. J. J., Liddell, I. J., Briggs, R. M. & Horrocks, J. L. 2001. Ages on weathered Plio-Pleistocene tephra sequences, western North Island, New Zealand. *Les Dossiers de l'Archeo-Logis*, **1**, 45–60.

Lowe, J. J., Blockley, S. P. E. *et al.* 2007. Age modelling of late Quaternary marine sequences in the Adriatic: towards improved precision and accuracy using volcanic event stratigraphy. *Continental Shelf Research*, **27**, 560–582.

Lowe, D. J., Shane, P. A. R., Alloway, B. V. & Newnham, R. M. 2008. Fingerprints and age models for widespread New Zealand tephra marker beds erupted since 30,000 years ago: a framework for NZ-INTIMATE. *Quaternary Science Reviews*, **27**, 95–126.

Lowe, D. J., Davies, S. M., Moriwaki, H., Pearce, N. J. G. & Suzuki, T. 2011. Enhancing tephrochronology and its application (INTREPID project): Hiroshi Machida commemorative volume. [Preface.] *Quaternary International*, **246**, 1–5.

Lowe, J. J., Barton, N. *et al.* 2012. Volcanic ash layers illuminate the resilience of Neanderthals and early modern humans to natural hazards. *Proceedings of the National Academy of Sciences of the United States of America*, **109**, 13 532–13 537.

Lowe, D. J., Blaauw, M., Hogg, A. G. & Newnham, R. M. 2013. Ages of 24 widespread tephras erupted since 30,000 years ago in New Zealand, with re-evaluation of the timing and palaeoclimatic implications of the Lateglacial cool episode recorded at Kaipo bog. *Quaternary Science Reviews*, **74**, 170–194.

Machida, H. & Arai, F. 1978. Akahoya Ash – a Holocene widespread tephra erupted from the Kikai caldera, south Kyushu. *The Quaternary Research (Japan)*, **17**, 143–163.

Manville, V. & Wilson, C. J. N. 2004. Vertical density currents: a review of their potential role in the deposition and interpretation of deep-sea ash layers. *Journal of the Geological Society, London*, **161**, 947–958.

Marcaida, M., Mangan, M. T., Vazquez, J. A., Bursik, M. & Lidzbarski, M. I. 2014. Geochemical fingerprinting of Wilson Creek formation tephra layers (Mono Basin, California) using titanomagnetite compositions. *Journal of Volcanology and Geothermal Research*, **273**, 1–14.

Marr, J. P., Carter, L., Bostock, H. C., Smith, E. & Bolton, A. 2013. Southwest Pacific Ocean response to a warming world: using Mg/Ca, Zn/Ca and Mn/Ca in foraminifera to track surface ocean water masses during the last deglaciation. *Paleoceanography*, **28**, 1–16.

Matthews, N. E., Smith, V. C., Costa, A., Durant, A. J., Pyle, D. M. & Pearce, N. J. G. 2012. Ultra-distal tephra deposits from super-eruptions: examples from Toba, Indonesia, and Taupo Volcanic Zone, New Zealand. *Quaternary International*, **258**, 54–79.

Matsu'ura, T., Ueno, T. & Furusawa, A. 2011. Characterization and correlation of cryptotephras using major-element analyses of melt inclusions preserved in quartz in last interglacial marine sediments, south-eastern Shikoku, Japan. *Quaternary International*, **246**, 48–56.

Matsu'ura, T., Furusawa, A. & Yanagida, M. 2012. Detection and correlation of widespread cryptotephras in middle Pleistocene loess in NE Japan using cummingtonite geochemistry. *Journal of Asian Earth Sciences*, **60**, 49–67.

Mew, G., Hunt, J. L., Froggatt, P. C., Eden, D. N. & Jackson, R. J. 1986. An occurrence of Kawakawa Tephra from the Grey River valley, South Island, New Zealand. *New Zealand Journal of Geology and Geophysics*, **29**, 315–322.

Naish, T. R. 2005. New Zealand's shallow-marine record of Pliocene–Pleistocene global sea-level and climate change. *Journal of the Royal Society of New Zealand*, **35**, 1–8.

Naish, T. R., Kamp, P. J. J., Alloway, B. V., Pillans, B. J., Wilson, G. S. & Westgate, J. A. 1996. An integrated tephrochronology and magnetostratigraphy for cyclothemic marine strata, Wanganui Basin: implications for the Pliocene–Pleistocene boundary in New Zealand. *Quaternary International*, **34–36**, 29–48.

Nelson, C. S. 1988. Revised age of a late Quaternary tephra at DSDP site 594 off eastern South Island and some implications for correlation. *Geological Society of New Zealand Newsletter*, **82**, 35–40.

Nelson, C. S., Hendy, C. H., Jarrett, G. R. & Cuthbertson, A. M. 1985*a*. Near-synchroneity of New Zealand alpine glaciations and Northern Hemisphere continental glaciations during the past 750 kyr. *Nature*, **318**, 361–363.

Nelson, C. S., Froggatt, P. C. & Gosson, G. J. 1985*b*. Nature, chemistry, and origin of late Cenozoic megascopic tephras in Leg 90 cores from the southwest Pacific. *In*: Kennett, J. P. & Von Der Borch, C. C. (eds) *Proceedings of the Ocean Drilling Program, Initial Reports*, 90. Ocean Drilling Program, Texas A & M University, College Station, TX, 1160–1173.

Nelson, C. S., Hendy, C. H., Cuthbertson, A. M. & Jarrett, G. R. 1986. Late Quaternary carbonate and isotope stratigraphy, subantarctic Site 594, southwest Pacific. *In*: Kennett, J. P. & Von Der Borch, C. C. (eds) *Proceedings of the Ocean Drilling Program, Initial Reports*, 90. Ocean Drilling Program, Texas A & M University, College Station, TX, 1425–1436.

Nelson, F. E., Wilson, G. S. & Neil, H. L. 2013. Marine magnetic signature of the Last Glacial Maximum and last deglaciation from the Southern Hemisphere mid-latitudes. *Marine Geology*, **346**, 246–255.

Newnham, R. M., Vandergoes, M. J. *et al.* 2012. Does the bipolar seesaw extend to the terrestrial southern mid-latitudes? *Quaternary Science Reviews*, **36**, 214–222.

Payne, R. J., Kilfeather, A. A., Van Der Meer, J. J. M. & Blackford, J. J. 2005. Experiments on the taphonomy of tephra in peat. *Suo*, **56**, 147–156.

Payne, R. J., Blackford, J. & Van Der Plicht, J. 2008. Using cryptotephras to extend regional tephrochronologies: an example from southeast Alaska and implications for hazard assessment. *Quaternary Research*, **69**, 42–55.

Pearce, N. J. G., Westgate, J. A., Perkins, W. T. & Preece, S. J. 2004. The application of ICP-MS methods to tephrochronological problems. *Applied Geochemistry*, **19**, 289–322.

Pearce, N. J. G., Denton, J. S., Perkins, W. T., Westgate, J. A. & Alloway, B. V. 2007. Correlation and characterisation of individual glass shards from tephra deposits using trace element laser ablation ICP-MS analyses: current status and future potential. *Journal of Quaternary Science*, **22**, 721–736.

Pearce, N. J. G., Westgate, J. A., Perkins, W. T. & Wade, S. C. 2011. Trace-element microanalysis by LA-ICP-MS: the quest for comprehensive chemical characterisation of single, sub-10 μm volcanic glass shards. *Quaternary International*, **246**, 57–81.

Pearce, N. J. G., Abbott, P. M. & Martin-Jones, C. 2014. Microbeam methods for the analysis of glass in fine-grained tephra deposits: a SMART perspective on current and future trends. *In*: Austin, W. E. N., Abbott, P. M., Davies, S. M., Pearce, N. J. G. & Wastegård, S. (eds) *Marine Tephrochronology*. Geological Society, London, Special Publications, **398**. First published online February 27, 2014. http://dx.doi.org/10.1144/SP398.1

Pillans, B. J. & Wright, I. 1992. Late Quaternary tephrostratigraphy from the southern Havre Trough-Bay of Plenty, northeast New Zealand. *New Zealand Journal of Geology and Geophysics*, **35**, 129–143.

Pillans, B. J., Kohn, B. P., Berger, G. W., Froggatt, P. C., Duller, G., Alloway, B. V. & Hesse, P. 1996. Multi-method dating comparison for mid-Pleistocene Rangitawa Tephra, New Zealand. *Quaternary Science Reviews*, **15**, 641–653.

Pillans, B. J., Alloway, B. V., Naish, T., Abbott, S. A., Westgate, J. A. & Palmer, A. 2005. The distal archive of North Island silicic volcanism recorded in Pleistocene shallow marine sediments of Wanganui Basin, New Zealand. *Journal of the Royal Society of New Zealand*, **35**, 43–90.

Ponomareva, V., Portnyagin, M. *et al.* 2013. Early Holocene M~6 explosive eruption from Plosky volcanic massif (Kamchatka) and its tephra as a link between terrestrial and marine paleoenvironmental records. *International Journal of Earth Sciences*, **102**, 1673–1699.

Preece, S. J., Westgate, J. A., Froese, D. G., Pearce, N. J. G. & Perkins, W. T. 2011. A catalogue of late Cenozoic tephra in the Klondike goldfields, Yukon. *Canadian Journal of Earth Sciences*, **48**, 1386–1418.

Pyne-O'Donnell, S. D. F. 2011. The taphonomy of Last Glacial–Interglacial Transition (LGIT) distal volcanic ash in small Scottish lakes. *Boreas*, **40**, 131–145.

Pyne-O'Donnell, S. D. F., Hughes, P. D. M. *et al.* 2012. High-precision ultra-distal Holocene tephrochronology in North America. *Quaternary Science Reviews*, **52**, 6–11.

Riede, F. & Thastrup, M. D. 2013. Tephra, tephrochronology and archaeology – a (re)view from northern Europe. *Heritage Science*, **1**, 1–17.

Robertson, S. M. & Mew, G. 1982. The presence of volcanic glass in soils on the West Coast, South Island, New Zealand. *New Zealand Journal of Geology and Geophysics*, **25**, 503–507.

Sandhu, A. S. & Westgate, J. A. 1995. The correlation between reduction in fission-track diameter and areal track density in volcanic glass shards and its application in dating tephra beds. *Earth and Planetary Science Letters*, **131**, 289–299.

Sarna-Wojcicki, A. 1976. Correlation of late Cenozoic tuffs in the central Coast Ranges of California by means of trace- and minor-element chemistry. US Geological Survey, Reston, VA, Professional Papers, **972**, 1–30.

Schwartz, M. A. 2008. The importance of stupidity in scientific research. *Journal of Cell Science*, **121**, 1771.

Self, S. & Sparks, R. S. J. (eds) 1981. *Tephra Studies*. Reidel, Dordrecht.

Shackleton, N. J. & Opdyke, N. D. 1973. Oxygen isotope and palaeomagnetic stratigraphy of equatorial Pacific core V28-238: oxygen isotope temperatures and ice volumes on a 10^5 and 10^6 year scale. *Quaternary Research*, **3**, 39–55.

Shane, P. A. R. 2005. Towards a comprehensive distal andesitic tephrostratigraphic framework for New Zealand based on eruptions from Egmont volcano. *Journal of Quaternary Science*, **20**, 45–57.

Shane, P., Black, T., Eggins, S. & Westgate, J. A. 1998. Late Miocene marine tephra beds: recorders of rhyolitic volcanism in North Island, New Zealand. *New Zealand Journal of Geology and Geophysics*, **41**, 165–178.

Shane, P. A. R., Sikes, E. L. & Guilderson, T. P. 2006. Tephra beds in deep-sea cores off northern New Zealand: implications for the history of Taupo Volcanic Zone, Mayor Island and White Island volcanoes. *Journal of Volcanology and Geothermal Research*, **154**, 276–290.

Shane, P. A. R., Nairn, I. A., Martin, S. B. & Smith, V. C. 2008. Compositional heterogeneity in tephra deposits resulting from the eruption of multiple magma bodies: implications for tephrochronology. *Quaternary International*, **178**, 44–53.

Smalley, I. J. 1980. Volcanic ash southern style. *Nature*, **286**, 841.

Smith, V. C. 2013. Volcanic markers for dating the onset of the Anthropocene. *In*: Waters, C. N., Zalasiewicz, J. A., Williams, M., Ellis, M. A. & Snelling, A. M. (eds) *A Stratigraphical Basis for the Anthropocene*. Geological Society, London, Special Publications, **395**. First published online November 21, 2013. http://dx.doi.org/10.1144/SP395.11

Smith, D. G. W. & Westgate, J. A. 1968. Electron probe technique for characterizing pyroclastic deposits. *Earth and Planetary Science Letters*, **5**, 313–319.

Smith, V. C., Shane, P. A. R. & Nairn, I. A. 2005. Trends in rhyolite geochemistry, mineralogy, and magma storage during the last 50 kyr at Okataina and Taupo volcanic centres, Taupo Volcanic Zone, New Zealand. *Journal of Volcanology and Geothermal Research*, **148**, 372–406.

Smith, V. C., Staff, R. A. *et al.* & SUIGETSU 2006 PROJECT MEMBERS 2013. Identification and correlation of visible tephras in the Lake Suigetsu SG06 sedimentary archive, Japan: chronostratigraphic markers for synchronising of east Asian/west Pacific palaeoclimatic records across the last 150 ka. *Quaternary Science Reviews*, **67**, 121–137.

Streeter, R., Dugmore, A. J. & Vésteinsson, O. 2012. Plague and landscape resilience in premodern Iceland. *Proceedings of the National Academy of Sciences of the United States of America*, **109**, 3664–3669.

Todd, J. A., Austin, W. E. & Abbott, P. E. 2014. Quantifying bioturbation of a simulated ash fall event. *In*: Austin, W. E. N., Abbott, P. M., Davies, S. M., Pearce, N. J. G. & Wastegård, S. (eds) *Marine Tephrochronology*. Geological Society, London, Special Publications, **398**. First published online February 5, 2014. http://dx.doi.org/10.1144/SP398.9

Tomlinson, E. L., Thordarson, T., Müller, W., Thirlwall, M. F. & Menzies, M. A. 2010. Micro analysis of tephra by LA-ICP-MS – strategies, advantages and limitations assessed using the Thorsmork ignimbrite (Southern Iceland). *Chemical Geology*, **279**, 73–89.

Topping, W. W. & Kohn, B. P. 1973. Rhyolitic tephra marker beds in the Tongariro area, North Island, New Zealand. *New Zealand Journal of Geology and Geophysics*, **16**, 375–395.

Turney, C. S. M. 1998. Extraction of rhyolitic component of Vedde microtephra from minerogenic lake sediments. *Journal of Paleolimnology*, **19**, 199–206.

Turney, C. S. M., Lowe, J. J. *et al.* 2004. Tephrochronology of Last Termination sequences in Europe: a protocol for improved analytical precision and robust correlation procedures (a joint SCOTAV–INTIMATE proposal). *Journal of Quaternary Science*, **19**, 111–120.

Vandergoes, M. J., Hogg, A. G. *et al.* 2013. A revised age for the Kawakawa/Oruanui tephra, a key marker for the Last Glacial Maximum in New Zealand. *Quaternary Science Reviews*, **74**, 195–201.

Walker, G. P. L. 1973. Explosive volcanic eruptions – a new classification scheme. *Geologische Rundschau*, **62**, 431–446.

Walker, G. P. L. 1983. Ignimbrite types and ignimbrite problems. *Journal of Volcanology and Geothermal Research*, **17**, 65–88.

Wastegård, S. & Davies, S. M. 2009. An overview of distal tephrochronology in northern Europe during the last 1000 years. *Journal of Quaternary Science*, **24**, 500–512.

Wastegård, S., Turney, C. S. M., Lowe, J. J. & Roberts, S. J. 2000. New discoveries of the Vedde Ash in southern Sweden and Scotland. *Boreas*, **29**, 72–78.

Walker, M. & Lowe, J. J. 2007. Quaternary science 2007: a 50-year retrospective. *Journal of the Geological Society, London*, **164**, 1073–1092.

Westgate, J. A. 1989. Isothermal plateau fission track ages of hydrated glass shards from silicic tephra beds. *Earth and Planetary Science Letters*, **95**, 226–234.

Westgate, J. A. & Fulton, R. J. 1975. Tephrostratigraphy of the Olympia interglacial sediments in southern British Columbia, Canada. *Canadian Journal of Earth Sciences*, **12**, 489–502.

Westgate, J. A. & Gorton, M. P. 1981. Correlational techniques in tephra studies. *In*: Self, S. & Sparks, R. S. J. (eds) *Tephra Studies*. Reidel, Dordrecht, 73–94.

Westgate, J. A., Perkins, W. T., Fuge, R., Pearce, N. J. G. & Wintle, A. G. 1994. Trace-element analysis of volcanic glass shards by laser ablation inductively coupled plasma mass spectrometry: application to tephrochronological studies. *Applied Geochemistry*, **9**, 323–335.

Westgate, J. A., Pearce, N. J. G., Perkins, W. T., Preece, S. J., Chesner, C. A. & Muhammad, R. F. 2013*a*. Tephrochronology of the Toba tuffs: four primary glass populations define the 75 ka Youngest Toba Tuff, northern Sumatra, Indonesia. *Journal of Quaternary Science*, **28**, 772–776.

Westgate, J. A., Naeser, N. D. & Alloway, B. V. 2013*b*. Fission-track dating. *In*: Elias, S. A. & Mock, C. J. (eds) *The Encyclopaedia of Quaternary Science*, 2nd edn. Elsevier, Amsterdam, **1**, 643–662.

Wilson, C. J. N. 1986. Pyroclastic flows and ignimbrites. *Science Progress (Oxford)*, **70**, 171–207.

Wilson, C. J. N. 1993. Stratigraphy, chronology, styles and dynamics of Late Quaternary eruptions from Taupo volcano, New Zealand. *Philosophical Transactions of the Royal Society of London Series A*, **343**, 205–306.

Wilson, C. J. N. & Walker, G. P. L. 1985. The Taupo eruption, New Zealand I. General aspects. *Philosophical Transactions of the Royal Society of London Series A*, **314**, 199–228.

Preparation of micro- and crypto-tephras for quantitative microbeam analysis

MIKE HALL* & CHRIS HAYWARD

School of Geosciences, University of Edinburgh, The King's Buildings, West Mains Road, Edinburgh EH9 3JW, UK

**Corresponding author (e-mail: mike.hall@ed.ac.uk)*

Abstract: The smallest tephra grains present special challenges in their preparation for quantitative microbeam chemical analysis by EPMA and LA-ICP-MS. High-quality polished surfaces are an essential pre-requisite for the collection of high quality quantitative chemical data, as these shards also present significant difficulties for analysis. A method for the preparation of tephra grains as small as 10–50 μm is described in detail. This method uses inexpensive, widely available materials and is easily implemented in facilities in which polished geological samples are prepared for microanalysis. Samples prepared in this way reduce or eliminate the difficulties in microbeam analysis associated with small grains, when appropriate analytical protocols are used.

High-quality chemical analysis of the smallest tephra grains (less than 50–100 μm across) is important in a wide range of palaeoclimate and environmental work (e.g. Abbott *et al.* 2012; Rea *et al.* 2012). These grains are sampled from marine and lake sediment cores, ice cores and peat bogs, and represent the most distal fallout from eruptions. In addition to their small size, they are usually present in very small quantities, sometimes fewer than 10 grains within a single sample. By virtue of their size and scarcity, such tephra samples present special challenges in their preparation for chemical analysis by the two methods most commonly used to identify specific tephra layers: electron probe microanalysis (EPMA) and laser ablation inductively-coupled plasma mass spectrometry (LA-ICP-MS). The finest-grained shards are non-trivial materials to analyse because of the very limited areas of glass they contain, through vesicularity and/or the presence of crystals, necessitating the development of specialized analytical protocols (Hayward 2012; Pearce *et al.* 2011. See also the reviews of Pearce *et al.* 2014). Without high-quality sample preparation, usable quantitative analyses cannot be obtained, and there is a significant risk of damage or loss of shards during the preparation process. For small populations of shards, it is essential to be able to analyse *every* shard in order to obtain the most statistically reliable representation of the individual layer for identification against existing geochemical data, requiring preparation methods that, ideally, eliminate the loss of material and ensure that all grains are exposed at the final polished surface. The following paper makes recommendations of ways of storing and preparing very fine-grained tephras for quantitative microbeam analysis by EPMA and LA-ICP-MS. The method uses inexpensive and widely available materials, and should be easily implemented within any facility in which thin sections or similar types of sample preparation are carried out.

Storage and transport

It is necessary to minimize the amount of handling of samples comprising very small and/or very limited numbers of grains, as there is risk of the loss of material each time the sample is manipulated, transferred between containers, etc. The best type of container for storage and transportation of fine-grained tephras is wide and shallow, because this best facilitates access to the shards for removal from the container. Dimensions similar to 25 mm in diameter and depth are suitable. Plastic containers have the advantage of leading to slight static charging of the interior surfaces, which causes the shards to become 'fixed' to these surfaces, thereby making it easier to remove the shards. It is best to transport samples in a dry condition. Where the sample comprises a small number of grains, say fewer than a few tens, that number should be indicated on the container or associated documentation. The range of grain sizes must be indicated, to enable grains to be placed in groups as close as possible in size. This maximizes, during preparation, the area of each shard that is exposed in the final polished surface. Samples should be washed in petroleum spirit in an ultrasonic bath for a minimum of 10 min to remove any dust or residue from any chemical separation process, which may reduce adhesion of the resin to the sample.

From: Austin, W. E. N., Abbott, P. M., Davies, S. M., Pearce, N. J. G. & Wastegård, S. (eds) 2014. *Marine Tephrochronology*. Geological Society, London, Special Publications, **398**, 21–28.
First published online February 27, 2014, updated August 29, 2014, http://dx.doi.org/10.1144/SP398.5

Preparation of polished samples

The method relies on the properties of silicone rubber sheet. Its flexibility, elasticity and relatively non-stick surface make it easy to remove the cured resin from the mould without damage, and its inherent flatness reduces the amount of grinding required, and critically keeps shards within a narrow range of depth so that most or all will appear at the final polished surface. Other workers have used extremely flat, graphite substrates to achieve the essential level surface (Kuehn & Froese 2010).

Materials used

(1) *Silicone rubber sheet.* The type used here is 1.5 mm thick, and cut to fit on a flat glass plate 100 mm square. The rubber sheet is attached to a glass plate with double-sided adhesive tape to provide support and maintain flatness, which is vital for keeping the maximum number of tephra grains within the plane of the final polished surface.

(2) *Double-sided adhesive tape.* Many types are available, the best for this method being a smooth non-reinforced tape with a thin, not too aggressive adhesive layer which is degradable in water, petroleum spirit or isopropanol. Tape of suitable qualities is available from hardware stores as smooth, double-sided carpet-floor tape.

(3) *Epoxy resin.* A low-viscosity epoxy resin such as Buehler 'Epo-Thin' and the corresponding hardener works very well. The low viscosity maximizes wetting of the complex tephra grain surfaces and helps reduce the frequency of air bubble formation. It is also less prone to forming a tacky surface when cured in air. The lack of a tacky surface is important for good adhesion between the resin holding the tephra onto the rubber sheet and the bulk of the resin block.

(4) *Ring moulds.* One-inch diameter moulds such as Buehler 'SampleKup' two-part moulds are used. The base of the mould is not used. The ring has a flanged base to give a large area to stick to the rubber sheet. The mould is used with Buehler release agent, which ensures easy release of the finally cured resin disc, without damage, from the mould. The imperial measurement mould is compatible with metric sample holders supplied by electron probe microanalyser manufacturers such as Cameca and Jeol, whose available standard sample holders include examples with circular spaces of 26 mm diameter.

(5) *Polishing cloth.* A medium hardness, non-woven cloth should be used, such as Kemet PSU-M. A cloth with a woven texture, or a visible pile or nap, or too soft a cloth is more liable to create relief during polishing, as will an unused non-woven cloth. A new cloth should be avoided as it will produce an excessive amount of relief. A new cloth should be broken in by polishing blank resin blocks or glass slides for 10–20 min with the same 0.3 μm aluminium oxide slurry as is used for polishing the samples.

The method can be broken down into three steps: (a) mounting the samples; (b) filling with resin and curing; and (c) grinding and polishing. A summarized flow diagram of the method is shown in Figure 1.

Step 1. Mounting the samples

(1) On a glass plate 100 × 100 mm, stick two strips of 50 mm-wide double-sided adhesive tape side by side. Avoid bubbles or gaps or overlaps between the two strips of tape (Fig. 2).

(2) Remove the backing paper from the tape. Place a sheet of silicon rubber 100 × 100 mm, smoothest side up, onto the tape. Start at an edge and carefully roll the silicon rubber sheet over the tape. Press down well. Mark out four or six circles 1 inch in diameter with a pencil or ballpoint pen using the ring mould as a guide, ensuring enough space is left between them for the flanges of the moulds to be stuck down later.

(3) Mix a small amount of epoxy resin. The useful working time is about 10 min. If more than one sample fraction is being mounted in the resin block, dividers cut from a bristle taken from a nylon scrubbing brush or similar can be stuck to the rubber sheet with a drop of resin at each end of the bristle, to divide the sample area into four or more sectors (Fig. 3). The dividers do not need to be stuck to the rubber along their full length as they will always be visible in the body of the resin. The nylon bristle polishes at the same rate as the epoxy resin, thus maintaining surface flatness.

(4) A tool for manipulating the tephra grains can be made from a single hair, taken from a fine watercolour paint brush and glued into a hole made in the end of a matchstick. After several uses the hair will become clogged with resin and should be discarded.

(5) Place a single drop of resin onto the rubber sheet, near the centre of the sample area, with a mounted needle or wooden toothpick, and spread out as thinly as possible to cover an

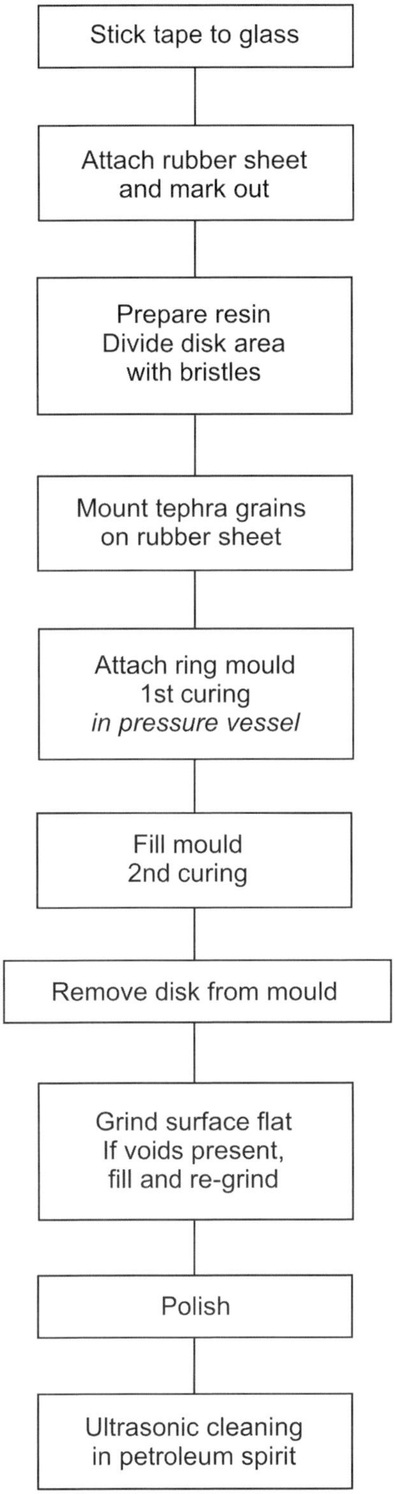

Fig. 1. Summary flow diagram of the method.

area of several square millimetres. Only a trace of resin is required, it should not be so deep as to cover the grains, just enough to hold them in place. Dip the end of the mounted hair in resin, and wipe most of it off, leaving just a trace on the hair. The hair should be just tacky enough to hold a tephra grain, without coating it in resin. Pick up a single tephra grain and transfer to the resin smear on the rubber sheet. Grains should be picked up using the very tip of the hair. Continue mounting further grains, refreshing the resin on the hair as required. Mark sample reference numbers on the rubber sheet near the samples, but outside the diameter of the ring mould. The orientation of each grain, if possible, should aim to maximize the area that will be exposed at the final polished surface. Arrangement of the grains in a compact, regular array will enable easy location of samples when observing under relatively restricted ranges of magnification during analysis (Fig. 3).

(6) When all the samples are in place, run a thin bead around the base of the mould (previously treated with mould release agent to ensure easy removal of the cured block) and position on the rubber sheet around the samples. The samples should be central within the mould ring (Fig. 4). Leave the resin to cure overnight. Curing in a pressure vessel at approximately 2 bar is essential to minimize the occurrence of air bubbles in the resin. If a pressure vessel is not available, bubbles may form on the rubber sheet or on the tephra grains, which will need to be filled after curing as described in step 3 paragraph 8. Any voids or bubbles breaking through the prepared surface will trap debris from grinding, which is difficult to remove, even with ultrasonic cleaning. This debris from preceding preparation stages will contaminate later stages, and may lead to deep scratches during polishing. Bubbles adjacent to tephra grains may also cause the grains to be plucked out of the resin, especially during earlier preparation stages, with the danger of loss of material – especially problematic where sample sizes are very small.

Step 2. Filling with resin and curing

(1) When the holding resin has cured, top up the mould with fresh resin. Using 4 g resin in a 1 inch mould gives a finished block thickness of around 7 mm. Using a disposable pipette, introduce the resin near the edge of the mould and let it flow over the mounted grains. Cure under 2 bar pressure overnight. When

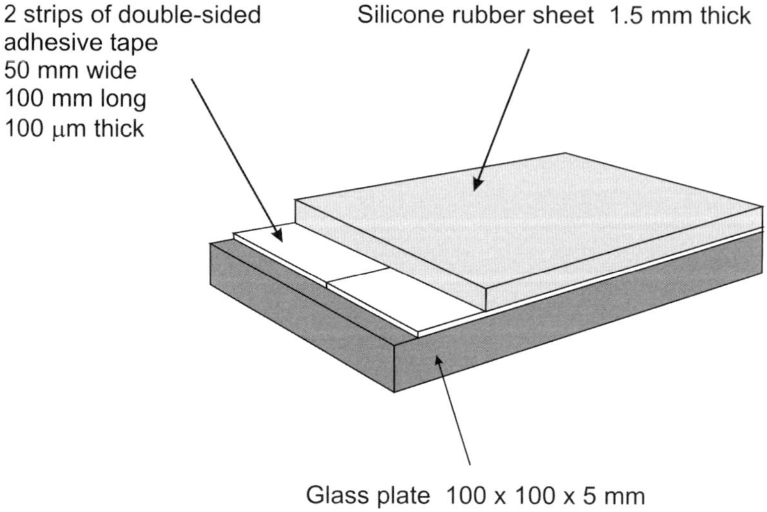

Fig. 2. Diagram showing the fixing with double-sided tape of the silicone rubber sheet onto the supporting glass plate.

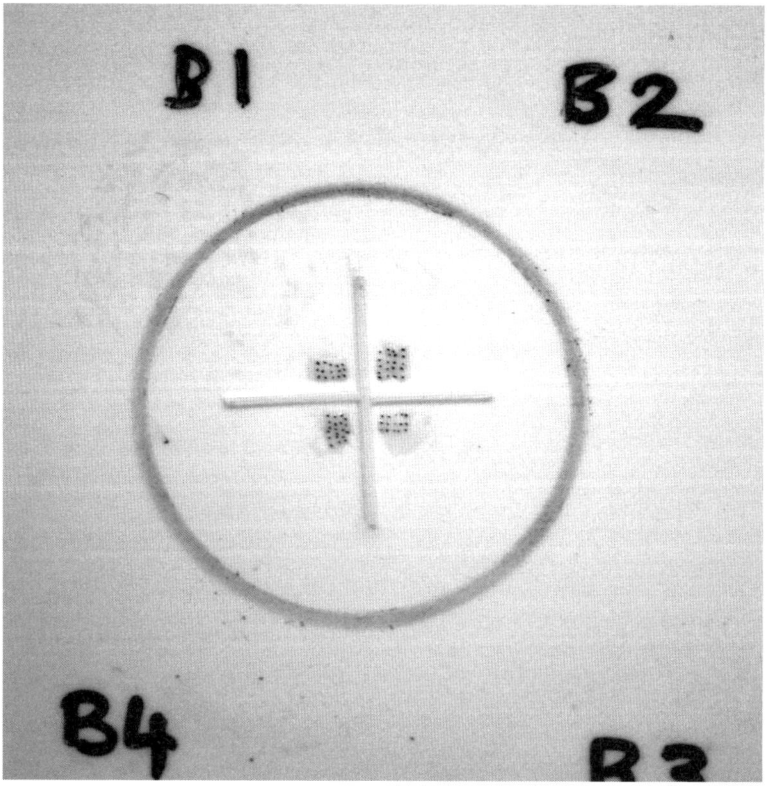

Fig. 3. Tephra shards placed in a regular array onto small areas of resin applied to the silicone rubber sheet. The four samples are separated by nylon bristles. Note that for illustrative purposes, relatively large (125–250 μm size fraction) are shown in all figures.

Fig. 4. The mould after covering of the shards with a small volume of epoxy resin.

cured, transfer the sample numbers to the top of the resin block using a permanent marker. To remove the blocks squeeze the sides of the mould to break the adhesion between the mould and the resin. This is easier if the mould ring is no more than half filled. Peel the rubber sheet away from the adhesive tape and glass plate at one corner. Pulling and stretching the rubber sheet will break the bond between the mould flange and the rubber sheet and release the mould ring, leaving the resin block attached to the rubber sheet. The rubber sheet will then separate easily from the adhesive tape and the resin block can be removed from the rubber sheet (Fig. 5). If the ring mould does not separate easily, soaking in warm water for 10–15 min or more will help soften the tape adhesive and aid removal. Alternatively, a little solvent such as petroleum spirit or isopropanol can be introduced between the silicone rubber sheet and the tape. Excessive force should not be applied to the glass plate during removal in case it breaks and causes injury. Once removed, leave the resin blocks to fully cure. Although the resin is hard enough to handle, full cure of epoxy resin takes about 72 h and no grinding or polishing should be attempted within this period. When fully cured, mark sample numbers on the block using an engraving tool or scriber. The appearance of the surface after removal of the resin disc from the mould is shown in Figure 6a.

Step 3. Grinding and polishing

(1) Grind the sharp edge of the meniscus from the back surface of the resin block using 800 or 1200 grit silicon carbide paper and water on a flat glass plate. Leave the edge of the block as a sharp right angle. Do not grind a chamfer – this is not necessary for sample preparation and will complicate placing of the finished disc in instrument sample holders.

Any surface voids can be filled by placing a drop of resin next to the hole using a mounted hair, and dragging the resin down one side of the void. Do not place a drop of resin directly over the void, as it will not be able to displace air beneath it. Leave slightly proud and leave to cure. As the volume of resin is so small, an overnight cure is sufficient. If the hole only just breaks the surface, or is so close to the surface that breakthrough seems inevitable during grinding, the entry to the hole can be enlarged by careful use of the point of a scalpel blade. The situation to avoid is where a spherical bubble is only just exposed, giving a relatively large volume to fill through a small aperture. Once cured, grinding for a few seconds on 2500 grit paper is usually enough to restore a plane surface to the block.

The analysis face of the block should need very little grinding. The amount of material to be removed is determined by the shape of the samples. Tephra fragments typically show conchoidal fracture and rest on the base of the mould on two or three points of contact; grinding slightly deeper than these exposed points will expose a larger area of the sample. The greater the range of grain size within each block is, the more difficult the optimal exposure of all the grains becomes. Grinding should be done with water on 2500 grit silicon carbide paper on a flat glass plate. Hold the block with a thumb and two fingers and move the block over the paper in a figure-of-eight or a circular motion, using the full surface area of the paper. The pressure applied should be equivalent to a load of 150–200 g. Initially make five or 10 circles or figure-of-eight movements of the sample over the abrasive paper, making sure to rotate the disc by approximately 90° every few movement cycles, so as to ensure maintenance of a level surface. Check the appearance under a binocular microscope with illumination at a low angle to the prepared surface. The resin should appear uniform and the tephra grains will be more reflective. Continue until the best exposure of grains is achieved. Any grains that remain below the surface can exposed at a later date, if necessary, once the currently exposed material has been analysed. The appearance of the surface after grinding is shown in Figure 6b.

(2) Polish with 0.3 μm aluminium oxide powder mixed 1:2 parts by volume with de-ionized water on a Kemet PSU-M cloth or equivalent. Wet the polishing cloth with water and leave to soak for 10 min. With the polisher running, apply five or six drops of the polishing slurry to the centre of the cloth. Stop the polisher and load the sample blocks. Apply a load of 500 g to each block and run at a speed of 200 rpm. During polishing add de-ionized water at a rate of three or four drops per minute to the cloth, and refresh the slurry by one to two drops per minute.

Fig. 5. The mould and sample after removal from the silicone rubber sheet.

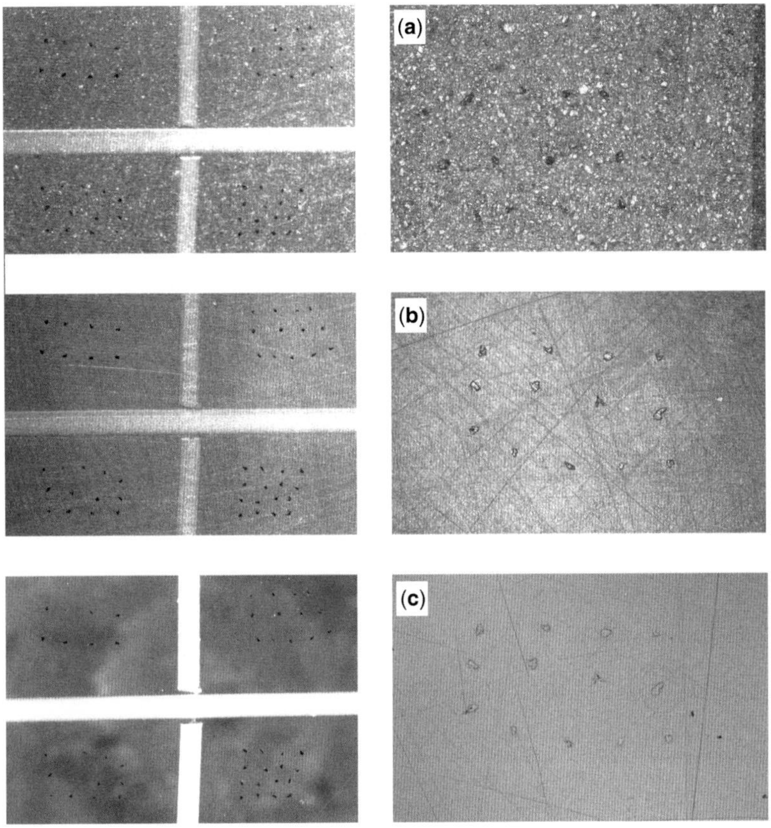

Fig. 6. (**a**) The surface after removal from the mould. (**b**) The surface after the first grinding stage. Note the slightly increased reflectivity of the shards relative to the epoxy resin. (**c**) The surface after final polishing.

Polish for 2 min, rinse with water, dry and examine. The final polished surface should be mirror-like across its entire area, and the shards in the same plane as the resin – that is, without significant relief. Repeat if required. It is very important to polish for only the minimum time needed to produce the required polished surface. Excessive polishing will cause relief on the surface of the tephra grains – rounding of the surfaces. Development of minor relief cannot always be prevented, but efforts to minimize it are essential, since EPMA requires flat, horizontal surfaces for quantitative analysis and rounding of grain surfaces effectively reduces the surface area available for analysis, potentially reducing the number of shards available for analysis. If excessive polishing relief is found, the blocks must be lightly re-ground on 2500 grit paper to remove all the relief before re-polishing. This is obviously possible only if grains are a suitable thickness beneath the polished surface. For the smallest tephra grains it is necessary to get it right first time.

Diamond polishing compounds can be used instead of aluminium oxide. Use the same type of cloth as for aluminium oxide polishing. A separate cloth is required for each grade of diamond compound. Polishing times and pressure are as for aluminium oxide polishing. Apply a few drops of polishing compound to the cloth at the start of polishing and add a few drops of water-based lubricant at intervals during polishing. The polishing cloth should be kept damp, but not overly wet. Only a little liquid should be flung off the edge of the rotating polishing plate. Successively finer grades are used, starting with 3 or 1 μm and finishing with 0.25 μm. Polish for 2–3 min at a time with each grade, checking frequently under a microscope, until the entire surface is well polished. The appearance of the final polished surface is shown in Figure 6c.

Extra care should be taken to thoroughly clean the samples and the polishing machine when changing diamond grades, to avoid contamination of the finer polishing cloths with coarser compounds.

Post-preparation

Abrasive particles always remain on the polished surface and within any pores exposed, such as bubbles in the resin and vesicles not impregnated by resin. Particles of alumina must be removed by ultrasonic cleaning in de-ionized water *immediately* after completion of this stage of polishing, or they can adhere to the surface and be impossible to remove other than through additional polishing, which may lead to the formation of excessive surface relief. Diamond abrasives, even most of those marketed as being aqueous, include an oil-based component. This will volatalize under electron bombardment during EPMA or SEM observation, obscuring the surface and probably leading to puncturing of the carbon coat and thus to charging of the surface and errors in chemical analysis. Oil-based abrasive remnants and associated lubricants are best removed by ultrasonic cleaning in AnalaR petroleum spirit for 10 min. This solvent causes no damage to fully cured epoxy resins. Ketone-based solvents such as acetone must not be used as they are solvents of epoxy resin and can cause swelling and gelatinization of the polished sample surface, rendering it useless for microbeam analysis. Ultrasonic cleaning causes mild heating of the petroleum spirit (it will feel slightly warm through nitrile protective gloves after 10 min of cleaning). The boiling range of petroleum spirit is 60–80 °C. It is *essential* that the ultrasonic cleaner is placed beneath a *metal* fume hood with suitable dimensions and extraction flow rate. *Under no circumstances* should any heater associated with the ultrasonic cleaner be activated during the cleaning as this will inevitably cause ignition of the solvent.

Conclusion

The method described above will enable tephra grains as small as 10–50 μm to be prepared with highly polished and low-relief surfaces. This will enable high-quality quantitative analysis by EPMA and LA-ICP-MS, and make accessible for chemical characterization even the most distal tephras, thereby maximizing the range of tephra layers that can be identified, and the range of depositional environments and locations from which tephras may be studied.

References

Abbott, P. M., Davies, S. M. *et al.* 2012. A detailed framework of Marine Isotope Stages 4 4 and 5 volcanic events recorded in two Greenland ice-cores. *Quaternary Science Reviews*, **36**, 59–77.

Hayward, C. L. 2012. High spatial resolution electron probe microanalysis of tephras and melt inclusions without beam-induced chemical modification. *The Holocene*, **22**, 119–125.

Kuehn, S. C. & Froese, D. G. 2010. Tephra from ice – a simple method to routinely mount, polish, and quantitatively analyze sparse fine particles. *Microscopy and Analysis*, **16**, 218–225.

Pearce, N. G., Perkins, W. T., Westgate, J. A. & Wade, S. C. 2011. Trace element microanalysis by LA-ICP-MS: the quest for comprehensive chemical characterisation of single, sub-10 μm volcanic glass shards. *Quaternary International*, **246**, 57–81.

Pearce, N. J. G., Abbott, P. M. & Martin-Jones, C. 2014. Microbeam methods used for the analysis of glass in tephra deposits: a SMART perspective on current and future trends. *In*: Austin, W. E. N., Abbott, P. M., Davies, S. M., Pearce, N. J. G. & Wastegård, S. (eds) *Marine Tephrochronology*. Geological Society, London, Special Publications, **398**. First published online February 27, 2014, http://dx.doi.org/10.1144.SP399.1

Rea, H. A., Swindles, G. T. & Roe, H. M. 2012. The Hekla 1947 tephra in the north of Ireland: regional distribution, concentration and geochemistry. *Journal of Quaternary Science*, **27**, 425–431.

Microbeam methods for the analysis of glass in fine-grained tephra deposits: a SMART perspective on current and future trends

NICHOLAS J. G. PEARCE[1]*, PETER M. ABBOTT[2] & CATHERINE MARTIN-JONES[1]

[1]*Department of Geography and Earth Sciences, Aberystwyth University, Aberystwyth SY23 3DB, UK*

[2]*Department of Geography, College of Science, Swansea University, Singleton Park, Swansea SA2 8PP, UK*

**Corresponding author (e-mail: Nick.Pearce@aber.ac.uk)*

Abstract: Correlation of tephra deposits frequently relies on the analysis of glass shards separated from their host. Small shards from distal deposits (marine-, ice- or lake-cores, peat bogs) are difficult to analyse. Here current methods for glass shard analysis from (particularly) marine tephra deposits are reviewed. These methods apply equally to other distal deposits, where linking repositories of climatic information is central to many research programmes. Electron probe microanalysis is used widely to determine the major element compositions of volcanic glass. However, electron beam irradiation causes permanent damage to glass, especially hydrated or silica- and alkali-rich compositions. Recent developments have shown that reliable data is obtained with beam diameters >3 μm on basaltic or moderately hydrated rhyolitic glass, provided low electron beam current densities are used. For robust correlation of tephra deposits from different environments, it is recommended that analytical data are normalized to an anhydrous basis. Glass trace element analysis is commonly performed by laser ablation inductively coupled plasma mass spectrometry, which determines approx. 30 elements at <1 ppm from ablation craters 10 μm diameter, and although element fractionation occurs, it can be corrected. Recent developments in both methods should facilitate analysis of smaller, more distal material, expanding the geographical range over which reliable tephra correlations can be achieved.

Attempts to integrate climate records from diverse environments have, in recent years, seen an expansion in the use of tephra horizons in stratigraphic correlation. Tephra, the product of explosive volcanic activity, forms essentially isochronous markers and these have become particularly important in many aspects of Quaternary research. In particular, projects such as INTIMATE (Integration of Ice-core Marine and Terrestrial records), SMART (Synchronizing Marine and Ice-core Records using Tephrochronology), RESET (Response of Humans to Abrupt Environmental Transitions) and TRACE (Tephra Constraints on Rapid Climate Events) have all sought to correlate disparate palaeoclimatic records using tephra horizons as independent chronostratigraphic tie-lines (see Blockley *et al.* 2012; Davies *et al.* 2012). In marine- and ice-core and terrestrial records, the presence of discrete tephra layers is invaluable in correlating and reconciling the chronologies of these records and these can, for example, also help in determining marine reservoir effects for radiocarbon age determinations (see Ascough *et al.* 2005; Lowe 2011) and reconstructing volcanic history as they may preserve a more complete record than proximal sequences.

The correlation of tephra deposits between disparate sequences relies heavily on the matching of chemical characteristics of individual phases, most notably glass shards, within deposits. In this contribution, the current analytical methods applicable to the compositional analysis (major and trace elements) of relatively fine-grained tephric material recovered from marine-cores are described, and possible future directions are highlighted. While tephras deposited in the marine environment provide some particular challenges for analysis, many of the approaches described here apply equally to fine-grained material recovered from tephra deposits in ice-cores or cryptotephra deposits in terrestrial sequences.

Marine tephra deposits

Tephra is typically recovered from the marine environment within sediment-cores. Tephra deposits may be present within these records as visible layers up to tens of centimetres in thickness in relatively proximal localities (Alloway *et al.* 2005; Albert *et al.* 2012; Salisbury *et al.* 2012) from which it is straightforward to separate material for analysis, even if the material is comparatively fine-grained. Salisbury *et al.* (2012), for example, describe a 5 cm-thick layer of tephra with an average

From: Austin, W. E. N., Abbott, P. M., Davies, S. M., Pearce, N. J. G. & Wastegård, S. (eds) 2014. *Marine Tephrochronology*. Geological Society, London, Special Publications, **398**, 29–46.
First published online February 27, 2014, http://dx.doi.org/10.1144/SP398.1

grain size of <30 μm in marine-cores from near to the Sunda Trench off Sumatra, which they correlate with the Youngest Toba Tuff. In such proximal settings, easily recognized deposits are likely to be relatively common, but with distance from volcanic sources, tephra deposits will become thinner and less obvious in marine-cores, as is exemplified by material from marine-cores in the North Atlantic, which store records of relatively small volume Icelandic volcanic activity, commonly as cryptotephras. Even tephra horizons containing coarse tephra particles (>150 μm) may not create a visible expression within a core, and many such deposits have been recognized within many marine studies/sequences, in particular in the North Atlantic (van Kreveld *et al.* 2000; Wastegård & Rasmussen 2001; Austin *et al.* 2004). Because shard concentrations are low in these deposits, they are classed as cryptotephras: their identification results from the sieving of marine sediments at >150 μm to isolate material for foraminifera and ice-rafted debris studies. These cryptotephras are difficult to recognize within the cores, containing only a small amount of material which may be dispersed and diluted over several centimetres of the core by processes such as bioturbation (see for example Todd *et al.* 2014) and bottom current reworking, and this problem is exacerbated away from source as grain sizes decrease. Thus the definition of these tephra horizons, and the separation of material from these for analysis, is both complex and time-consuming. Coarse tephra shards can be picked out of the sieved fraction prior to mounting for geochemical analysis (see below).

The recognition of many fine-grained cryptotephras within terrestrial sequences (e.g. Davies *et al.* 2005; Pyne-O'Donnell *et al.* 2008; Lane *et al.* 2012*a*) has led to the application of additional techniques to identify further cryptotephras within marine sequences. This has been achieved both through the inspection and quantification of less coarse fractions (>63 μm) during the sieving of samples (e.g. Brendryen *et al.* 2010) and the application of the density separation technique (Turney 1998; Blockley *et al.* 2005). Within the latter method the heavy liquid sodium polytungstate has been used to isolate fine-grained tephra material (25–80 μm) within specific density fractions (e.g. Lowe *et al.* 2007; Bourne *et al.* 2010; Abbott *et al.* 2011; Abbott *et al.* 2013). In both instances tephra concentrations are typically expressed as shards per gram dry weight of sediment (g.d.w.) with concentrations within a gram of sediment ranging from thousands to tens of shards. Average grain sizes from these marine cryptotephra horizons are small, with some only composed of material with diameters <80 μm. From this it can be calculated that a deposit containing 1000 shards/g.d.w. of 80 μm diameter would yield far less than 500 μg of glass/g.d.w. sample. In most sediment-cores, there is insufficient material available to yield enough sample for bulk analysis, which requires some 2–4 orders of magnitude more mass than is present in 1 g.d.w. of these cryptotephras. Even if this material were present, to fully isolate a sufficient mass of a 'pure' glass separate for bulk analysis would be prohibitively and excessively time-consuming, and essentially impossible for many tephras which contain significant proportions of crystals. Typical glass fragments recovered from such marine-cores, and for comparison, some from Greenlandic ice-cores are illustrated in Figure 1, showing the range of sizes of material which need to be prepared for microbeam analysis.

Because of the small amounts of material in marine cryptotephra deposits, routine bulk analysis (e.g. X-ray fluorescence (XRF) and solution inductively coupled plasma (ICP) methods, see Pearce *et al.* 2004) would be inappropriate in most cases. That is not to say that bulk methods cannot be applied to these sorts of deposits, and some studies use bulk methods for cryptotephra analysis. Lim *et al.* (2008), for example, used instrumental neutron activation analysis to detect and then subsequently analyse picked material from cryptotephras within marine-cores from the Japan Sea, but this application was made easier by the alkaline nature of many of the target tephras (from Ulleung Island), providing a strong contrast in chemistry from the host sediments. Similarly, a strong compositional contrast has allowed XRF core scanning to be applied successfully to the recognition of basaltic cryptotephra layers (<1000 shards cm^{-3}) in gyttja-rich sediments in the Faroe Islands (Kylander *et al.* 2012). It is, however, unlikely that these chemical applications would be successful in locating diffuse basaltic or rhyolitic cryptotephra (a few hundred shards per cubic centimetre) in silicicalstic sediments such as those found in the marine realm, because of the lack of compositional contrast and the dilution of the tephra component. Their recognition often relies on the labour-intensive density separation methods described above.

Because of the small volume of material often recovered from distal marine-cores (and equally ice-core or terrestrial cryptotephra deposits such as peat-core sequences), coupled with its typically fine grain size, microbeam analytical methods which offer grain-specific analyses at high spatial resolution are the most appropriate for the analyses of these small volume samples. Below, methods for the preparation of glass microshards from marine cryptotephra for microbeam analysis are described, before the current status of major and trace element analytical methods are reviewed.

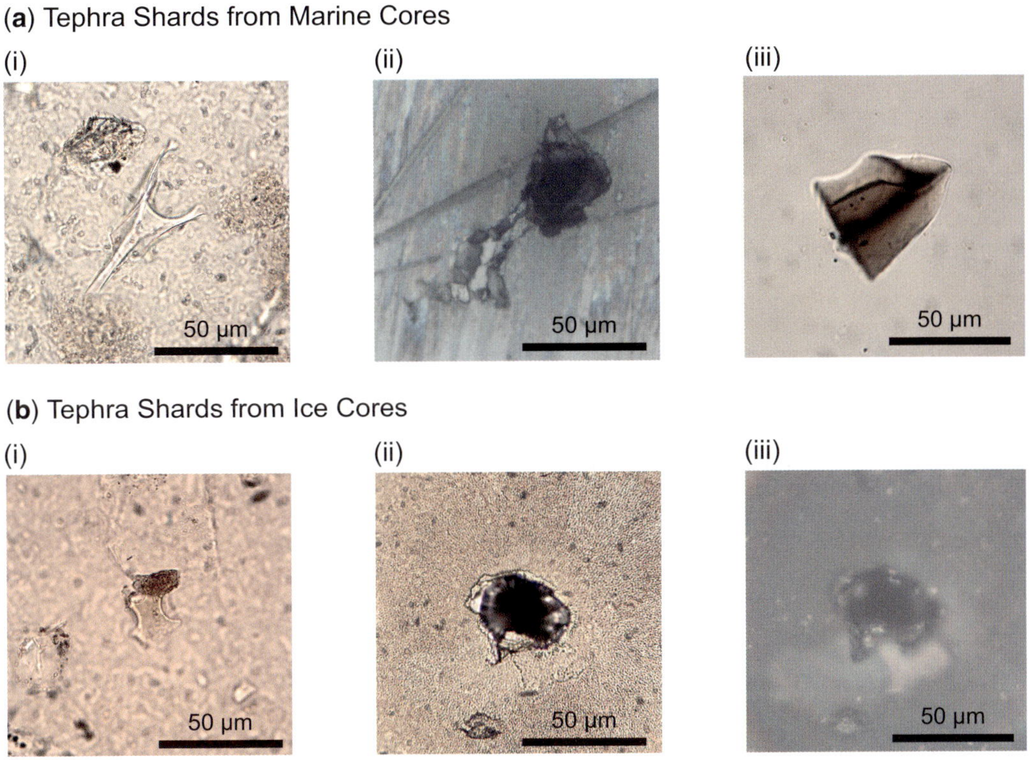

Fig. 1. Examples of tephra shards from within North Atlantic marine and Greenland ice-core cryptotephra horizons. (**a**) (i) rhyolitic shard; (ii) LA-ICP-MS analysis crater in a rhyolitic shard; (iii) basaltic shard. (**b**) (i) basaltic shard; (ii) LA-ICP-MS crater in a basaltic shard (transmitted light); and (iii) LA-ICP-MS crater in a basaltic shard (reflected light).

Preparation of cryptotephras for microbeam analysis

The requirement for a flat sample for electron probe microanalysis (EPMA) poses some challenges for tephra studies in the preparation of material. Where there is an abundance of glass shards from even relatively modest tephra deposits, preparation of grain mounts encapsulated in resin blocks and their subsequent polishing is relatively straightforward (see Lowe 2011, p. 124). However, where smaller amounts of material are available, particularly where the glass is present as microshards from cryptotephras, extreme care is required to produce a mount which is appropriate for analysis, and various methods of sample preparation can be employed. For example, in the EPMA analysis of tephra separated from ice-cores, Kuehn & Froese (2010) recently described a method to produce resin blocks on polished graphite planchettes, which required only a very gentle polish, ensuring fine shards were not ground or polished away. Hall & Hayward (2014) also describe methods for the preparation of glass derived from cryptotephra deposits prior to EPMA analysis.

In the authors' laboratories fine-grained shards from cryptotephra deposits are prepared for geochemical analysis on 28 by 48 mm frosted microscope slides. Density or grain size fractions of interest for tephra analysis are directly evaporated onto the microscope slides prior to being covered in epoxy resin. This does not produce a pure sample of tephra shards and some minerogenic and biogenic material may be present; however, in most circumstances this provides an ample number of glass shards, which can be discriminated from any contaminants, to gain a robust chemical characterization of horizons. When tephra shards are only present in very low concentrations tephra shards can be picked out of suspension in distilled water prior to mounting on the slides using either a micromanipulator or a fine needle or a hair. To achieve thin sections of the shards the resin overlying the shards is first ground using decreasing grades of silicon carbide paper until, depending on the size of the glass shards, the resin is between approx. 60

and 30 μm thick and shards have been sectioned. To ensure that a flat surface is achieved during grinding, the thickness of the epoxy resin is checked at various points across the resin surface using a digital micrometer. Once sections through a sufficient number of shards are acquired, the shard and resin surfaces are polished using 6 and then 1 μm diamond suspension until a mirror finish is achieved. This polishing can be undertaken using a polishing machine such as the Logitech CL50 or by hand. Following polishing, slides are washed in distilled water and cleaned by immersing them in petroleum sprit and placing them in an ultrasonic bath for 10 min. The microscope slides can be substituted for cylindrical blocks of epoxy resin or scanning electron microscopy (SEM) stubs; however, the same grinding and polishing procedures are still required to achieve thin sections. This method has been successfully utilized to gain characterizations of material from a North Atlantic marine-core (Abbott *et al.* 2011, 2013) and cryptotephras from the Greenland ice-cores (e.g. Davies *et al.* 2010; Bourne *et al.* 2013).

Electron probe microanalysis

It is over 60 years since the first experiments to generate X-rays from a solid sample by focussing an electron beam on the sample surface were performed (Castaing 1951). Since that time the basic geometry of electron probe instruments has remained essentially unchanged with a 0.5–3 μm electron beam typically being focussed on the sample surface, generating fluorescent X-rays which are picked up by some form of detector. Modern electron probe microanalysers may have between one and five wavelength dispersive spectrometers and/or a solid-state, energy dispersive detector. Samples for EPMA are prepared as polished thin sections or blocks, with the EPMA requiring a flat sample geometry to account for absorption and secondary fluorescence effects related to the passage of the primary X-rays out of the sample (Reed 1995, 2005). These effects are corrected during calculation of concentrations by instrument software, with various approaches being adopted by different manufacturers (see Reed 2005 for more details).

Electron probe microanalysis was first applied to the analysis of tephras by Smith & Westgate (1968), who used glass major element compositions of single grains to correlate and discriminate between deposits. This approach has since been adopted by the tephrochronology community and now forms the foundation of the majority of tephra studies (Lowe 2008, 2011); however, electron probe analysis of glass from tephra deposits is not without problems, and care needs to be exercised during analysis. Electron beam irradiation causes permanent damage to glass, especially to hydrated glasses and silica- and alkali-rich compositions (e.g. Humphreys *et al.* 2006). In particular, the mobilization of Na under the electron beam can be significant, especially in rhyolitic glasses, which may in turn lead to the problem of overestimation of Si and Al, and this is typically overcome by use of relatively low beam currents with a defocused (approx. 10 μm diameter) or rastered electron beam. Determination of Na must be made at the start of an analysis to reduce or eliminate the effects of its mobilization, and the order in which elements are determined on multispectrometer instruments also helps to optimize the data acquisition. These aspects of EPMA analysis of glass have been the subject of numerous publications (including among others Smith & Westgate 1968; Froggatt 1992; Hunt & Hill 1993, 2001 Morgan & London 1996, 2005; Kuehn *et al.* 2011; Hayward 2012).

In addition to intrinsic analytical issues, such as Na mobility, which need to be overcome in EPMA, different laboratories operating different instruments may employ different calibration and data reduction procedures (to correct for X-ray absorption and fluorescence effects, for example), and these differences can conspire to create subtle inter-laboratory variations even for the analysis of the same materials. This has led the EPMA glass analysis and tephra community over the last 30-plus years to embark on a series of inter-laboratory comparisons, in an attempt to promote best practice and ensure robust correlations can be made between data produced on different instruments (Jarosewich *et al.* 1979; Hunt & Hill 1996; Hunt *et al.* 1998; Potts *et al.* 2002). With the expansion in recent years in the number of laboratories undertaking tephra analysis by EPMA, the most extensive of these inter-laboratory comparisons was undertaken between 2009 and 2010 (Kuehn *et al.* 2011), which compiled the results from 24 separate laboratories, including most of the then currently active tephra analysis laboratories worldwide. The materials analysed included a range of compositions (basaltic, rhyolitic and phonolitic with high Na samples). Despite a wide range of instruments, calibration strategies, beam conditions (diameter, raster and current) and software, the overwhelming majority of laboratories involved in this study produced data for both the circulated tephras and reference materials, which compared very favourably. This inter-laboratory comparison study concluded with a series of recommendations that will help make the routine analyses of tephras more consistent, not least the recommendation for the community to consistently publish analyses of reference materials (Kuehn *et al.* 2011). This provides a definitive monitor of data quality. Despite the good inter-comparison between many EPMA laboratories and the willingness of the community to publish reference material analyses, day-to-day variations in calibration may still however generate very subtle differences between the analyses of the same material. To overcome these slight variations and particularly to test a proposed correlation, some studies have reanalysed materials during the same analytical run (Jensen *et al.* 2008; Preece *et al.* 2011). This 'side-by-side' analysis of closely similar materials has been advocated recently as a means to achieve the most reliable results (Lowe 2011) as it eliminates any potential differences between instruments, calibration strategies etc (e.g. Bourne *et al.* 2013). This may not always be feasible within cryptotephra studies if horizons are being correlated between disparate sequences being worked on by different research groups. It is hoped that closer collaboration between groups when key correlations are being tested could result in 'side-by-side' analyses, and a consistent use of the same suite of secondary standards will help increase the robustness of correlations.

Another of the key recommendations in Kuehn *et al.* (2011) is to use a beam diameter between 5 and 20 μm, or to raster the beam across a similar area, to minimize Na migration in the sample. This may be problematic in the analysis of small samples, such as the smaller grain sizes (e.g. microshards, <32 μm) which are often present in marine cryptotephras, or microcryst-rich shards, in which crystal-free areas of glass are small compared with ideal EPMA beam diameters. Recently, robust methods for the analysis of glass from tephra deposits have been described with beams diameters as small as 3 μm (Hayward 2012). The methods described by Hayward (2012) require the use of low beam currents (0.5–2 nA, giving current densities <0.1 nA μm^{-2}) in moderately hydrated samples to avoid Na migration. However, Hayward (2012) also suggests that, at higher currents, slightly hydrated or anhydrous basaltic glasses may withstand beam-induced damage. In hydrated samples, particularly rhyolites, however, analyses using a small electron beam diameters (<5 μm) do suffer from Na migration and their compositions cannot be accurately determined in these conditions (Hayward 2012). Nonetheless, this provides a significant development in the analysis of distal microshards, particularly basaltic material, and moderately hydrated (<2 wt% H_2O) silicic material.

Following the acquisition of EPMA data, consideration has to be given to the way in which glass shard analyses are presented, and this to some extent depends on how the data is to be used. The choice is whether or not to normalize the analytical data to 100% total oxides, and this issue of normalization divides tephrochronologists. In no circumstances, however, should reference material data be normalized, as this provides a check on analytical quality. Some practitioners advocate the use of raw glass analyses with analytical totals >95% (e.g. Hunt & Hill 1993; Pollard *et al.* 2006), and argue that normalization might mask the quality of data and potentially lead to false correlations. Here reference material analyses can assist the reader in assessing analytical quality. Elsewhere, others contend that the difference from 100% is dominantly controlled by the water content of the glass, being composed of both primary magmatic water and water from post-depositional hydration. Thus for comparing tephra deposited in different environments, data should be normalized and presented on volatile-free (i.e. anhydrous) basis (e.g. WoldeGabriel *et al.* 2005; Pearce *et al.* 2008*b*) for totals >90%, provided two criteria can be met: (a) that analytical quality (based on published reference material data) can be assured; and (b) that there has been no alteration of the sample, particularly alkali exchange or mobility, during any post depositional hydration of the glass (see Pearce *et al.* 2008*b* for discussion of hydration in glasses). It appears that this alkali loss does not occur for most rhyolitic glasses with totals >90% (e.g. WoldeGabriel *et al.* 2005; Pearce *et al.* 2008*b*), although this needs to be assessed for each study. Less commonly, other circumstances may result in low analytical totals, and these include the omission of elements from the analytical suite (often Mn, Cl, F are not determined) or an inappropriate assignment of the oxidation state of Fe (to either FeO or Fe_2O_3). Where reference material data indicate good analytical quality, but analytical totals are very low (<90%), other factors such as glass vesicularity or the very small size of the shards (e.g. in ice-core studies) may be responsible. Here it may be difficult or impossible to place even a 5 μm diameter electron beam onto a large enough area of glass, and thus the electron beam can irradiate some of the mounting epoxy (e.g. Dunbar *et al.* 2003). Normalization here may be the only practical solution to enable data comparisons, but it must be remembered that some loss in data quality (e.g. in mass absorption or fluorescence corrections) is likely, and the data should be treated accordingly.

Figures 2 & 3 illustrate the positive impact of normalization on a dataset of shard analyses of the widespread Vedde Ash from material extracted from a range of environments (see Lane *et al.* 2012*b* for a summary regarding this horizon). Figure 2 highlights how normalizing data removes the linear relationship observed between particular unnormalized oxide values and the total oxide sum for the analysis. In some cases these trends (introduced by dilution during hydration) can be

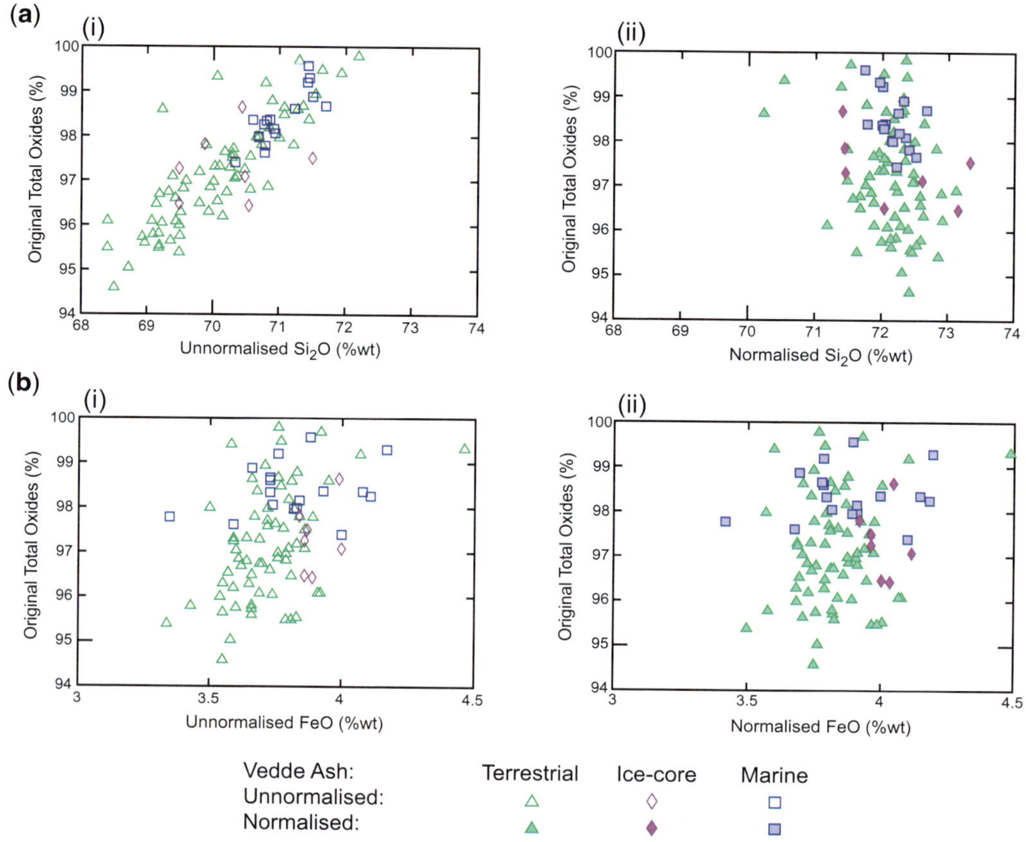

Fig. 2. Comparison of oxide values for specific elements with total oxide values for single shard analyses from the Vedde Ash. (**a**) (i) unnormalized SiO_2; (ii) normalized SiO_2. (**b**) (i) unnormalized FeO; (ii) normalized FeO. Vedde Ash data from: terrestrial – Turney *et al.* (1997), Wastegård *et al.* (2000), Blockley *et al.* (2007), Lane *et al.* (2012*a*, *b*); marine – Thornalley *et al.* (2011); ice-core – Mortensen *et al.* (2005).

misinterpreted as magmatic trends. Figure 3 highlights how the spread of data is reduced following normalization with tighter clustering of analyses observed on Figure 3(aii and bii). Outliers are still apparent following normalization and have not been masked by the correction, furthermore, on Figure 3(aii) two outliers can be observed that are only identified following data normalization and would be incorporated in the characterization of this horizon if only unnormalized oxide values were considered.

Normalization is essential for marine tephra horizons, especially for the correlation of rhyolites, as they typically display lower total oxide values which can be attributed to increased hydration resulting from their depositional environment (e.g. Wallrabe-Adams & Lackschewitz 2003; Austin *et al.* 2004; Brendryen *et al.* 2010). Recent studies of marine tephra where analyses with total oxides of between 93 and 95% have been gained have advocated the normalization of data to 100% total oxides to remove the influence of hydration and permit comparisons with deposits preserved in other environments (e.g. Abbott *et al.* 2011; Albert *et al.* 2012). Despite this advocacy of normalization for use in data correlation, it is vital that data should be presented in such a way that it can be recalculated to either a normalized or raw state (i.e. by presenting original analytical totals), alongside analyses of reference materials, which will allow other researchers to fully assess the data quality.

Trace element analytical methods

While the application of EPMA revolutionized the major element analysis of juvenile material from tephra deposits from the 1970s onwards, the

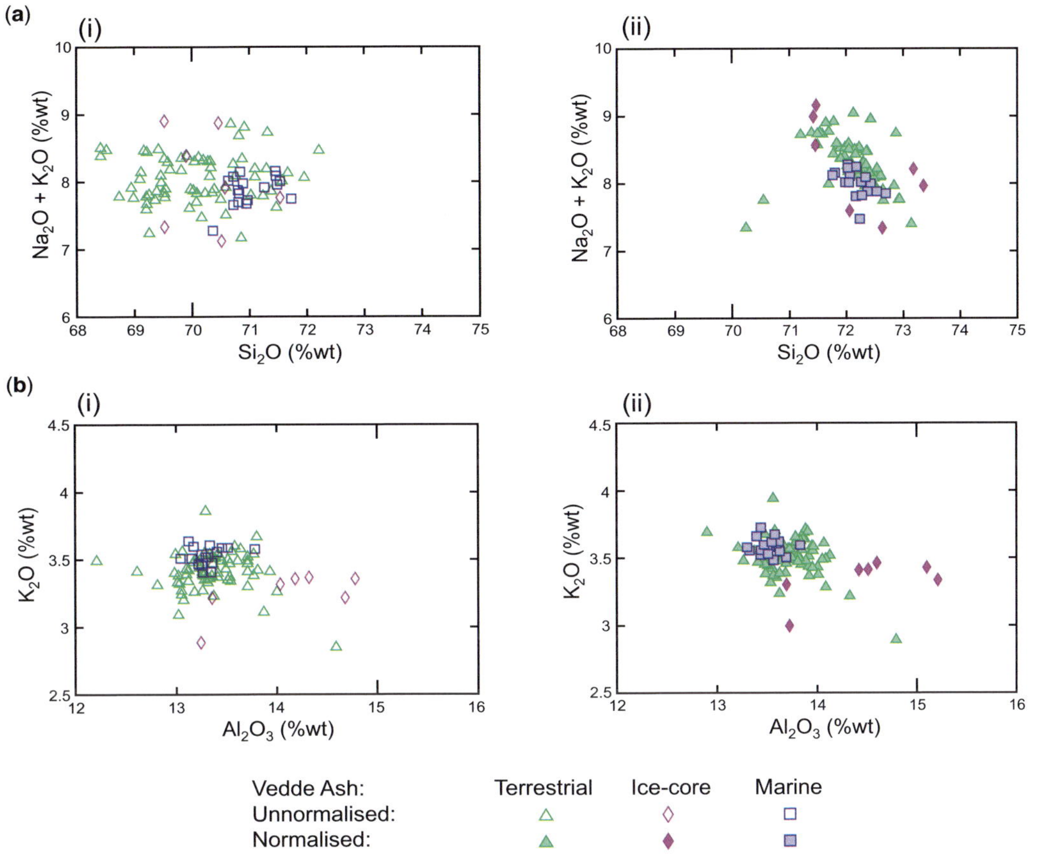

Fig. 3. Compositional variation diagrams of single shard analyses of the Vedde Ash from a range of environments. (**a**) $Na_2O + K_2O$ v. SiO_2: (i) unnormalized; (ii) normalized. (**b**) K_2O v. Al_2O_3: (i) unnormalized; (ii) normalized. Analyses are derived from the sources as for Figure 2.

fine-grained and diffuse nature of tephric material in distal deposits was suitable for trace element analysis only by bulk methods at this time (see e.g. Westgate & Gorton 1981). However, grain-specific trace element techniques with beam diameters below 1 μm were available by the early 1980s, such as proton induced X-ray emission, PIXE or ion microprobe analysis (Fraser 1995; Hinton 1995), but these had not been applied to the determination of trace elements in tephras, presumably because these methods are very expensive and relatively slow. Unfortunately, because of the phase controls on magmatic evolution (Pearce *et al.* 2008*b*), most erupted magmas are either cotectic (common for basalts) or minimum/eutectic compositions (e.g. granites or rhyolites), and thus many widespread tephra deposits (particularly from the same source) can be very similar in composition, often being indistinguishable in terms of their major element chemistry (see e.g. Westgate *et al.* 1994, 2013). It was not until the development of laser ablation inductively coupled plasma mass spectrometry (LA-ICP-MS) (Gray 1985; Arrowsmith 1987) that a rapid and relatively inexpensive technique with the capability for high spatial resolution trace element analysis became widely available.

The potential for the analysis of small glass separates from tephra was first recognized by Westgate *et al.* (1994), who performed the first analyses using an infra red (IR) laser to ablate material from approx. 10–20 shards held in a small mound by cyano-acrylate glue. In this study, Westgate *et al.* (1994) used Ce as the internal standard for calibration and produced data from very small amounts of material (approx. 80 μg). While this demonstrated the potential of LA-ICP-MS, the use of a separate trace element technique for the determination of the internal standard was a drawback. Pearce *et al.* (1996) overcame this issue by using a minor isotope of a major elements (i.e. Si, Fe or

Ca) as the internal standard, these being determined by EPMA in the analysed glasses. The development of UV lasers (frequency modified IR lasers operating at 266 or 213 nm, and now commonly 193 nm Excimer lasers), which offer good coupling to glass and spatial resolutions below 10 μm, led to the analysis of individual glass shards in the 40–100 μm size being performed in the late 1990s (Bryant *et al.* 1999; Pearce *et al.* 1999), with many single-shard trace element analysis being performed for petrogenetic or stratigraphic correlation studies since (Bryant *et al.* 1999; Pearce *et al.* 2002, 2008*a*, 2011; Ukstins-Peate *et al.* 2003, 2008; Harangi *et al.* 2005; Narcisi *et al.* 2006; Jordan *et al.* 2007; Aksu *et al.* 2008; Allan *et al.* 2008; Brendryen *et al.* 2010; Tomlinson *et al.* 2010; Abbott *et al.* 2011; 2012 ; Smith *et al.* 2011*a*, *b*; Salisbury *et al.* 2012; Tamburrino *et al.* 2012; Bramham-Law *et al.* 2013). A smaller number of studies, using other microbeam techniques, have been applied to the trace element analysis of tephra, such as ion probe (SIMS) analysis (Clift & Fitton 1998; Clift & Blusztajn 1999; Beget & Keskinen 2003; Dunbar *et al.* 2003; Hammer *et al.* 2003; Clift *et al.* 2005; Denton & Pearce 2008), but while these methods provide high-quality (precise and accurate) analyses, they are not widely used because they are generally slow and thus relatively expensive per analysis (Hinton 1995; Perkins & Pearce 1995; Pearce *et al.* 2007).

The development of ICP-MS methodology, with particular reference to LA-ICP-MS, for the analysis of glass from tephra deposits has been reviewed by Pearce *et al.* (2004, 2007, 2011) and by Tomlinson *et al.* (2010). Driving the development of LA-ICP-MS applications to the analysis of tephra has been the desire to analyse ever smaller grains of glass from more distal locations, such as very fine-grained tephra deposited thousands of kilometres from source, including cryptotephra in peat and marine- or ice-cores. Glass from these cryptotephras can be as small as 10 μm in diameter and often only ranges in size up to between 20 and 40 μm, and thus crater diameters in the approx. 10 μm size range are required for trace element analysis of typical distal marine-/ice-core-derived glass shards (see Fig. 1). Trace element analysis at these spatial resolutions presents particular problems.

In LA-ICP-MS analyses it is typical to calibrate the instrument with a synthetic reference material such as the NIST 610 or NIST 612 reference glasses (Pearce *et al.* 1997), which contain about 60 trace elements at approx. 450 and 40 ppm, respectively. For analyses of similar matrices, calibration using these glasses can generate high-quality data (see e.g. Eder *et al.* 2013); however, matrix effects can become apparent when material with different major element compositions is analysed. This matrix effect can lead to 'element fractionation', which is the non-stoichiometric removal of material from the sample (Günther & Hattendorf 2005; Hattendorf *et al.* 2005). The effects of this fractionation have been reduced by decreases in laser wavelength, which improves absorption of the laser radiation by the sample (Gonzalez *et al.* 2002) and generates a better aerosol for transport to the plasma (Kuhn & Gunther 2004; Weis *et al.* 2005; Günther & Koch 2008). However, as the laser crater diameter drops below 20 μm (using a Coherent GeoLas 193 nm Excimer laser ablation system), ablation crater morphology changes such that the surface area of the crater produced in relation to its volume increases, and the effects of element fractionation become worse (Pearce *et al.* 2011). Observed under an SEM, it is clear that Excimer laser ablation systems produce craters with a melt film on the crater walls, typically of the order of 0.5 μm thick, and this is illustrated in Figure 4, which shows 6 and 4 μm ablation craters in shards of rhyolitic glass (see also Eggins *et al.* 1998; Pearce *et al.* 2011). It has been suggested (Pearce *et al.* 2011) that element retention or volatilization from this melt film is the cause of fractionation, and Pearce *et al.* (2011) related this to the structure of the molten glass. Martin-Jones (2012) studied fractionation in a wider range of glass compositions than had been studied by Pearce *et al.* (2011), and showed similar degrees of element fractionation, which related strongly to crater diameter but showed no clear relationship with glass composition. As ablation diameter decreases, the volume of ablated material also decreases and the detectable concentration of an element, recorded as the LLD, lower limit of detection, increases (Perkins & Pearce 1995). Figure 5 shows the typical LLD for the typical range of elements determined in the authors' laboratories for rhyolitic glasses.

Fractionation can, however, be corrected for by the application of a fractionation correction factor, and craters as small as 10 μm can be used for the routine analysis of around 25 trace elements in glass shards from tephra deposits (Pearce *et al.* 2011). For analyses performed using both 20 and 10 μm ablation craters, most elements are detectable at above 1 ppm with only a few elements (i.e. the alkali and alkaline earth metals Rb, Cs, Sr and Ba, and the rare earth element Gd, which have either elevated instrument blanks or are relatively low abundance isotopes) having LLDs above 1 ppm. The relative ability of LA-ICP-MS to detect particular elements in particular glass compositions is illustrated further in Figure 6, which shows whether the concentrations in a range of reference materials exceed the LLD (calculated from 3σ on the blank) or the LLQ – the lower limit of

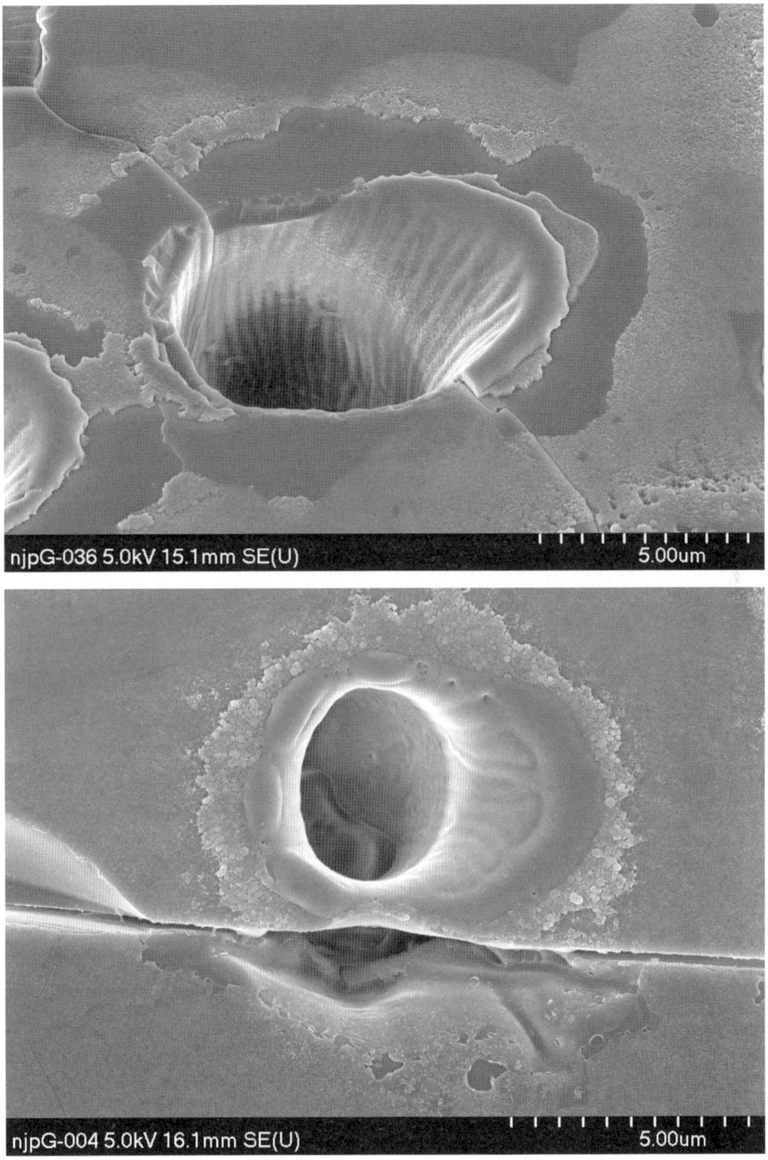

Fig. 4. SEM images of ablation craters in rhyolitic glass, of 6 μm (upper image) and 4 μm (lower image) diameters. Note in each image the formation of a melt lip which is thicker on the right-hand side of the crater, the direction in which the Ar gas flows through the ablation cell of the LA system.

quantitation (calculated from 10σ on the blank). For most materials, analysed with ablation craters 20 μm in diameter, the 25 petrogenetically significant trace elements which are routinely determined in the authors' laboratories are above the LLQ, with the exception of the LREE (light rare earth elements), some LILE (large-ion lithophile elements) and HFSE (high field-strength elements) in the ultrabasic reference glass GOR-132G, which is LREE depleted. As the ablation crater size decreases, the amount of material ablated and transported to the plasma decreases, causing LLDs to worsen. Thus the number of elements which fall below the LLD increases with decreasing ablation crater diameter. However, even from 4 μm craters, some of the more abundant trace elements (e.g. LREE, Ba, Zr, Nb) and those elements where instrumental blanks are particularly low (Th and U)

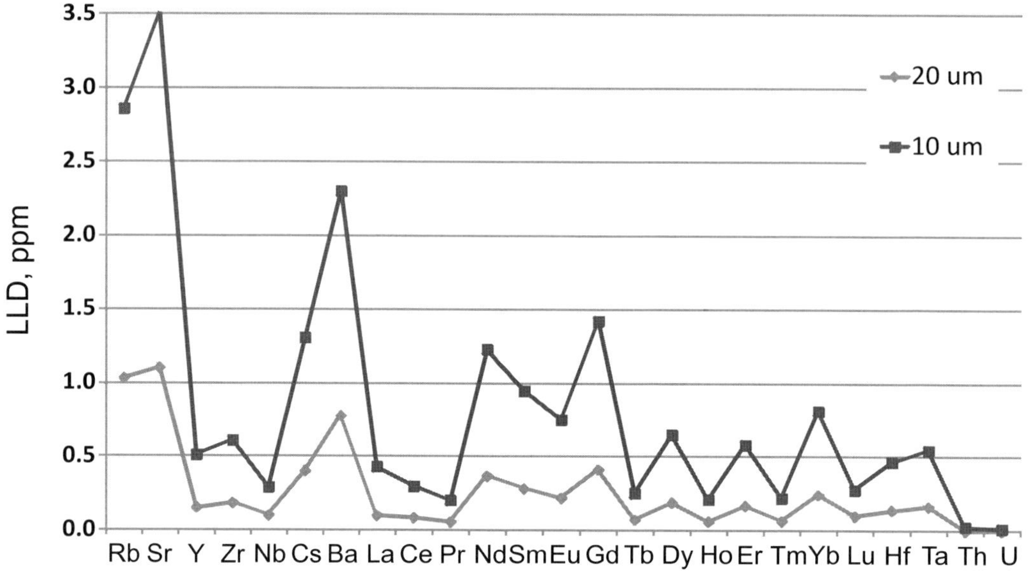

Fig. 5. Typical lower limit of detection (LLD, 3σ) for 25 trace elements routinely determined by LA-ICP-MS. Data are an average of 12 separate LLDs calculated over 1 week from individual calibrations from NIST 612.

remain detectable across a range of compositions. Figure 6 also shows that at these small crater diameters there are limits on the possible internal standards which can be employed. SiO_2 remains above the LLD at all crater diameters, and above the LLQ at 6 μm and above, making this a useable internal standard, although analyses at 4 μm will have a bigger associated error because of the lower signal intensity. CaO (measured from either ^{43}Ca or ^{44}Ca) remains above the LLQ for almost all compositions except rhyolites for 20 μm analyses, and is above the LLD for non-rhyolitic materials at 10 μm, but falls below the LLD for 6 and 4 μm analyses, meaning that either Ca isotope cannot be used as an internal standard, even for basaltic compositions, at smaller crater diameters.

To date, the majority of studies of marine tephra which have used LA-ICP-MS have used ablation craters (where specified by the authors) 25–100 μm in diameter (e.g. Bryant *et al.* 1999; Ukstins-Peate *et al.* 2003; Jordan *et al.* 2007; Aksu *et al.* 2008; Allan *et al.* 2008; Brendryen *et al.* 2010; Tomlinson *et al.* 2010; Albert *et al.* 2012), although some studies have worked with smaller crater diameters at between 10 and 14 μm (Abbott *et al.* 2011, 2013; Tamburrino *et al.* 2012). In all of these studies, the trace element data are used to confirm, refine or refute correlations based on major element data from the same materials. For example, in Davies *et al.* (2012) the previously defined ice–marine tie-line of the Faroe Marine Ash Zone II (FMAZ II) and the ice–marine–terrestrial tie-line of the Saksunarvatn Ash were tested using trace element analysis. This supported the correlation of the FMAZ II, but trace element differences between marine- and ice-core occurrence of the Saksunarvatn Ash hinted that they might be related to different events or different phases of the same eruption (Fig. 7). This issue was subsequently further investigated in Bramham-Law *et al.* (2013), who analysed the major and trace element composition of material from the Saksunarvatn-type site on the Faroe Islands and an in-filled lake sequence from Germany. They demonstrated that trace element analysis supports the correlation between the Faroe Islands and German occurrences and that they both have trace element affinities to the marine occurrence characterized in Davies *et al.* (2012). Albert *et al.* (2012) utilized LA-ICP-MS analysis of material from a Tyrrhenian Sea core to correlate three distal marine horizons to proximal Italian terrestrial deposits and highlighted how the trace element analysis was crucial in providing more extensive and diagnostic geochemical fingerprints for the horizons. Single tephra shards from a series of tephra horizons deposited during MIS 6-4 in a North East Atlantic marine-core (MD04-2822) have been analysed for their trace element composition (Abbott *et al.* 2011, 2013). At present a limited number of eruptions from the time period of interest have been characterized, which precludes their use in testing correlations. However, Abbott *et al.* (2013) used the trace element characterizations to provide additional support for tentative correlations between the distal marine horizons and Icelandic proximal deposits suggested by major element comparisons.

It is clear that a combination of major and trace element analysis provides a powerful tool for the analysis and correlation of cryptotephras from marine-cores with their terrestrial- and ice-core counterparts, with current technology enabling analyses of shards in the 10–20 μm range. At present, the trace element database for particular eruptions is small in comparison to the amount of major element data which is available, but with the continued adoption of LA-ICP-MS in tephra studies, this situation will gradually improve.

Future directions for the analysis of cryptotephra

Current technologies allow quantitative major element analysis from approx. 3 μm diameter spots by EPMA (Hayward 2012), and trace element analyses by ICP-MS from approx. 10 μm laser ablation craters (Pearce *et al.* 2011). There may be potential for improvements in the size of material which can be analysed.

Alkali mobility is a significant problem in the EPMA of glasses with even moderate beam currents. Nonetheless, careful optimization of beam currents and diameters has enabled the high-quality analysis of extremely small fragments of basaltic glass using a 3 μm diameter electron beam, rather than the more widely used, defocused 10 μm beam (Kuehn *et al.* 2011; Hayward 2012). The effectiveness of the 3 μm electron beam analyses relies on not allowing the current density to increase sufficiently to cause alkali mobility in the sample, thus focussing the beam to 2 or 1 μm would require concomitant decreases in beam current, so as not to exceed the suggested 0.1 nA μm^{-2} limit to allow analysis of a wide compositional spectrum of glasses (Hayward 2012). In some EPMA instruments this may be impossible because of an inability to regulate low beam currents, decreasing analytical quality (Hayward pers. comm. 2013).

The more widespread recent availability of high-brightness electron beam sources attached to electron probe microanalysers, namely the replacement of the W filament by LaB_6 or CeB_6 sources (Reed 2005), gives an order of magnitude increase in beam current and can allow smaller diameter beams (of the order of 100 nm) to be used for analysis. Initial studies of these so called field emission (FE) electron probes for the major element analysis of glass shards from tephra have been successfully conducted (Coulter *et al.* 2010) at operating conditions similar to conventional EPMA analysis (cf. Hayward 2012). These FE electron probes have not yet been used at their highest resolutions for the analysis of volcanic glass, but, for major element analysis, increasing beam intensity to achieve better spatial resolution is unlikely to be achieved without significant alkali mobility. Nonetheless, FE probes with high beam currents may offer some interesting possibilities for analysis at small beam diameters, which may have some tephra applications. For some robust materials, which can withstand high beam currents and current densities, it may be possible to determine a few abundant trace elements relatively quickly, which may be significant in tephra studies (e.g. Ba in the Toba tephras as described by Westgate *et al.* 2013). Improved spatial resolution may allow some analytical data to be produced from very small shards of material, or may have the potential to navigate safely around microlite-infested shards, as is typical of the many andesitic tephra deposits. Whether volcanic glass survives the potentially high current densities which can be produced by FE probes, and what the impact of FE electron probes on tephra studies will be, remain to be seen. The work of Hayward (2012) suggests that, even using very low beam currents, it is unlikely that analyses at beam diameters below 1 μm could be performed on relatively stable (anhydrous, basaltic) glasses, and for sensitive glasses (such as hydrated rhyolites or alkali-rich glass) beam-induced element mobility or damage is even more likely.

Since the first introduction of ICP-MS in the late 1980s, every few years have seen modifications which have improved instrument sensitivity, from early basic quadrupole instruments to double-focussing instruments with high mass resolutions and single or multiple collectors, and instruments with reaction or collision cells. The potential for many of these ICP-MS instruments has not been explored for tephra studies, for example, the use of collision cell instruments for the determination of low concentrations of first-row transition metals in rhyolites, which have been difficult to determine to date because of high blanks in this region of the mass spectrum. Currently, very few cryptotephra studies use trace element analyses at resolution of 10 μm (e.g. Abbott *et al.* 2011, 2013), most being performed on material with >25 μm analyses. The data presented in Figure 6 (from Martin-Jones 2012) suggests that analyses at 6 μm could provide data for about 18 elements, and at 4 μm, data may be obtained for perhaps six to eight trace elements, once the degree of fractionation has been established at these crater diameters. Careful setting up of the ICP-MS scan parameters could improve the counts for those elements determined by removing analytes which would be below LLDs from the acquired spectrum, and thus improve

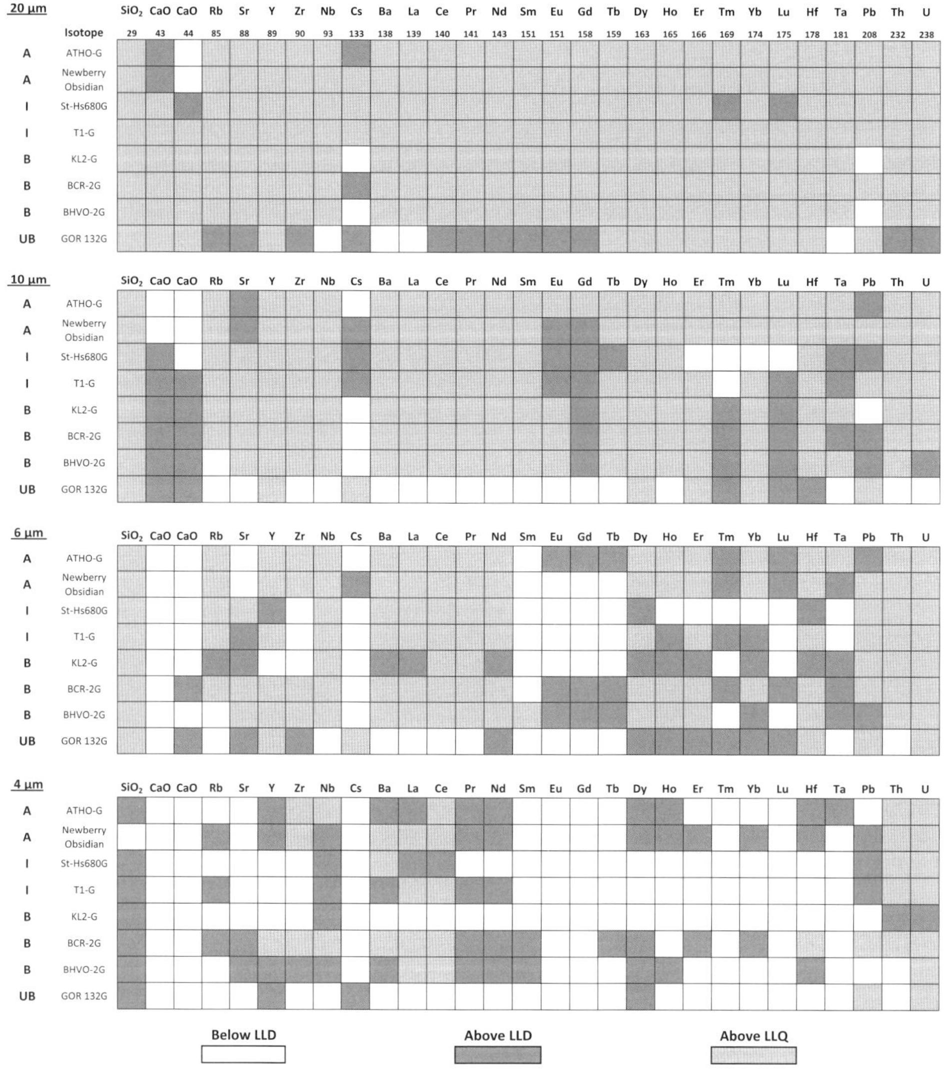

	SiO_2	CaO	CaO	Rb	Sr	Y	Zr	Nb	Cs	Ba	La	Ce	Pr	Nd	Sm	Eu	Gd	Tb	Dy	Ho	Er	Tm	Yb	Lu	Hf	Ta	Pb	Th	U
	wt%	wt%	wt%	ppm	ppm	ppm	ppm	ppm	ppm	ppm	ppm	ppm	ppm	ppm	ppm	ppm	ppm	ppm	ppm	ppm	ppm	ppm	ppm	ppm	ppm	ppm	ppm	ppm	ppm
ATHO-G	75.6	1.70	1.70	65.3	94.1	94.5	512	62.4	1.08	547	55.6	121	14.6	60.9	14.2	2.76	15.3	2.51	16.2	3.43	10.3	1.52	10.5	1.54	13.7	3.9	5.67	7.4	2.37
Newberry Obsidian	73.1	0.81	0.81	113	44.7	44.3	356	24.9	4.66	758	31	68.6	7.4	28.4	6.66	0.87	6.64	1.1	7.18	1.48	4.5	0.72	5.43	0.81	9.63	1.85	18.5	12.1	5.05
St-Hs680G	63.7	5.28	5.28	30.7	482	11.4	118	6.94	1.75	298	12	26.1	3.2	13	2.78	0.95	2.59	0.37	2.22	0.42	1.18	0.17	1.13	0.17	3.07	0.42	10.3	2.28	1.01
T1-G	58.6	7.10	7.10	79.7	284	23.9	144	8.87	2.69	388	70.4	127	12.4	41.4	6.57	1.21	5.31	0.77	4.5	0.86	2.49	0.35	2.38	0.35	3.88	0.46	11.6	31.3	1.71
KL2-G	50.3	10.9	10.9	8.7	356	25.4	152	15	0.12	123	13.1	32.4	4.6	21.6	5.54	1.92	5.92	0.89	5.22	0.96	2.54	0.33	2.1	0.29	3.93	0.96	2.07	1.02	0.55
BCR-2G	54.4	7.06	7.06	47	342	35	184	12.5	1.16	683	24.7	53.3	6.7	28.9	6.59	1.97	6.71	1.02	6.44	1.27	3.7	0.51	3.39	0.5	4.84	0.78	11	5.9	1.69
BHVO-2G	49.3	11.4	11.4	9.2	396	26	170	18.3	0.1	131	15.2	37.6	5.35	24.5	6.1	2.07	6.16	0.92	5.28	0.98	2.56	0.34	2.01	0.28	4.32	1.15	1.7	1.22	0.4
GOR 132G	45.5	8.45	8.45	2.1	15.3	12.9	9.9	0.07	7.45	0.82	0.08	0.39	0.09	0.69	0.51	0.26	1.19	0.27	2.15	0.51	1.56	0.23	1.61	0.24	0.36	0.03	19.5	0.01	0.05

Fig. 6. Graphical representation of the change in an elements 'detectability' and 'quantifiability' (i.e. concentrations $>3\sigma$ are above the LLD and $>10\sigma$ are above the lower limit of quantitation, LLQ, on the blank) by LA-ICP-MS at differing ablation crater diameters. At 20 µm most elements are detected in almost all compositions of glass, except for the LREE, some LILE and HFSE in the ultrabasic reference glass GOR-132G, which is LREE depleted. As crater sizes decrease, the number of elements falling below the LLD increases, but even from 4 µm craters, some of the more abundant trace elements (e.g. LREE, Ba, Zr and Nb) and those elements where instrumental blanks are particularly low (Th and U) remain detectable across a range of compositions. Data for the Newberry Obsidian from Martin-Jones (2012) determined by LA-ICP-MS calibrated using major element data from Higgins (1973); all other reference glasses from GeoReM (Jochum *et al.* 2005, 2006; Jochum & Stoll 2008).

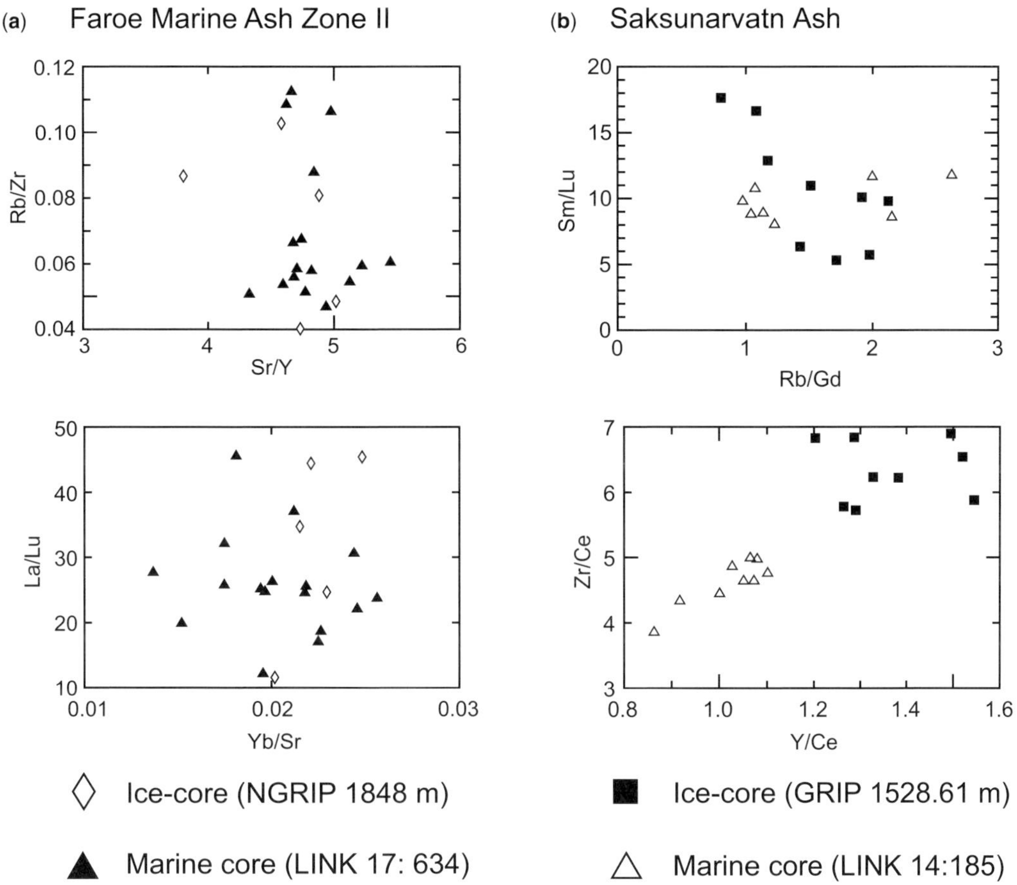

Fig. 7. Comparisons of single grain LA-ICP-MS characterizations of tephra horizons previously correlated between ice and marine sequences using major element characterizations. (**a**) Characterizations support the correlation of the Faroe Marine Ash Zone II occurrences. (**b**) Characterizations suggest differences between the Saksunarvatn Ash occurrences (adapted from Davies *et al.* 2012).

analytical quality at these smaller spot sizes. Again, analyses from ablation craters this small have yet to be attempted. In the trace element analysis of larger glass shards from tephra deposits, using ablation craters >30 μm in diameter, elemental fractionation issues are less problematic than for analyses at crater diameters <20 μm, a result of the changing crater geometry and the reduction in the volume of the melt film relative to the total volume ablated. The use of very short pulse laser ablation systems offers some promise of overcoming some of the fractionation issues; these femtosecond (fs) lasers have pulses which are of the order of 50–200 fs long, compared with the 10–20 ns pulse typical of solid-state or ArF Excimer laser systems, and this short pulse causes less peripheral heating around the ablation site (Gonzalez *et al.* 2006; Pisonero & Gunther 2008; Claverie *et al.* 2009; Borisova *et al.* 2010; Shaheen *et al.* 2012). As yet these systems have been used for melt inclusion analysis (Borisova *et al.* 2008), but not for glass analysis from tephras.

A very limited number of tephra studies have used single-grain Pb isotopic analyses by LA-ICP-MS to help provenance tephras (Ukstins-Peate *et al.* 2003; Westgate *et al.* 2011), which can be helpful in recognizing possible regional sources. This has yet to be taken up by the tephra community, but with tephra deposits containing larger shards where 20–30 μm craters could be used for analysis, Pb isotopic data from single grains of glass can be produced and may be useful in discriminating between sources (Ukstins-Peate *et al.* 2003). The increase in the number of multicollector ICP-MS instruments in research institutions should facilitate this.

Conclusions

Many of the issues discussed in this paper for the analysis and correlation of marine tephra deposits relate to the wider area of cryptotephra analysis. The separation and preparation for analysis of glass shards from marine cryptotephra deposits is a painstaking and time-consuming process, and is possibly one area where much change cannot be readily foreseen. Analytical protocols for the major and trace element analysis of glass fragments, once prepared, can produce data from 3–5 μm diameter spots for major elements. Trace elements are presently determined in some laboratories routinely from 10 μm diameter LA-ICP-MS analyses, but most laboratories do not determine trace elements from craters less than approx. 25 μm diameter spots. Potential exists to reduce the size of the ablated area to around 6 μm, but to maintain data quality, because of lower signal intensities, the number of trace elements determined needs to be reduced. However, with smaller crater diameters, problems of element fractionation increase in LA-ICP-MS, although progress is being made to overcome these issues. New, short-pulse (fs) lasers may also overcome these fractionation issues, but have as yet to be used in tephra studies. Data interpretation relies on comparisons of major and/or trace element data, and to facilitate this correlation between different sedimentary environments, it is vital that data is normalized to an anhydrous basis.

Present methodologies and instrumentation (dominantly EPMA and LA-ICP-MS) continue to be extremely successful in the compositional analysis of small fragments of glass from cryptotephra deposits in marine, ice and terrestrial deposits and is at the heart of many research programmes (e.g. INTIMATE, TRACE). These methods have evolved in recent years, having improved spatial resolution, and the potential exists for their use on finer-grained, more distal material than has been previously analysed. These developments should expand the geographical range over which tephra correlations can be made, and will no doubt be crucial to future tephra studies.

N.J.G.P. and P.M.A. acknowledge financial support from NERC through the SMART project (NE/F020600/1, NE/F02116X/1, NE/F021445/1) which has contributed to this research described in this paper. P.M.A. also acknowledges the support of the Climate Change Consortium of Wales (C3W) and of the European Research Council (TRACE project: 259253). N.J.G.P. is grateful to Dr Richard Bevins, Keeper of Geology at the National Museum of Wales in Cardiff, for providing space and facilities during a period of research leave in 2013. C.M.J. would like to thank her family for financial support during the course of her MPhil. This paper contributes to the INTREPID project (Enhancing tephrochronology as a global research tool through improved fingerprinting and correlation techniques and uncertainty modelling – an INQUA INTAV-led project; International Focus Group on Tephrochronology and Volcanism, no. 0907). We are grateful for the helpful and constructive reviews of C. Hayward and H. Haflidason which have improved this contribution.

References

Abbott, P. M., Davies, S. M., Austin, W. E. N., Pearce, N. J. G. & Hibbert, F. D. 2011. Identification of cryptotephra horizons in a North East Atlantic marine record spanning marine isotope stages 4 and 5a (~60,000–82,000 a b2k). *Quaternary International*, **246**, 177–189.

Abbott, P. M., Davies, S. M. *et al.* 2012. A detailed framework of marine isotope stages 4 and 5 volcanic events recorded in two Greenland ice-cores. *Quaternary Science Reviews*, **36**, 59–77.

Abbott, P. M., Austin, W. E. N., Davies, S. M., Pearce, N. J. G. & Hibbert, F. D. 2013. Crypto tephrochronology of a North East Atlantic marine sequence over Termination II, the Eemian and the last interglacial-glacial transition. *Journal of Quaternary Science*, **28**, 501–514.

Aksu, A. E., Jenner, G., Hiscott, R. N. & Isler, E. B. 2008. Occurrence, stratigraphy and geochemistry of Late Quaternary tephra layers in the Aegean Sea and the Marmara Sea. *Marine Geology*, **252**, 174–192.

Albert, P. G., Tomlinson, E. L. *et al.* 2012. Marine-continental tephra correlations: volcanic glass geochemistry from the Marsili Basin and the Aeolian Islands, Southern Tyrrhenian Sea, Italy. *Journal of Volcanology and Geothermal Research*, **229–230**, 74–94.

Allan, A. S. R., Baker, J. A., Carter, L. & Wyscozanksi, R. J. 2008. Reconstructing the Quaternary evolution of the world's most active silicic volcanic system: insights from a 1.65 Ma deep ocean tephra record sourced from the Taupo Volcanic Zone, New Zealand. *Quaternary Science Reviews*, **27**, 2341–2360.

Alloway, B. V., Pillans, B. J., Carter, L., Naish, T. R. & Westgate, J. A. 2005. Onshore–offshore correlation of Pleistocene rhyolitic eruptions from New Zealand: implications for TVZ eruptive history and palaeoenvironmental construction. *Quaternary Science Reviews*, **24**, 1601–1622.

Arrowsmith, P. 1987. Laser ablation of solids for elemental analysis by inductively coupled plasma mass spectrometry. *Analytical Chemistry*, **59**, 1437–1444.

Ascough, P., Cook, G. & Dugmore, A. 2005. Methodological approaches to determining the marine radiocarbon reservoir effect. *Progress in Physical Geography*, **29**, 532–547.

Austin, W. E. N., Wilson, L. J. & Hunt, J. B. 2004. The age and chronostratigraphical significance of North Atlantic Ash Zone II. *Journal of Quaternary Science*, **19**, 137–146.

Beget, J. E. & Keskinen, M. J. 2003. Trace element geochemistry of individual glass shards of the Old

Crow tephra and age of the Delta glaciation, central Alaska. *Quaternary Research*, **60**, 63–69.

Blockley, S. P. E., Pyne-O'Donnell, S. D. F. *et al.* 2005. A new and less destructive laboratory procedure for the physical separation of distal glass tephra shards from sediments. *Quaternary Science Reviews*, **16–17**, 1952–1960.

Blockley, S. P. E., Lane, C. S., Lotter, A. F. & Pollard, A. M. 2007. Evidence for the presence of the Vedde Ash in Central Europe. *Quaternary Science Reviews*, **26**, 3030–3036.

Blockley, S. P. E., Lane, C. S., Bronk Ramsey, C. & Turney, C. S. M. 2012. The integration of ice core, marine and terrestrial records of the last termination (INTIMATE) 60,000 to 8000 BP, *Quaternary Science Reviews*, **36**, 232pp.

Borisova, A. Y., Freydier, R., Polve, M., Salvi, S., Candaudap, F. & Aigouy, T. 2008. In situ multi-element analysis of the Mount Pinatubo quartz-hosted melt inclusions by NIR femtosecond laser ablation inductively coupled plasma mass spectrometry. *Geostandards and Geoanalytical Research*, **32**, 209–229.

Borisova, A. Y., Freydier, R., Polve, M., Jochum, K. P. & Candaudap, F. 2010. Multi-element analysis of ATHO-G rhyolitic glass (MPI-DING reference material) by femtosecond and nanosecond LA-ICP-MS: evidence for significant heterogeneity of B, V, Zn, Mo, Sn, Sb, Cs, W, Pt and Pb at the millimetre scale. *Geostandards and Geoanalytical Research*, **34**, 245–255.

Bourne, A. J., Lowe, J. J. *et al.* 2010. Distal tephra record for the last 105,000 years from PRAD 1–2 in the central Adriatic Sea: implications for marine tephrostratigraphy. *Quaternary Science Reviews*, **29**, 3079–3094.

Bourne, A. J., Davies, S. M., Abbott, P. M., Rasmussen, S. O., Steffensen, J. P. & Svensson, A. 2013. Revisiting the Faroe Marinse Ash Zone III in two Greenland ice cores: implications for marine-ice-core correlations. *Journal of Quaternary Science*, **28**, 641–646.

Bramham-Law, C. W. F., Theuerkauf, M., Lane, C. S. & Mangerud, J. 2013. New findings regarding the Sakunarvatn Ash in Germany. *Journal of Quaternary Science*, **28**, 248–257.

Brendryen, J., Haflidason, H. & Sejrup, H. P. 2010. Norwegian Sea tephrostratigraphy of marine isotope stages 4 and 5: prospects and problems for tephrochronology in the North Atlantic region. *Quaternary Science Reviews*, **29**, 847–864.

Bryant, C. J., Arculus, R. J. & Eggins, S. M. 1999. Laser ablation inductively coupled plasma mass spectrometry and tephras: a new approach to understanding arc magmatism. *Geology*, **27**, 1119–1122.

Castaing, R. 1951. *Application of electron beams to a method of local chemical and crystallographic analysis*. PhD, University of Paris.

Claverie, F., Fernandez, B., Pecheyran, C., Alexis, J. & Donard, O. F. X. 2009. Elemental fractionation effects in high repetition rate IR femtosecond laser ablation ICP-MS analysis of glasses. *Journal of Analytical Atomic Spectrometry*, **24**, 891–902.

Clift, P. D. & Blusztajn, J. 1999. The trace element characteristics of Aegean and Aeolian volcanic arc marine tephra. *Journal of Volcanology and Geothermal Research*, **92**, 321–347.

Clift, P. D. & Fitton, J. G. 1998. Trace and rare earth element chemistry of volcanic ashes from sites 918 and 919: implications for Icelandic volcanism. *In*: Saunders, A. D., Larsen, H. C. & Wise, S. W., Jr. (eds) *Proceedings of the Ocean Drilling Program, Scientific Results*, **152**. Ocean Drilling Program, Texas A & M University, College Station, TX.

Clift, P. D., Chan, L. H., Blusztajn, J., Layne, G. D., Kastner, M. & Kelly, R. K. 2005. Pulsed subduction accretion and tectonic erosion reconstructed since 2.5 Ma from the tephra record offshore Costa Rica. *Geochemistry, Geophysics, Geosystems*, **6**, Q09016.

Coulter, S. E., Pilcher, J. R., Hall, V. A., Plunkett, G. & Davies, S. M. 2010. Testing the reliability of the JEOL FEGSEM 6500F electron microprobe for quantitative major element analysis of glass shards from rhyolitic tephra. *Boreas*, **39**, 163–169.

Davies, S. M., Hoek, W. Z., Bohncke, J. P., Lowe, J. J., Pyne-O'Donnell, S. D. F. & Turney, C. S. M. 2005. Detection of lateglacial distal tephra layers in the Netherlands. *Boreas*, **34**, 123–135.

Davies, S. M., Wastegård, S. *et al.* 2010. Tracing volcanic events in the NGRIP ice-core and synchronising North Atlantic marine records during the last glacial period. *Earth and Planetary Science Letters*, **294**, 69–79.

Davies, S. M., Abbott, P. M., Pearce, N. J. G., Wastegård, S. & Blockley, S. P. E. 2012. Integrating the INTIMATE records using tephrochronology: rising to the challenge. *Quaternary Science Reviews*, **36**, 11–27.

Denton, J. S. & Pearce, N. J. G. 2008. Comment on Vinther *et al.* 'A synchronized dating of three Greenland ice cores throughout the Holocene': no Minoan tephra in the 1642 B.C. layer of the GRIP ice core. *Journal of Geophysical Research (Atmospheres)*, **113**, D04303.

Dunbar, N. W., Zielinski, G. A. & Voisins, D. T. 2003. Tephra layers in the Siple Dome and Taylor Dome ice cores, Antarctica: sources and correlations. *Journal of Geophysical Research*, **108**, 2374–2385.

Eder, F. M., Neelmeijer, C., Pearce, N. J. G., Bichler, M., Sterba, J. H., Ntaflos, T. & Merchel, S. 2013. Volcanic glass under fire – a comparison of three complimentary analytical techniques. *X-ray Spectrometry*, **6**, 412–422.

Eggins, S. M., Kinsley, L. P. J. & Shelley, J. M. G. 1998. Deposition and element fractionation processes during atmospheric pressure laser sampling for analysis by ICP-MS. *Applied Surface Science*, **127–129**, 278–286.

Fraser, D. G. 1995. The nuclear microprobe – PIXE, PIGE, RBS, NRA and ERDA. *In*: Potts, P. J., Bowles, F. W., Reed, S. J. B. & Cave, M. R. (eds) *Microprobe Techniques in the Earth Sciences*. The Mineralogical Society, London.

Froggatt, P. C. 1992. Standardization of the chemical analysis of tephra deposits. Report of the ICCT working group. *Quaternary International*, **13–14**, 93–96.

Gonzalez, J., Mao, X. L., Roy, J., Mao, S. S. & Russo, R. E. 2002. Comparison of 193, 213 and 266 nm laser ablation ICP-MS. *Journal of Analytical Atomic Spectrometry*, **17**, 1108–1113.

GONZALEZ, J., DUNDAS, S. H., LIU, C. Y., MAO, X. & RUSSO, R. E. 2006. UV-femtosecond and nanosecond laser ablation-ICP-MS: internal and external repeatability. *Journal of Analytical Atomic Spectrometry*, **21**, 778–784.

GRAY, A. L. 1985. Solid sample introduction by laser ablation inductively coupled plasma mass spectrometry. *Analyst*, **61**, 551–556.

GÜNTHER, D. & HATTENDORF, B. 2005. Solid sample analysis using laser ablation inductively coupled plasma mass spectrometry. *Trends in Analytical Chemistry*, **24**, 255–265.

GÜNTHER, D. & KOCH, J. 2008. Formation of aerosols generated by laser ablation and their impact on element fractionation in LA-ICP-MS. *In*: SYLVESTER, P. (ed.) *Laser ablation ICP-MS in the Earth Sciences: Current Practices and Outstanding Issues, 40*. Mineralogical Association of Canada, Vancouver, Short Course Series.

HALL, M. & HAYWARD, C. L. 2014. Preparation of micro- and crypto-tephras for quantitative microbeam analysis. *In*: AUSTIN, W. E. N., ABBOTT, P. M., DAVIES, S. M., PEARCE, N. J. G. & WASTEGÅRD, S. (eds) *Marine Tephrochronology*. Geological Society, London, Special Publications, **398**. First published online February 27, 2014, http://dx.doi.org/10.1144/SP398.5

HAMMER, C. U., KURAT, G., HOPPE, P., GRUM, W. & CLAUSEN, H. B. 2003. Thera eruption date 1645 BC confirmed by new ice core data? *Proceedings of SCIEM 2000 – EuroConference*, 2–7 May 2001, Vienna, SCIEM2000, 87–94.

HARANGI, S., MASON, P. R. D. & LUKACS, R. 2005. Correlation and petrogenesis of silicic pyroclastic rocks in the Northern Pannonian Basin, Eastern-Central Europe: in situ trace element data of glass shards and mineral chemical constraints. *Journal of Volcanology and Geothermal Research*, **143**, 237–257.

HATTENDORF, B., LATKOCZY, C. & GÜNTHER, D. 2005. Laser ablation ICP-MS. *Analytical Chemistry*, **75**, 341A–347A.

HAYWARD, C. L. 2012. High spatial resolution electron micorporbe analysis of tephras and melt inclusions without beam-induced chemical fractionation. *The Holocene*, **22**, 119–125.

HIGGINS, M. W. 1973. Petrology of the Newberry Volcano, Central Oregon. *Bulletin of the Geological Society of America*, **84**, 455–488.

HINTON, R. W. 1995. Ion microprobe analysis in geology. *In*: POTTS, P. J., BOWLES, J. F. W., REED, S. J. B. & CAVE, M. R. (eds) *Microprobe Techniques in the Earth Sciences*. Chapman and Hall, London, 235–289.

HUMPHREYS, M. C. S., STUART, K. & BLUNDY, J. D. 2006. SIMS investigation of electron-beam damage to hydrous, rhyolitic glasses: implications for melt inclusion analysis. *American Mineralogist*, **91**, 667–679.

HUNT, J. B. & HILL, P. G. 1993. Tephra geochemistry: a discussion of some persistent analytical problems. *The Holocene*, **3**, 271–278.

HUNT, J. B. & HILL, P. G. 1996. An inter laboratory comparison of the electron probe microanalysis of glass geochemistry. *Quaternary International*, **34–36**, 229–241.

HUNT, J. B. & HILL, P. G. 2001. Tephrological implications of beam size – sample size effects in electron microprobe analyses of glass shards. *Journal of Quaternary Science*, **16**, 105–117.

HUNT, J. B., CLIFT, P. D., LACASSE, C., VALLIER, T. L. & WERNER, R. 1998. Interlaboratory comparison of electron probe microanalysis of glass geochemistry. *In*: SAUNDERS, A. D., LARSEN, H. C. & WISE, S. W., JR (eds) *Proceedings of the Ocean Drilling Program, Scientific Results*, **152**. Ocean Drilling Program, Texas A & M University, College Station, TX, 85–91.

JAROSEWICH, E., PARKES, A. S. & WIGGINS, L. B. 1979. Microprobe analyses of four natural glasses and one mineral: an interlaboratory study of precision and accuracy. *Smithsonian Contributions to the Earth Sciences*, **22**, 53–67.

JENSEN, B. J. L., FROESE, D. G., PREECE, S. J., WESTGATE, J. A. & STACHEL, T. 2008. An extensive middle to late Pleistocene tephrochronologic record from east-central Alaska. *Quaternary Science Reviews*, **27**, 411–427.

JOCHUM, K.-P. & STOLL, B. 2008. Reference materials for elemental and isotopic analysis by LA-(MC)-ICP-MS: successes and outstanding needs. *In*: SYLVESTER, P. (ed.) *Laser Ablation-ICP-MS in the Earth Sciences, Current Practices and Outstanding Issues*. Mineralogical Association of Canada, Vancouver, Short Course Series, 147–168.

JOCHUM, K.-P., NOHL, U., HERWIG, K., LAMMEL, E., STOLL, B. & HOFMANN, A. W. 2005. GeoReM: a new feochemical database for reference materials and isotopic standards. *Geostandards and Geoanalytical Research*, **29**, 333–338.

JOCHUM, K.-P., STOLL, B. ET AL. 2006. MPI-DING reference glasses for in situ microanalysis: new reference values for element concentrations and isotope ratios. *Geochemistry, Geophysics, Geosystems*, **7**, Q02008, doi: 02010.01029/02005GC001060

JORDAN, B. R., SIGURDSSON, H., CAREY, S. N., ROGERS, R. & EHRENBORG, J. 2007. Geochemical variation along and across the Central American Miocene paleoarc in Honduras and Nicaragua. *Geochimica et Cosmochimica Acta*, **71**, 3581–3591.

KUEHN, S. C. & FROESE, D. G. 2010. Tephra from ice – a simple method to routinely mount, polish, and quantitatively analyze sparse fine particles. *Microscopy and Microanalysis*, **16**, 218–225.

KUEHN, S. C., FROESE, D. G. & SHANE, P. A. 2011. The INTAV intercomparison of electron-beam microanalysis of glass by tephrochronology laboratories, results and recommendations. *Quaternary International*, **246**, 19–47.

KUHN, H. R. & GUNTHER, D. 2004. Laser ablation-ICP-MS: particle size dependent elemental composition studies on filter-collected and online measured aerosols from glass. *Journal of Analytical Atomic Spectrometry*, **19**, 1158–1164.

KYLANDER, M. E., LIND, E. M., WASTEGÅRD, S. & LOWEMARK, L. 2012. Recommendations for using an XRF core scanner as a toop in tephrochronology. *The Holocene*, **22**, 371–375.

LANE, C. S., BLOCKLEY, S. P. E., LOTTER, A. F., FINSINGER, W., FILIPPI, M. L. & MATTHEWS, I. P. 2012*a*. A regional tephrostratigraphic framework for central and southern European climate archives during the Last Glacial to Interglacial transition: comparisons north and south of the Alps. *Quaternary Science Reviews*, **36**, 50–58.

Lane, C. S., Blockley, S. P. E. *et al.* 2012*b*. Was the 12.1 ka Icelandic Vedde Ash one of a kind? *Quaternary Science Reviews*, **33**, 87–99.

Lim, C., Ikehara, K. & Toyoda, K. 2008. Cryptotephra detection using high-resolution trace element analysis of Holocene marine sediments, southwest Japan. *Geochimica et Cosmochimica Acta*, **72**, 5022–5036.

Lowe, D. J. 2008. Globalization of tephrochronology: new views from Australasia. *Progress in Physical Geography*, **32**, 311–335.

Lowe, D. J. 2011. Tephrochronology and its application: a review. *Quaternary Geochronology*, **6**, 107–153.

Lowe, J. J., Blockley, S. P. E. *et al.* 2007. Age modelling of late Quaternary marine sequences in the Adriatic: towards improved precision and accuracy using volcanic event stratigraphy. *Continental Shelf Research*, **27**, 560–582.

Martin-Jones, C. 2012. *Defining fractionation in LA-ICP-MS analysis of volcanic glass shards and its application to the correlation of tephra deposits from Ulleungdo, Korea.* MPhil thesis, Aberystwyth University.

Morgan, G. B. & London, D. 1996. Optimising the electron microprobe analysis of hydrous alkali aluminosilicate glasses. *American Mineralogist American Mineralogist*, **81**, 1176–1185.

Morgan, G. B. & London, D. 2005. Effect of current density on the electron microprobe analysis of alkali aluminosilicate glasses. *American Mineralogist*, **90**, 1131–1138.

Mortensen, A. K., Bigler, M., Gronvold, K., Steffensen, J. P. & Johnsen, S. J. 2005. Volcanic ash layers from the Last Glacial termination in the NGRIP ice core. *Journal of Quaternary Science*, **20**, 209–219.

Narcisi, B., Petit, J. R. & Tiepolo, M. 2006. A volcanic marker (92 ka) for dating deep Antarctic ice cores. *Quaternary Science Reviews*, **25**, 2682–2687.

Pearce, N. J. G., Westgate, J. A. & Perkins, W. T. 1996. Developments in the analysis of volcanic glass shards by laser ablation ICP-MS: quantitative and single internal standard-multi-element methods. *Quaternary International*, **34–36**, 213–227.

Pearce, N. J. G., Perkins, W. T., Westgate, J. A., Gorton, M. P., Jackson, S. E., Neal, C. R. & Chenery, S. P. 1997. A compilation of new and published major and trace element data for NIST SRM 610 and NIST SRM 612 glass reference materials. *Geostandards Newsletter*, **21**, 115–144.

Pearce, N. J. G., Westgate, J. A., Perkins, W. T., Eastwood, W. J. & Shane, P. A. R. 1999. The application of laser ablation ICP-MS to the analysis of volcanic glass shards from tephra deposits: bulk glass and single shard analysis. *Global and Planetary Change*, **21**, 151–171.

Pearce, N. J. G., Eastwood, W. J., Westgate, J. A. & Perkins, W. T. 2002. The composition of juvenile volcanic glass from the c. 3,600 B.P. Minoan eruption of Santorini (Thera). *Journal of the Geological Society*, **159**, 545–556.

Pearce, N. J. G., Westgate, J. A., Perkins, W. T. & Preece, S. J. 2004. The application of ICP-MS methods to tephrochronological problems. *Applied Geochemistry*, **19**, 289–322.

Pearce, N. J. G., Denton, J. S., Perkins, W. T., Westgate, J. A. & Alloway, B. V. 2007. Correlation and characterisation of individual glass shards from tephra deposits using trace element laser ablation ICP-MS analyses: current status and future potential. *Journal of Quaternary Science*, **22**, 721–236.

Pearce, N. J. G., Alloway, B. V. & Westgate, J. A. 2008*a*. Mid-Pleistocene silicic tephra beds in the Auckland region, New Zealand: their correlation and origins based on the trace element analyses of single glass shards. *Quaternary International*, **178**, 16–43.

Pearce, N. J. G., Bendall, C. A. & Westgate, J. A. 2008*b*. Comment on 'Some numerical considerations in the geochemical analysis of distal microtephra' by A.M. Pollard, S.P.E. Blockley and C.S. Lane. *Applied Geochemistry*, **23**, 1353–1364.

Pearce, N. J. G., Perkins, W. T., Westgate, J. A. & Wade, S. C. 2011. Trace element analysis by laser ablation ICP-MS: the quest for comprehensive chemical characterisation of single sub-10um volcanic glass shards. *Quaternary International*, **246**, 57–81.

Perkins, W. T. & Pearce, N. J. G. 1995. Mineral microanalysis by laserprobe inductively coupled plasma mass spectrometry. *In*: Potts, P. J., Bowles, J. F. W.,, Reed, S. J. B. & Cave, M. R. (eds) *Microprobe Techniques in the Earth Sciences*. The Mineralogical Society, London.

Pisonero, J. & Gunther, D. 2008. Femtosecond laser ablation inductively coupled plasma mass spectrometry: fundamentals and capabilities for depth profiling analysis. *Mass Spectrometry*, **27**, 609–623.

Pollard, A. M., Blockley, S. P. E. & Lane, C. S. 2006. Some numerical considerations in the geochemical analysis of distal microtephra. *Applied Geochemistry*, **21**, 1692–1714.

Potts, P. J., Thompson, M. & Wilson, S. 2002. G-Probe-1 – an international proficiency test for microprobe laboratories – report on round 1: february 2002 (TB-1 basaltic glass). *Journal of Geostandards and Geoanalysis*, **26**, 197–235.

Preece, S. J., Pearce, N. J. G., Westgate, J. A., Froese, D. G., Jensen, B. J. L. & Perkins, W. T. 2011. Old Crow tephra across eastern Beringia: a single cataclysmic eruption at the close of Marine Isotope Stage 6. *Quaternary Science Reviews*, **30**, 2069–2090.

Pyne-O'Donnell, S. D. F., Blockley, S. P. E., Turney, C. S. M. & Lowe, J. J. 2008. Distal volcanic ash layers in the Lateglacial Interstadial (GI-1): problems of stratigraphic discrimination. *Quaternary Science Reviews*, **27**, 72–84.

Reed, S. J. B. 1995. Electron probe microanalysis. *In*: Potts, P. J., Bowles, J. F. W., Reed, S. J. B. & Cave, M. R. (eds) *Microprobe Techniques in the Earth Sciences*. Chapman and Hall, London, 49–89.

Reed, S. J. B. 2005. *Electron Microprobe Analysis and Scanning Electron Microscopy in Geology*. Cambridge University Press, Cambridge.

Salisbury, M. J., Patton, J. R., Kent, A. J. R., Goldfinger, C., Djadjadihardja, Y. & Hanifa, U. 2012. Deep-sea ash layers reveal evidence foe large, late-Pleistocene and Holocene explosive activity from

Sumatra, Indonesia. *Journal of Volcanology and Geothermal Research*, **231–232**, 61–71.

Shaheen, M. E., Gagnon, J. E. & Fryer, B. J. 2012. Femtosecond (fs) lasers coupled with modern ICP-MS instruments provide new and improved potential for in situ elemental and isotopic analyses in the geosciences. *Chemical Geology*, **330–331**, 267–273.

Smith, D. G. W. & Westgate, J. A. 1968. Electron probe technique for characterising pyroclastic deposits. *Earth and Planetary Science Letters*, **5**, 313–319.

Smith, V. C., Isaia, R. & Pearce, N. J. G. 2011*a*. Tephrostratigraphy of post-15 kyr Campi Flegrei eruptions: implications for eruption history and chronostratigraphic markers. *Quaternary Science Reviews*, **30**, 3638–3660.

Smith, V. C., Pearce, N. J. G., Matthews, N. E., Westgate, J. A., Petraglia, M. D., Haslam, M., Lane, C. S., Korisettar, R. & Pal, J. N. 2011*b*. Geochemical fingerprinting the widespread Toba tephra using biotite compositions. *Quaternary International*, **246**, 97–104.

Tamburrino, S., Insinga, D. D., Sprovieri, M., Petrosino, P. & Tiepolo, M. 2012. Major and trace element characterisation of tephra layers offshore Pantelleria Island: insights into the last 200 ka of volcanic activity and contribution to the Miditerranean tephrochronology. *Journal of Quaternary Science*, **27**, 129–140.

Thornalley, D. J. R., McCave, I. M. & Elderfield, H. 2011. Tephra in deglacial ocean sediments south of Iceland:stratigraphy,geochemistryandoceanicreservoir ages. *Journal of Quaternary Science*, **26**, 190–198.

Todd, J. A., Austin, W. E. N. & Abbott, P. M. 2014. Quantifying bioturbation of a simulated ash fall event. *In*: Austin, W. E. N., Abbott, P. M., Davies, S. M., Pearce, N. J. G. & Wastegård, S. (eds) *Marine Tephrochronology*. Geological Society, London, Special Publications, **398**. First published online February 5, 2014, http://dx.doi.org/10.1144/SP398.9

Tomlinson, E. L., Thordarson, T., Muller, W., Thirlwall, M. & Menzies, M. 2010. Microanalysis of tephra by LA-ICP-MS – strategies, advantages and limitations assessed using the Thorsmork ignimbrite (Southern Iceland). *Chemical Geology*, **279**, 73–89.

Turney, C. S. M. 1998. Extraction of rhyolitic ash from minerogenic lake sediments. *Journal of Paleolimnology*, **19**, 199–206.

Turney, C. S. M., Harkness, D. D. & Lowe, J. J. 1997. The use of microtephra horizons to correlate lateglacial lake sediment successions in Scotland. *Journal of Quaternary Science*, **12**, 525–531.

Ukstins-Peate, I., Baker, J. A., Kent, A. J. R., Al-Kadasi, M., Al-Subbary, A., Ayalew, D. & Menzies, M. 2003. Correlation of Indian Ocean tephra to individual Oligocene silicic eruptions from Afro-Arabian flood volcanism. *Earth and Planetary Science Letters*, **211**, 311–327.

Ukstins-Peate, I., Kent, A. J. R., Baker, J. A. & Menzies, M. 2008. Extreme geochemical heterogeneity in Afro-Arabian Oligocene tephras: preserving the fractional crystallisation and mafic recharge processes in silicic magma chambers. *Lithos*, **102**, 260–278.

van Kreveld, S., Sarnthein, M. *et al.* 2000. Potential links between surging ice sheets, circulation changes, and the Dansgaard–Oeschger cycles in the Irminger Sea, 60–18 kyr. *Paleoceanography*, **15**, 425–442.

Wallrabe-Adams, H.-J. & Lackschewitz, K. S. 2003. Chemical composition, distribution, and origin of silicic volcanic ashes in the Greenland–Iceland–Norwegian Sea: explosive volcanism from 10 to 300 ka as recorded in deep-sea sediments. *Marine Geology*, **193**, 273–293.

Wastegård, S. & Rasmussen, T. L. 2001. New tephra horizons from Oxygen Isotope 5 in the North Atlantic: correlation potential for terrestrial, marine and ice-core archives. *Quaternary Science Reviews*, **20**, 1587–1593.

Wastegård, S., Wohlfarth, B., Subetto, D. A. & Sapelko, T. V. 2000. Extending the known distribution of the Younger Dryas Vedde Ash into northwestern Russia. *Journal of Quaternary Science*, **15**, 581–586.

Weis, P., Beck, H. & Gunther, D. 2005. Characterizing ablation and aerosol generation during elemental fractionation on absorption modified lithium tetraborate glasses using LA-ICP-MS. *Analytical and Bioanalytical Chemistry*, **381**, 212–224.

Westgate, J. A. & Gorton, M. P. 1981. Correlation techniques in tephra studies. *In*: Self, S. & Sparks, R. S. J. (eds) *Tephra Studies*. Reidel, Dordrecht, 73–94.

Westgate, J. A., Perkins, W. T., Fuge, R., Pearce, N. J. G. & Wintle, A. G. 1994. Trace element analysis of volcanic glass shards by laser ablation inductively coupled plasma mass spectrometry: application to Quaternary tephrochronological studies. *Applied Geochemistry*, **9**, 323–355.

Westgate, J. A., Pearce, N. J. G., Perkins, W. T., Shane, P. A. R. & Preece, S. J. 2011. Lead isotope ratios of volcanic glass by laser ablation inductively-coupled plasma mass spectrometry: application to Miocene tephra beds in Montana, USA and adjacent areas. *Quaternary International*, **246**, 82–96.

Westgate, J. A., Pearce, N. J. G., Perkins, W. T., Preece, S. J., Chesner, C. A. & Muhammad, R. F. 2013. Tephrochronology of the Toba Tuffs; four primary glass populations define the 75-ka Youngest Toba Tuff, northern Sumatra, Indonesia. *Journal of Quaternary Science*, doi:10.1002/jqs.2672.

WoldeGabriel, G., Hart, W. K. & Heiken, G. 2005. Innovative tephra studies in the East African Rift System Eos. *Transactions of the American Geophysical Union*, **86**, 255–256.

Physical characteristics of tephra layers in the deep sea realm: the Campanian Ignimbrite eruption

S. L. ENGWELL[1,2]*, R. S. J. SPARKS[1] & S. CAREY[3]

[1]*Department of Earth Science, University of Bristol, Will's Memorial Building, Queen's Road, Bristol BS8 1RJ, UK*

[2]*Present address: Istituto Nazionale di Geofisica e Vulcanologia, Sezione di Pisa, Via della Faggiola 32, 56126 Pisa, Italy*

[3]*Graduate School of Earth Science, University of Rhode Island, South Ferry Road, Narragansett, RI 02882, USA*

**Corresponding author (e-mail: samantha.engwell@pi.ingv.it)*

Abstract: Tephra deposits in the deep sea can survive undisturbed for long periods of time and, on regional scales, tend to be much better preserved than their subaerial counterparts. In this study, grain size distributions and thicknesses of tephra deposits from the Campanian Ignimbrite (CI) eruption (39 000 yr BP; magnitude *c.* 7.7) preserved in thirty-three deep sea cores are analysed to infer key eruption parameters. Distal deep sea tephra thickness data show an exponential decrease with distance from source. Such trends are difficult to identify in distal subaerial data owing to reworking and limited exposure. We find that tephra grain size distributions are much less affected by depositional environment than thickness, with trends that are consistent across distal subaerial, lacustrine and deep sea environments. The CI layer exhibits bimodal grain size distributions to distances of *c.* 1000 km, after which it becomes unimodal. Such trends can be related to different mechanisms of tephra transport in the atmosphere, whereby at proximal to medial distances the Plinian and co-ignimbrite phases produce distinct plumes. Within 150 and 900 km from source, Plinian tephra constitutes 40 ± 5% of the deposit volume. Beyond this region where coarse particles are deposited, the plumes merge and fines derived largely from co-ignimbrite elutriation are spread in the atmosphere at velocities greater than the settling velocities of the particles.

Understanding the physical characteristics of tephra deposits, such as componentry, grain size and thickness, is key to understanding the dynamics of volcanic plumes, and in particular how tephra is transported over great distances (e.g. Carey & Sparks 1986). Few large explosive eruptions (>10 km^3 dense rock equivalent) have occurred since there has been the technological ability to make key observations regarding tephra transport and deposition. To understand these processes, there is reliance on theoretical modelling and analysis of deposits (e.g. Bursik *et al.* 1992; Carey & Sparks 1986; Baines *et al.* 2008). As such, much is known about the characteristics of proximal and medial deposits, especially for small-to-moderate sized eruptions (e.g. Biass & Bonadonna 2010; Bonadonna & Houghton 2005; Watt *et al.* 2009). In comparison, there have been few systematic studies of more distal deposits. This is partially related to a lack of deposits and preferential preservation in a small number of environments, predominantly valleys (e.g. Lewis *et al.* 2012). Further complications arise owing to re-working of these deposits. However, recent studies have shown that it is possible to determine reworked tephra from primary tephra fall by analysis of stratigraphic features of the deposit (Matthews *et al.* 2012; Lewis *et al.* 2012).

The low frequency of large eruption events means that our understanding of these eruptions and their plumes is extrapolated from our knowledge of much smaller eruptions. Thus, less is known about how tephra is transported to distances of thousands of kilometres from source, especially for very large magnitude eruptions. Deposits from large explosive eruptions are commonly preserved in deep sea cores, which provide a high-resolution record of volcanism (e.g. Keller *et al.* 1978; Paterne *et al.* 1988, 1990, 2008). In addition, there is increasing potential for understanding distal dynamics using cryptotephra (e.g. Reide *et al.* 2011). The challenge is to find methods to utilize data from distal deposits, specifically their grain size and thickness, to determine key characteristics about exceptionally large explosive eruptions.

This study investigates marine, lacustrine and subaerial deposits of the Campanian Ignimbrite

From: AUSTIN, W. E. N., ABBOTT, P. M., DAVIES, S. M., PEARCE, N. J. G. & WASTEGÅRD, S. (eds) 2014. *Marine Tephrochronology*. Geological Society, London, Special Publications, **398**, 47–64.
First published online February 6, 2014, http://dx.doi.org/10.1144/SP398.7

(CI) eruption at medial to very distal locations (130 to >2000 km distance). The two types of pyroclastic deposit of importance with regards to this eruption are Plinian fall deposits and co-ignimbrite fall deposits. For Plinian eruptions the column height and the velocity and direction of atmospheric winds control the geometry and size of a deposit (e.g. Walker 1981; Pyle 1989; Sparks *et al.* 1997; Biass & Bonadonna 2010). Studies (e.g. Carey & Sigurdsson 1982; Hildreth & Drake 1992; Rose & Durant 2009; Rust & Cashman 2011) have shown that a large proportion of ejecta from Plinian eruptions is fine tephra, particularly glass shards, which can produce extensive distal deposits. Co-ignimbrite tephra is carried into the atmosphere either as a plume above the region of column collapse or as material transported within pyroclastic density currents becomes buoyant and rises above the flow (Bursik & Woods 1996). Once in the atmosphere, the resultant plume is dispersed downwind in the same manner as a Plinian plume. In both cases, deposition of fine material is enhanced by aggregation. The CI Plinian fall and co-ignimbrite fall deposits can be identified to distances of over 2000 km, allowing analysis and interpretation of deposits with regards to eruption dynamics and tephra dispersion.

The Campanian Ignimbrite (CI) eruption

The CI eruption is the largest known Quaternary eruption in Europe (Ninkovich & Heezen 1967; Barberi *et al.* 1978). Analysis of grain size and thickness of the proximal deposits indicate the source of the eruption is located within the Phlegraean fields area of southern Italy (Perrotta & Scarpati 2003; Fig. 1). Both caldera and fissure origins have been suggested, although the exact location of the vent is unknown. The eruption is dated as 39 ± 0.4 ka BP (De Vivo *et al.* 2001; Ton-That *et al.* 2001).

The eruption produced a sequence of deposits, with two proximal pumice fall units and a major ignimbrite that can be divided into numerous stratigraphic subunits (Rosi *et al.* 1999; Perrotta & Scarpati 2003). Geochemically, the deposits are heterogeneous with a compositional range from phonolite to trachyte. Pyle *et al.* (2006) suggested that the eruption evacuated a chemically zoned magma body with a volumetrically dominant and chemically evolved phonotrachyte tapped throughout the eruption, bounded by a less-evolved trachytic layer at the beginning and end. Plume heights in the Plinian phase reached 44–45 km (Pyle *et al.* 2006), dispersing tephra with a particularly high content of fine bubble-wall shards (Pyle *et al.* 2006; Sulpizio *et al.* 2010). These glass shards are distinctive with long, straight to slightly curved plates (Pyle *et al.* 2006). Such aerodynamic shapes are thought to have aided tephra dispersion to great distances.

Tephra deposits correlated with the CI eruption have been discovered in the mainland of Central Italy (Narcisi 1996; Frezzotti & Narcisi 1996), in Northern Italy (Chiesa *et al.* 1990), Bulgaria (Harkovska *et al.* 1990), Albania (Wagner *et al.* 2008; Sulpizio *et al.* 2010), Greece (Vitaliano *et al.* 1981; St. Seymour & Christanis 1995), Russia (Melekestsev *et al.* 1984; Pyle *et al.* 2006), Romania (Constantin *et al.* 2012; Veres *et al.* 2013; Fitzsimmons *et al.* 2013) and Montenegro (Morley & Woodward 2011) covering an area of at least 2×10^6 km^2 (Fedele *et al.* 2002; Pyle *et al.* 2006; Giaccio *et al.* 2008). Difficulties in volume estimation of the CI deposits arise owing to the unknown depositional limits and correlation of the Plinian and ignimbrite/co-ignimbrite phases, particularly at distance. As such, volume estimates for the CI vary by a factor of 2 and have increased, as more tephra deposit sites have been identified. An overview of estimated volumes is provided by Pyle *et al.* (2006), who estimated a total volume dense rock equivalent (DRE) for both the Plinian and the ignimbrite phase of between 105 and 210 km^3.

The CI tephra layer in deep sea cores

Within deep sea deposits from the Mediterranean Sea, the CI deposit is known as the Y5 (e.g. Ton-That *et al.* 2001) or C13 (Thunell *et al.* 1979) horizon. The Y5 is the most widespread tephra of the Eastern and Central Mediterranean region (Keller *et al.* 1978; Gallo *et al.* 1980; McCoy 1981; Vinci 1985; Paterne *et al.* 1986, 1988, 1990; Cramp *et al.* 1989; Vezzoli 1991). In early studies, the CI tephra in deep sea cores was labelled the ‘Lower’ tephra layer by Ninkovich & Heezen (1967). The layer was geochemically correlated with the 25 kyr eruption of Ischia Islands using samples from Salina Island, and was stratigraphically correlated throughout the Mediterranean Sea. Thunell *et al.* (1979) re-analysed the tephra layer and compared the geochemical characteristics with those from proximal deposits of the Campanian Ignimbrite eruption and found a much better correlation. Therefore, the ‘Lower’ tephra layer was instead attributed to the CI eruption. Reanalysis of the geochemical characteristics of ash layers in the Mediterranean has shown that glass chemistry is often not unique to an eruption (Turney *et al.* 2008; Smith *et al.* 2011). Therefore other characteristics, for example the stratigraphy, chronology of the core and the magnitude of the eruption, are taken into account when assessing tephra provenance (Turney *et al.* 2008; Smith *et al.* 2011; Tomlinson *et al.* 2012).

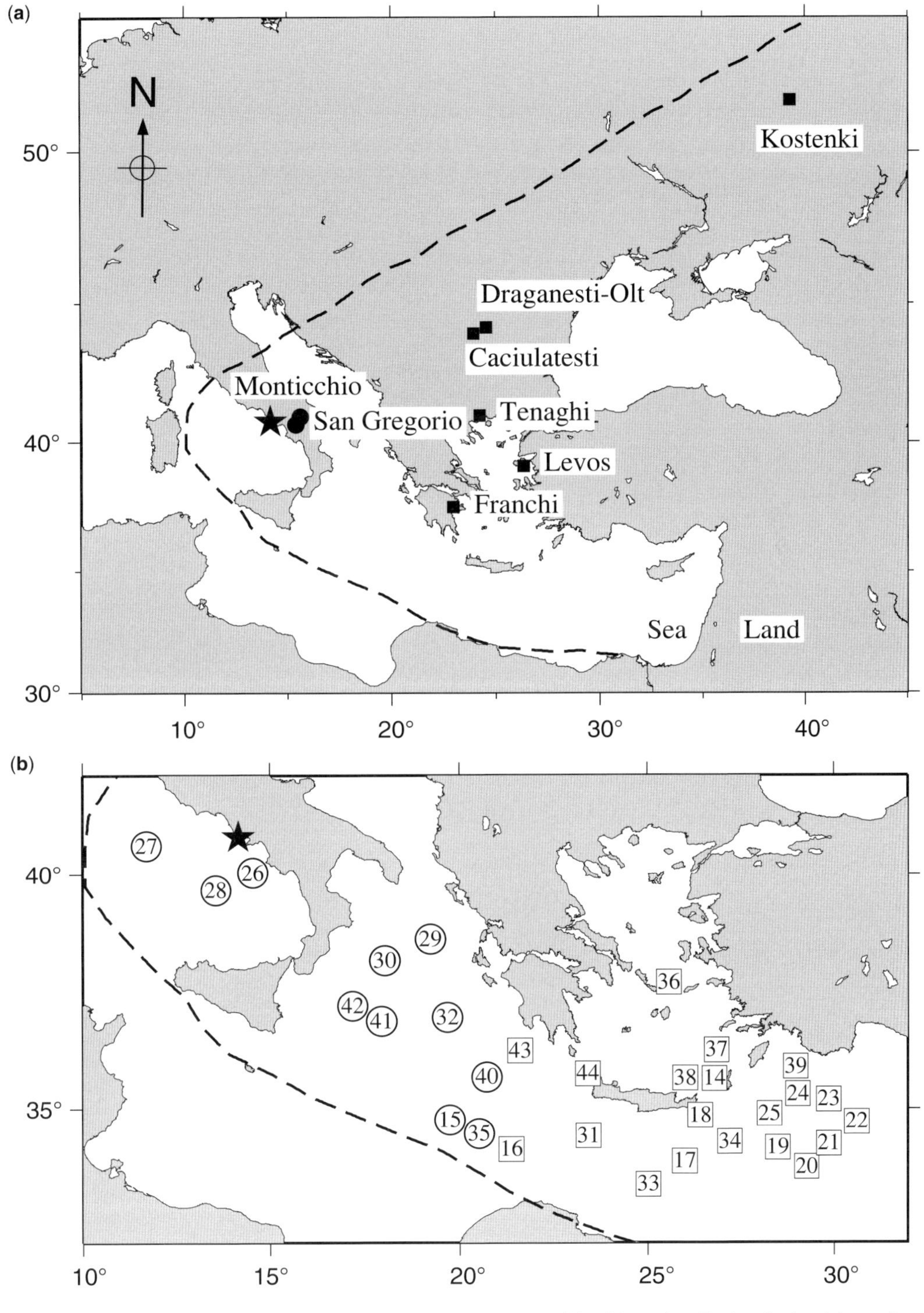

Fig. 1. (**a**) Map of the Mediterranean showing the approximate source of the CI eruption (Star) and subaerial sampling locations. The dashed line encloses the known extent of the eruption (modified from Giaccio *et al.* 2008 and Pyle *et al.* 2006). (**b**) Locations of the deep sea cores analysed in this study (see Table 1 for further information). Circles represent locations where the deposit is bimodal while square symbols identify locations where the deposit is unimodal.

Previous grain size analysis of deposits from deep sea cores was conducted by Sparks & Huang (1980) and Cornell *et al.* (1983). Both studies sieved the coarse fraction of the deposits. For the fine fraction, Cornell *et al.* (1983) used a Particle Data electro-resistance size analyser, whereas Sparks & Huang (1980) used a Coulter counter. Both methods rely on the electrical resistivity of particles. These studies noted a distinct bimodal character to deposits less than 800 km from source. Sparks & Huang (1980) commented that deposits at 440 and 630 km can be separated into a coarse lower unit dominated by elongate pumice separated by a sharp boundary from a fine-grained volcanic glass upper unit. Based on these physical differences the coarse pumice layer was attributed to the Plinian phase of the eruption and the fine-grained upper unit to the co-ignimbrite phase. With distance from source they describe mixing of these units and therefore difficulty distinguishing the separate units. In addition, both studies note that the coarse component decreases in size away from source, whereas the fine mode shows little change in median diameter with distance, residing at about 13 μm (Cornell *et al.* 1983). The studies showed the distal (>800 km from source) deposits were unimodal.

Both studies interpreted the coarse mode as being from the Plinian phase of the eruption, while the fine mode represents tephra elutriated from large-scale pyroclastic density currents during the co-ignimbrite phase of the eruption. Some studies (e.g. Brazier *et al.* 1983) attribute the bimodal grain size character of such deposits to aggregation, where the fine mode represents particles that are deposited much closer to source than would be expected if individual particulate settling was dominant. More recent studies have shown that the physical origin of CI tephra can be determined using glass chemistry (Tomlinson *et al.* 2012). However, it is not yet possible to determine whether a deposit is co-ignimbrite or Plinian by analysing physical characteristics such as grain size and morphology alone.

Methodology

Thirty-three deep sea cores from cruises MD81, TR171, TR172, Vema 10, Vema 14, Robert Conrad (RC) 9 and KET were analysed (Table 1). These cores sample the dispersal area across the Mediterranean Sea (Fig. 1). The cores were inspected to identify the CI tephra. Subaerial samples from Lago Grande di Monticchio (Wulf *et al.* 2004), San Gregorio di Magno (Munno & Petrosino 2007), Franchi Cave (Vitaliano *et al.* 1981), Megali Limni on the Island of Levos (Margari *et al.* 2007), Tenaghi Philippon peat bog (Müller *et al.* 2011), loess from Romania (Veres *et al.* 2013) and re-worked samples from Russia (Pyle *et al.* 2006) were also analysed.

The thickness of the deposit was measured in each core and the amount of bioturbation was estimated. Where possible, for each of the deep sea cores, samples were taken at 1 cm intervals through the CI layer. Carbonate material was removed by covering the samples with 1 M acetic acid for 24 h. Organic material was removed from subaerial samples by placing each sample in 30% hydrogen peroxide for 72 h. Samples were then cleaned three times using a centrifuge at 4000 rps for 20 min.

Grain size was measured using the Malvern Mastersizer 2000 at the Graduate School of Oceanography, Rhode Island University. The method measures grain size by laser diffraction whereby the angle of refraction of light off a particle is inversely proportional to its grain size. The samples were suspended in deionized water with sodium hexametaphosphate and were placed in an ultrasonic bath to discourage formation of aggregates. Three aliquots were taken for each sample and the average is reported. Results are presented in the form of percentage volume of sample, for example, if a result indicates that 25% of the distribution is within a certain size category, this means that the total volume of all particles with diameters in this ranges represents 25% of the total volume of all particles in the distribution.

The deposits within Lago Grande di Monticchio cores J10 and O10 (Fig. 2) were sampled at 1 cm intervals except where changes in stratigraphy could be identified. Within this core, the deposit can be split into two units. Wulf *et al.* (2004) attributed the lowermost layer to the Plinian phase of the eruption, and the overlying layer to the co-ignimbrite phase of the eruption. Samples from the Plinian fall deposit at Lago Grande di Monticchio were too coarse to measure using the Malvern Mastersizer. These samples were sieved at half-phi intervals down to 90 μm. The material less than 500 μm was also measured using the Malvern Mastersizer and the two datasets were weighted and combined for each of the samples following Eychenne *et al.* (2012).

Results

Within the Lago Grande di Monticchio cores, the tephra is easily separated into Plinian and co-ignimbrite fall deposits separated by a sharp boundary (Fig. 2). The Plinian deposit contains highly vesiculated and elongate buff coloured pumice with abundant lithics and crystals of sanidine, pyroxene and biotite. The co-ignimbrite fall deposit is much finer and is completely composed of glass shards.

The CI layer at medial distances (400–800 km) is composed of elongate pumice fragments, biotite

Table 1. *List of locations of samples of the CI tephra layer that have been analysed in this study; asterisks denote reworked deposits; numbers of deep sea cores correlate with those in **Figure 1***

	Sample\core	N	E	Depositional environment	Water depth	Depth in core (base; m)	Distance from source (km)	Thickness (cm)
1	LGM J10 P	40.93	15.61	Lake	Not known	–	130	16.5
	LGM J10 co-I	40.93	15.61	Lake	Not known	–	130	13
	LGM J10 T	40.93	15.61	Lake	Not known	25	130	29.5
2	LGM O10 P	40.93	15.61	Lake	Not known	–	130	17.5
	LGM O10 co-I	40.93	15.61	Lake	Not known	–	130	*c.* 23.5
	LGM O10 T	40.93	15.61	Lake	Not known	25	130	*c.* 41
3	SGM	40.65	15.43	Lake	Not known	8.5	110	60
4	ML00	39.1	26.33	Lake	Not known	7.6	1060	16
5	ML01	39.1	26.33	Lake	Not known	5.63	1060	8
6	ML2A	39.1	26.33	Lake	Not known	5.54	1060	37
7	ML2B	39.1	26.33	Lake	Not known	5	1060	12
8	Franchi Cave	37.51	22.96	Cave	NA	NA	850	–
9	Phillipon Tenaghi	40.97	24.22	Peat bog	NA	12.81	850	19
10	Dragnesti-OLT	44.16	24.54	Loess	NA	NA	940	<60*
11	Caciulastesti	43.94	23.94	Loess	NA	NA	880	<60*
12	Boschevo	51.4	39	Loess	NA	NA	2300	1–2
13	Kostenki	51.4	39	Loess	NA	NA	2300	<10
14	LC21	35.66	26.58	Deep sea	1522	4.8	1234	10
15	TR171-21	34.45	20.13	Deep sea	2785	0.25	900	1.5
16	TR171-22	34.1	21.36	Deep sea	2380	0.38	997	3.5
17	TR171-27	33.83	25.99	Deep sea	2680	1.13	1312	1.5
18	TR171-28	34.85	26.39	Deep sea	2075	1.27	1273	0.2
19	TR172-11	34.14	28.98	Deep sea	2585	1.03	1510	1.5
20	TR172-12	33.9	29.26	Deep sea	2930	5.52	1550	0.8
21	TR172-18	34.22	29.56	Deep sea	3116	0.39	1550	1
22	TR172-19	34.71	30.15	Deep sea	2354	1.28	1570	1
23	TR172-21	35.21	29.76	Deep sea	2650	0.83	1509	1.7
24	TR172-22	35.32	29.02	Deep sea	3150	1.45	1445	2
25	TR172-24	34.88	28.46	Deep sea	2600	1.53	1427	1
26	TR172-42	40.03	14.55	Deep sea	728	4.05	118	21
27	KET 8022	40.58	11.7	Deep sea	2430	3.2	300	12
28	DED 8708	39.67	13.57	Deep sea		6.18	200	35
29	RC9 190	38.65	19.23	Deep sea	1712	1.31	497	4.5
30	RC9 191	38.2	18.03	Deep sea	2345	3.1	445	4
31	RC9 183	34.5	23.42	Deep sea	2684	1.22	1066	3
32	RC9 189	36.98	19.68	Deep sea	3378	1.38	645	9
33	RC9 181	33.42	25.02	Deep sea	2286	1.03	1264	1.5
34	RC9 179	34.27	27.18	Deep sea	2604	1.42	1360	1
35	RC9 185	34.45	20.12	Deep sea	2858	1.15	890	3
36	V10 48	37.73	25.58	Deep sea	833	3.61	1050	4
37	V10 49	36.08	26.83	Deep sea	1170	2.2	1200	10
38	V10 58	35.67	26.3	Deep sea	2283	6.8	1205	10
39	V10 60	35.82	28.97	Deep sea	4081	4	1400	1.8
40	V10 67	35.7	20.72	Deep sea	2904	1.3	810	4.5
41	V10 68	36.9	17.95	Deep sea	3455	1.84	550	9
42	V10 69	37.23	17.28	Deep sea	3156	1.8	490	3
43	V10 26	36.27	21.6	Deep sea	3515	0.96	830	2
44	V14 132	35.77	23.4	Deep sea	2750	1.2	1100	0.1

Abbreviations: LGM, Lago Grande di Monticchio where P is Plinian; co-I, co-ignimbrite, T, total; SGM, San Gregorio di Magno; ML, Megali Limni.

and vitrious glass. At distances greater than 800 km the pumice component is no longer present and the deposit is made up of glass shards (Fig. 3). The deposit is often pale yellow/grey in colour (Munsell colour 2.5Y/8.1). Many of the deposits show a bioturbated top such as LC21 (Fig. 3).

Fig. 2. The Campanian Ignimbrite (CI) within cores (**a**) J10 and (**b**) O10 from Lago Grande di Monticchio (130 km from source). The Plinian deposit is separated from the overlying co-ignimbrite ash by a sharp boundary; images courtesy of S. Wulf.

Fig. 3. (**a**) The *c.* 15 cm thick CI layer within core LC21 from cruise MD81, located *c.* 1230 km from source. (**b**) Scanning electron microscope image of the CI tephra. (**c**) Grain size distributions from the base and top of the deposit. There is a distinct increase in fines towards the top.

Thickness

The thickness at each location is plotted alongside previously published data (Fig. 4). The plot shows a large range in thickness of the Plinian fall deposit close to source (<50 km). No measurements of co-ignimbrite thickness at these proximities are published. The Lago Grande di Monticchio (130 km from source) cores contain both Plinian fall and co-ignimbrite fall deposits, but owing to their distance from source and elevation are not thought to contain primary ignimbrite. Measurements for these thicknesses are plotted separately. For the more distal deposits it was not possible to separate the Plinian and co-ignimbrite layers, and therefore total thickness is plotted.

There is a large range in thickness of the deposit in distal subaerial environments, with most deposits typically much thicker than for equivalent deep sea deposits. This variability is attributed to reworking (e.g. Pyle *et al.* 2006) and, for this reason, regression was applied to the deep sea thickness measurements only. The thickness of the tephra layer is consistent with an exponential decrease with distance from source (Fig. 4). Prediction of the deposit thickness at 130 km using this exponential trend gives a thickness that is less than that seen at Lago Grande di Monticchio but within 95th percentile confidence levels determined by the deep sea deposits.

Grain size

Grain size distributions were obtained for each sample. At Lago Grande di Monticchio, 130 km from source, the deposit can be separated into a lower coarse-grained bimodal (modes between 620 and 590 μm) Plinian deposit overlain by a fine (25–30 μm) unimodal co-ignimbrite fall deposit. Deposits between 400 and 1000 km from source (e.g. RC9–190, 490 km; Fig. 5) cannot be divided into separate units but show distinct grain size bimodality. The bimodality varies with depth in each deposit. The coarse mode is dominant at the base

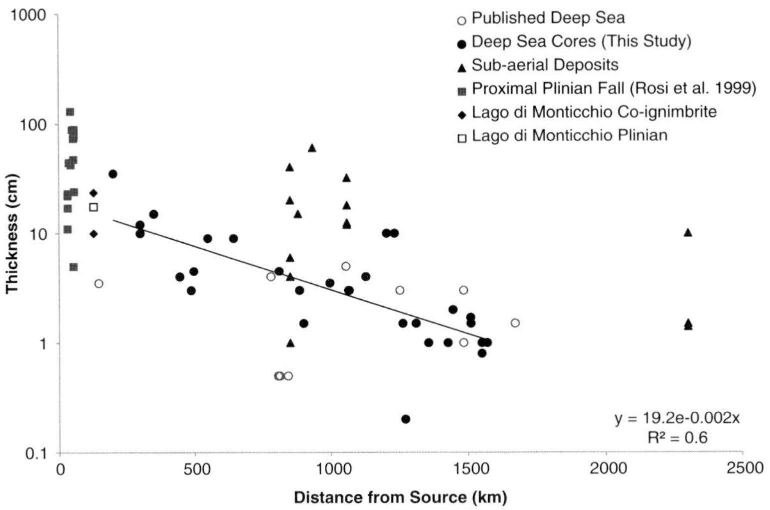

Fig. 4. Thickness of CI deposits with distance from source. The thickness in deep sea cores can be described by an exponential trend. Subaerial thickness is highly variable and is not included in the regression.

of the deposit, with the fine mode becoming more prevalent at the top. In addition, the coarse mode becomes finer with decreasing depth in the deposit, whereas the fine mode does not change with depth. Distal deposits (>850 km) are unimodal (Fig. 5), and deposits show very similar grain-size distributions independent of depositional environment. The exception to this is that of Franchi Cave, which is much coarser than all other deposits at similar distances from the source. This is attributed to the removal of fines by wind. The results obtained from analysis of subaerial deposits from Russia are identical to those shown by Pyle *et al.* (2006).

The grain size modes of the deposits are shown in Figure 6. Where multiple samples were taken from a single deposit, the weighted individual sample grain size distributions were averaged to produce a single size distribution. The fine and coarse modes for bimodal samples are plotted separately. In agreement with results of previous studies, the coarse mode becomes finer with distance from source, and this can be described by an exponential function. However, somewhat counter-intuitively, the fine mode becomes coarser with distance from source. This is highlighted when grain size distributions of the co-ignimbrite deposit at Lago Grande di Monticchio and the fine mode at distance are compared. By correlating the deposits at Lago Grande di Monticchio with more distal deposits, it is possible to determine the physical provenance of the distal subpopulations. Based on the above trends, we correlate the coarse mode with the Plinian and the fine mode with the co-ignimbrite phase of the eruption.

The deposit becomes unimodal at *c.* 850 km from the source. The mode of different deposits varies between 45 and 80 μm, much coarser than the fine mode of 13 μm measured by Cornell *et al.* (1983). This can be attributed to the different methods by which results are recorded and presented. The previously published data are based on the number of particles of a given grain size rather than the volume of the particles. Fitting a regression to the unimodal deposits suggests a slight decrease in grain size mode with distance from source. The very distal subaerial deposits are coarser than would be expected at such distances based on regression using the deep sea tephra samples, but are within uncertainty limits.

The equations generated by fitting the coarse and fine mode data of the bimodal deposits indicate convergence at a distance of 1400 km from source. However, deposits at distances of 850 km from source are unimodal. This indicates that between 850 and 1400 km the difference between the two grain size subpopulations which comprise the total grain size distribution is so minor that the deposit appears unimodal. Analysis of multiple samples from a single unimodal deposit shows that, between 850 and 1500 km from source, there is a distinct change in mode from the base of the deposit to the top. Tephra at the base are much coarser than at the top of the deposit. This trend is clearly illustrated when comparing grain size distributions from the top and base of the deposit in deep sea core LC21 (Fig. 3c). Between 850 and 1400 km, the grain size at the base of the deposit (the coarser mode) decreases with distance from source, while the

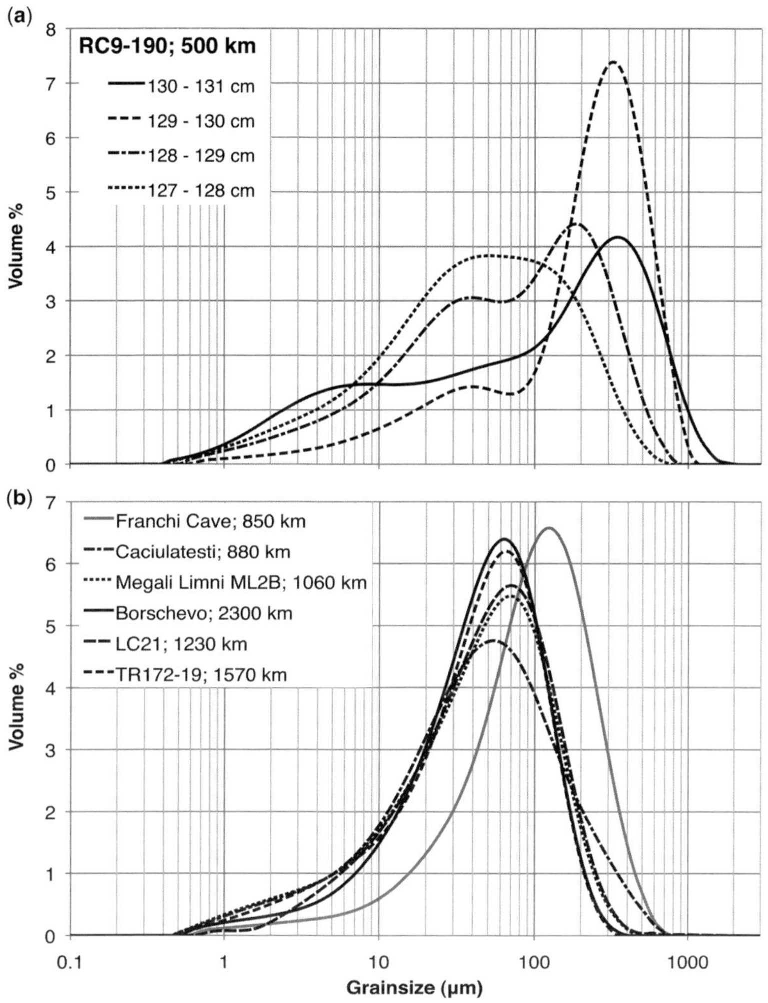

Fig. 5. (**a**) Grain size results from deep sea core RC9-190, 500 km from source, showing upward fining. The coarse mode and volumetric fraction of the coarse population decrease throughout the deposit whereas the fine mode stays constant. (**b**) A representative sample of grain size distributions from deep sea cores and subaerial environments at different distances from source, showing a striking similarity in grain size distributions, despite different depositional environments. The exception is tephra from Franchi Cave, Greece, whose coarse distribution is attributed to winnowing removing fines.

grain size at the top of the deposit (the finer mode) increases. These trends closely follow those of the separate modes from the bimodal deposits, showing that the coarse Plinian component is present to distances of at least 1425 km.

The grain size distributions were converted to cumulative distributions and the 95th percentile was used to represent the maximum grain size and show trends in the coarse tail of the grain size distribution with distance from source (Fig. 7). Maximum grain size shows a decrease with distance from source and can be well represented by an exponential trend. Even at great distance, where deposits are unimodal and where there is little change in mode, there is still a decrease in maximum grain size. The median grain size trend follows the trends in the 95th percentile closely, but there is much more scatter in these data.

The Kware Geological Software SFT application (sequential fragmentation/transport; Wohletz *et al.* 1989) was used to deconvolve the grain size distributions of each of the bimodal deposits into a number of separate subpopulations. In applying SFT, it is assumed that the component

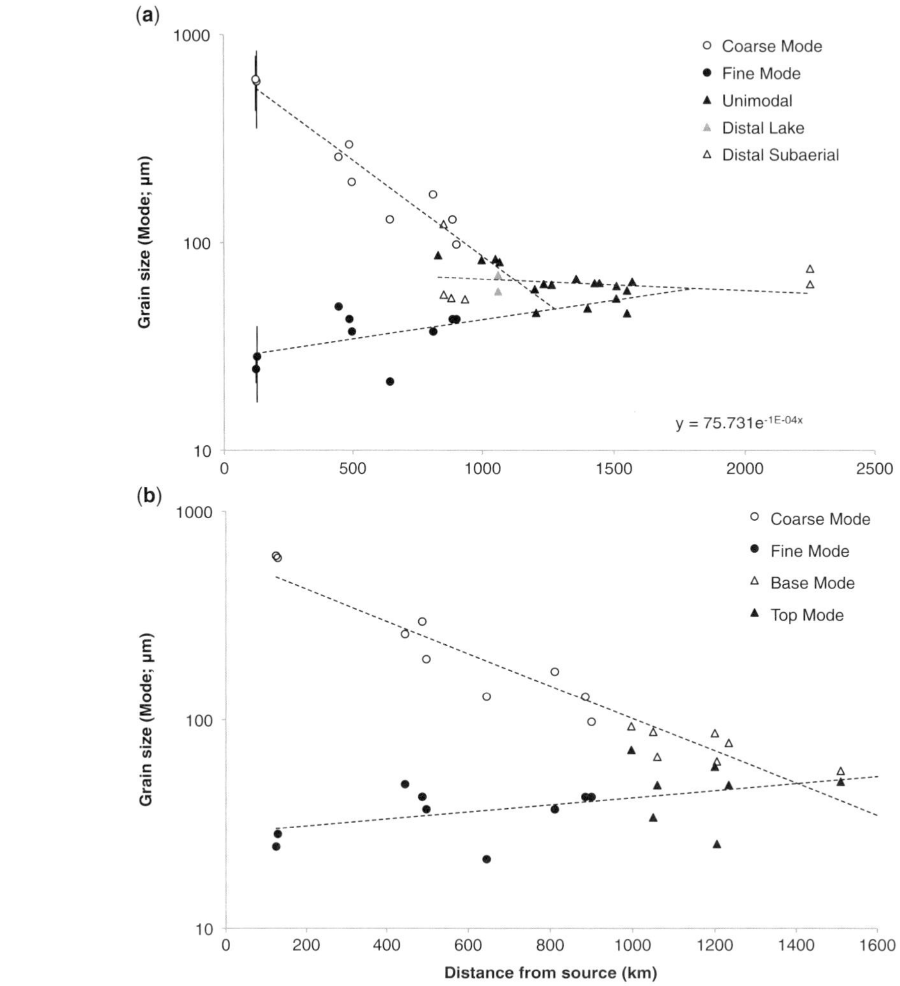

Fig. 6. (**a**) Grain size mode with distance from source. The coarse mode is described by an exponential decrease with distance from source ($y = 721.0e^{-0.002x}$; $R^2 = 0.9$). The fine ($y = 27.7e^{-0.0004x}$; $R^2 = 0.2$) and unimodal ($y = 75.7e^{-1E-4x}$; $R^2 = 0.1$) modes describe an exponential increase and slight exponential decrease, respectively. (**b**) Mode of deposit extremities. The solid symbols represent the coarse and fine modes of the bimodal samples. The open symbols represent samples from unimodal deposits displaying a change in mode from the base to the top of the deposit. The coarse mode depicts an exponential decrease ($y = 603.6e^{-0.002x}$; $R^2 = 0.9$), while the fine mode again describes an exponential increase with distance from source ($y = 28.6e^{0.004x}$; $R^2 = 0.2$).

subpopulation distributions follow a log–normal distribution. Similar analyses of deposits from Mt St Helens (Rose & Durant 2009) and Volcan de Colima (Evans *et al.* 2009) have enabled identification of subpopulations, especially of fine ash, and consequently implications for atmospheric dispersal. Here, a maximum number of three subpopulations were used such that the integrated residuals between the data and modelled subpopulation equaled less than 5%.

The deposit at Lago Grande di Monticchio can be separated into three subpopulations. The bimodal Plinian fall deposit was separated into two constituent distributions and these modes are plotted separately as subpopulations 1A and 1B. The grain size of the co-ignimbrite fall varies little throughout the deposit and is unimodal. This is plotted as subpopulation 2.

The best fit to the bimodal deep sea tephra was achieved using three subpopulations (Fig. 8a, Table 2). The coarse subpopulation (population 1) mode varies between 300 and 140 µm and the data can be described by an exponential decrease with distance from source. The middle subpopulation

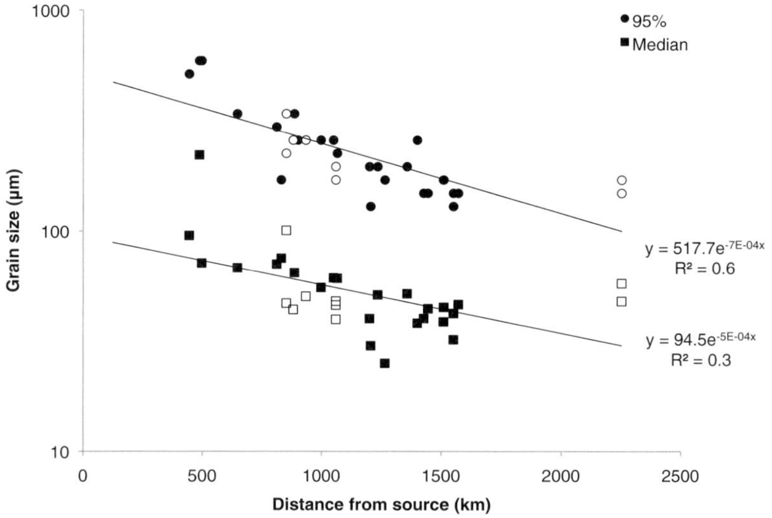

Fig. 7. The 95th percentile, calculated using cumulative grain size distributions, is used as measure of maximum grain size. The median grain size of each sample is also plotted for comparison. Deep sea deposits are shown by solid symbols and subaerial deposits by open symbols. The data can be described by an exponential trend with distance from source. Again, there is little difference between the deep sea and subaerial deposits.

(population 2) mode varies in size between 48 and 30 μm but shows no systematic change with distance from source. The third subpopulation is considerably finer than the other subpopulations with a mode of 3–4 μm. As with subpopulation 2, there is no systematic change in the mode of subpopulation 3 with distance from source. There is a clear correlation between the co-ignimbrite subpopulation in Lago Grande di Monticchio and the distal subpopulation 2. It is more difficult to correlate the coarse distal subpopulation with either one of the Plinian subpopulations (1A and 1B) at Lago Grande di Monticchio.

The volumetric fraction of each subpopulation relative to the total sample is illustrated in Figure 8b. The fraction of the deposit composed of subpopulation 1 and 2 varies significantly with distance from source. Between 400 and 700 km there is no clear trend in the data. At distances greater than 800 km, subpopulation 1, the population associated with the Plinian phase of the eruption, is still present but increasingly becomes the minor fraction. There is very little change in subpopulation 3 with distance from source. Assuming that the coarse mode represents the Plinian phase of the eruption and the fine mode the co-ignimbrite phase, it is possible to calculate the relative proportions of the separate phases. Integration of data shown in Figure 8 between 130 and 900 km show that the Plinian material constitutes 40 ± 5% of the tephra while co-ignimbrite material comprises 60 ± 6%. To date, there have been no studies that have quantified the proportion of eruptive phases within distal deposits and as such there are no figures with which to compare these results. Previous studies have highlighted the difficulties in attributing distal deposits to either Plinian or co-ignimbrite phases, for example, deposits from the 1815 eruption Tambora (Self *et al.* 2004; Sigurdsson & Carey 1989). Here we are able to show that the Plinian ash constitutes a great proportion of the CI deposit at distance with significant implications for ash dispersal. It is not possible to quantitatively determine the relative proportions of tephra from the different phases at distances greater than 900 km owing to the unimodal nature of the deposits. The rapid decrease in Plinian component between 800 and 900 km indicate that it is minor.

Discussion

Changes in the physical characteristics of tephra with distance from source can be used to infer key eruptive parameters and constrain how tephra is transported through the atmosphere and ocean. Analysis of deposits from a range of environments has shown that those in the deep sea are more reliable than those on land with regards to capturing trends, particularly the thickness characteristics. This is despite the potentially large influence of bioturbation, slumping and effects of currents on deep sea deposits (McCoy 1981). The grain size distributions of the CI deposits reflect the plume

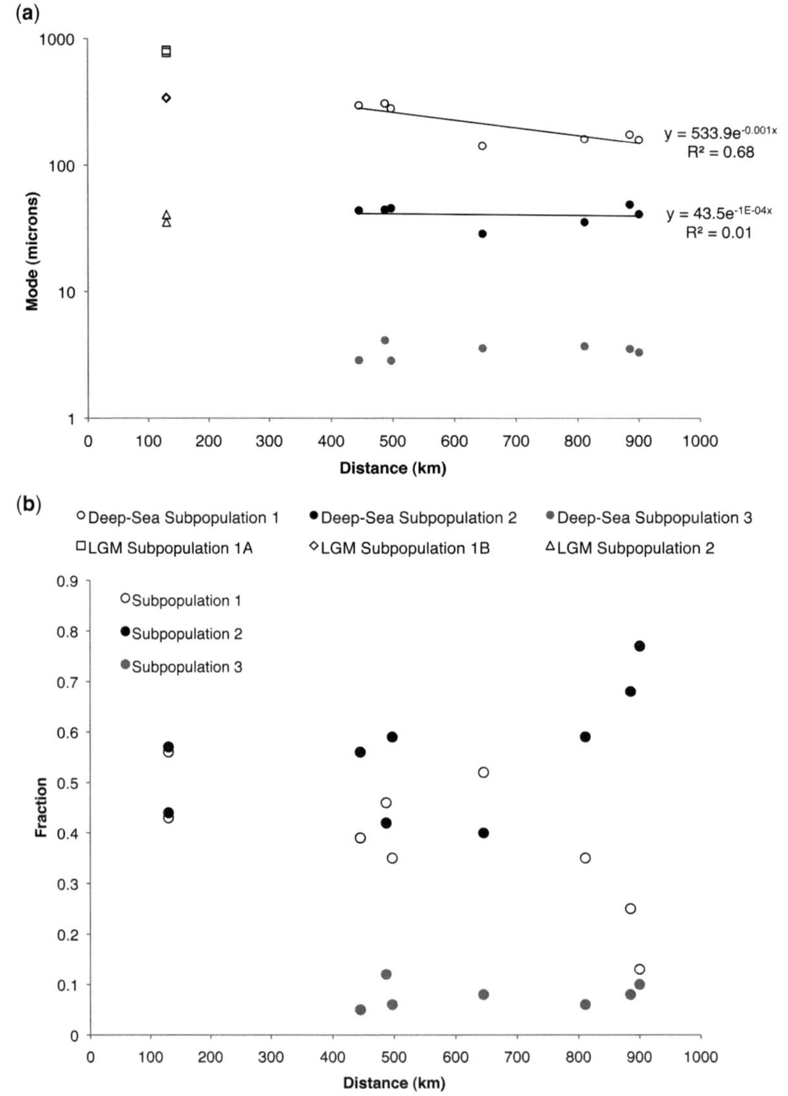

Fig. 8. Kware sequential fragmentation/transport (SFT) analysis results for the bimodal samples. Each of the sample grain size distributions was separated into three log–normal distributions. (**a**) The change in mode of the distributions with distance from source. (**b**) Fraction of subpopulations in each sample with distance from source.

dynamics of the Plinian and co-ignimbrite phases. Analysis of the CI deposits at Lago Grande di Monticchio, in addition to deconvolution of distal grain size distributions, allow the more distal coarse mode in deep sea cores to be attributed to the Plinian phase of the eruption while the fine mode is attributed to the co-ignimbrite phase. Boundaries between the Plinian and co-ignimbrite deposits are used to infer the relative timing of the phases while grain size trends provide information regarding tephra dispersion and sedimentation.

At a given distance from source, the grain size distributions in the deep sea and subaerial environments are virtually indistinguishable. For deposition in the deep sea, tephra is transported not only through the atmosphere but also through thousands of metres of water to the sea floor. Given that the settling velocity in water is significantly slower than in air (e.g. Cashman & Fiske 1991), an efficient depositional process is required to preserve the original grain size distributions. Carey (1997) and Manville & Wilson (2004) have shown that tephra

Table 2. *Kware sequential fragmentation/transport (SFT) results for bimodal samples, with their subpopulations (μm) and associated fractions*

Sample	RC9-185	RC9-189	RC9-190	RC9-191	TR171-21	V10-67	V10-69
Distance (km)	885	645	497	445	900	811	487
Thickness (cm)	3	10	4	4	2.5	4.5	0.5
Subpopulation 1 (μm)	174	142	281	297	158	160	308
Subpopulation 1 fraction	0.25	0.52	0.35	0.39	0.13	0.35	0.46
Subpopulation 2 (μm)	49	29	46	44	41	35	45
Subpopulation 2 fraction	0.68	0.4	0.59	0.56	0.77	0.59	0.42
Subpopulation 3 (μm)	4	4	3	3	3	4	4
Subpopulation 3 fraction	0.08	0.08	0.06	0.05	0.1	0.06	0.12

in the deep sea environment is deposited as vertical density currents. Tephra accumulates at the air–sea interface owing to the differential settling velocity until Rayleigh–Taylor instabilities develop owing to local reversals in density stratification. Vertically descending plumes then deliver the tephra to the bottom waters. By this process, the tephra is deposited much more quickly than when only Stoke's law settling is considered, enabling fines to be preserved in the deposits. Further evidence is provided by analysis of settling rates of ash in the South China Sea after the Pinatubo eruption (Weisner *et al.* 1995).

Thickness measurements of the CI deposits in the deep sea realm can be described by exponential thinning. Such trends are a common observation in volcanic fall deposits (e.g. Thorarinsson 1954; Pyle 1989). It is difficult to reconcile thickness data from subaerial deposits with those from deep sea deposits. We have shown that, except with regards to constraining the extent of dispersal, there are no extra benefits to focusing on subaerial deposits. In addition, the limited number of distal subaerial locations where the tephra is found makes drawing isopach maps difficult and hampers estimation of erupted volume. Recent volume estimates of the CI eruptions (e.g. Costa *et al.* 2012) are based on the fit of an advection–diffusion tephra dispersion model to thickness data. However, large uncertainties remain when fitting such models to distal subaerial deposits and therefore volume estimates contain large uncertainties. In addition, such models do not account for the different phases of the eruption and assume that the distal deposits originate from a single phase. We have shown that Plinian tephra is present in deposits to distances of at least 1400 km and between 130 and 900 km constitutes 40 ± 5% of the deposit. The dispersion of Plinian products to great distance may have been aided by the elongate nature of the pumice (Pyle *et al.* 2006). The proportion of Plinian material decreases with distance from source and therefore it is expected that the actual proportion of Plinian material is less but is difficult to quantify. Although the co-ignimbrite tephra is volumetrically dominant, it may not be feasible to fit distal deposits assuming only a single component and plume.

The grain size distributions can be interpreted to infer information regarding eruption dynamics. The rate of change in the coarse mode can be attributed to a number of different factors. These include volumetric flux, plume height, clast shape and the influence of wind. The presence of bimodal distributions is commonly attributed to aggregation of fine ash (e.g. Brazier *et al.* 1983) whereby the fine population is deposited as particle aggregates and the coarse population is due to the fallout of individual particles. Comparison of the distal deposits with those at Lago Grande di Monticchio shows that aggregation alone cannot explain bimodality in medial deposits. The presence of a distinct co-ignimbrite layer at Lago Grande di Monticchio suggests that fine ash from this eruption was deposited after the Plinian deposit. The correlation of the distal fine mode with the co-ignimbrite deposit at Lago Grande di Monticchio indicates that, for this eruption, the bimodal distribution represents the separate phases of the eruption. The concentration of fines so close to source indicates that aggregation played a significant role in deposition of the fine population.

Another interesting observation is the increase in fine mode with distance from source. The deposit at Lago Grande di Monticchio (*c.* 30 μm) is considerably finer than the more distal ash (50–80 μm). With greater distance, the fine mode becomes coarser still, as shown from the top of the unimodal deposits between 850 and 1400 km. This could be attributed to two processes; (1) mixing of coarse Plinian ash with co-ignimbrite ash in the atmosphere at distance; and (2) the full grain size distribution of the co-ignimbrite phase is deposited at Lago Grande di Monticchio owing to proximity to source and very high concentrations of tephra in the atmosphere at these distances. At greater distances, the coarser material (>40 μm) becomes concentrated at the base of the plume and is deposited as aggregates while finer tephra is retained and dispersed further downwind and deposited to produce cryptotephra. It is likely that both processes contribute to the coarsening trend.

The CI eruption timeline is relatively simple in comparison to other eruptions. Proximal deposits show that the eruption produced a high-intensity Plinian phase followed by multiple large pyroclastic density currents that produced voluminous co-ignimbrite clouds. Deposits give little information regarding the mechanism of eruption evolution from Plinian to ignimbrite forming, for example whether flows were produced by column collapse or were emitted from large fissures (e.g. De Vivo *et al.* 2001; Rolandi *et al.* 2003; Costa *et al.* 2011). The sharp boundary between the Plinian and co-ignimbrite deposits at Lago Grande di Monticchio indicates that the transition was rather abrupt, but with enough time to allow Plinian fall deposits to settle through the water column before co-ignimbrite ash was deposited. We therefore consider the eruption as forming two distinct plumes, the first being the Plinian plume and the second the co-ignimbrite plume (Fig. 9). The Plinian plume is thought to have reached 40–44 km in height (Pyle *et al.* 2006), injecting material with a wide grain size distribution into the atmosphere. The tephra elutriated from pyroclastic density currents formed co-ignimbrite plumes rising 37–40 km (Costa *et al.* 2012) with grain size distributions much smaller and finer than those from the Plinian phase. At distances of 400–630 km from source, Sparks & Huang (1980) were able to identify Plinian deposits with a sharp upper contact separating them from an overlying co-ignimbrite deposit. At distances greater than 630 km, this contact becomes indistinct as the tephra from the two phases became mixed in the atmosphere. This indicates that the co-ignimbrite plume propagated through the atmosphere at greater speeds than the Plinian plume.

Initially both the Plinian and co-ignimbrite plumes can be modelled as intrusive gravity currents

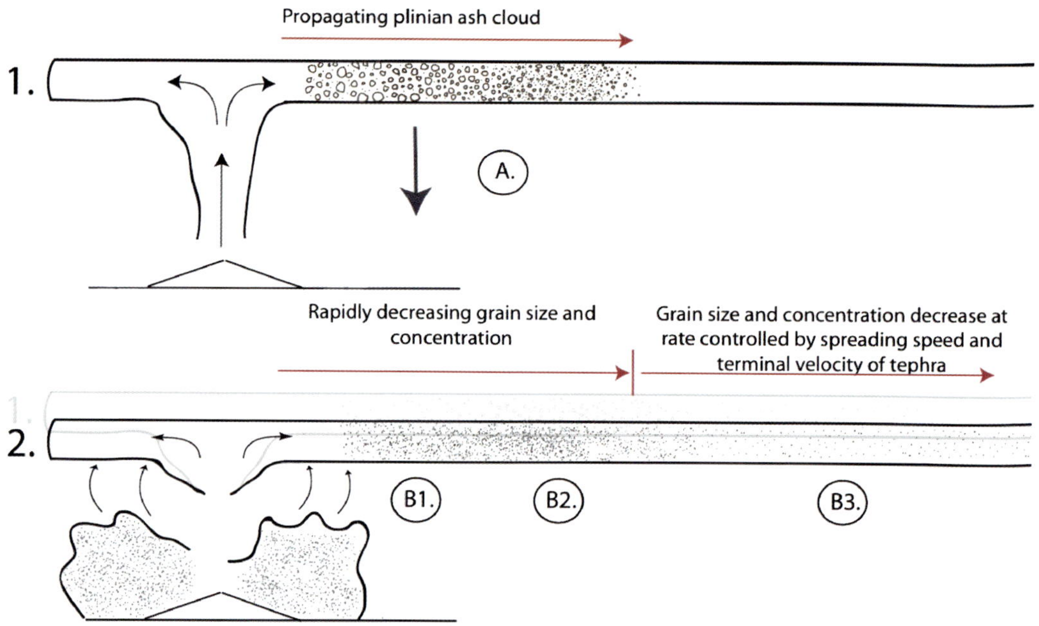

Fig. 9. Schematic model to show dispersion of tephra through the atmosphere. The evolution of the eruption plume can be separated into two main phases. 1, The Plinian phase where at proximal–medial distances (<500 km) the Plinian tephra is deposited (A) before (B) the co-ignimbrite cloud arrives. Close to source (B1), the co-ignimbrite tephra is deposited overlying the Plinian deposit forming strongly stratified deposits. At distances of 400–500 km from source there is still an identifiable Plinian deposit with a sharp contact before the overlying co-ignimbrite deposit. With further distance from source this contact becomes less defined as the tephra from the separate phases becomes increasingly mixed in the atmosphere (B2). At great distances (>1400 km) the tephra is completely mixed in the atmosphere and is transported downwind at velocities greater than the velocities of individual particles shown by similar grain size distributions over great areas (B3).

(e.g. Bursik *et al.* 1992 and Baines 2013), whereby gravitationally induced current velocities near source control the volcanic current spreading at neutral buoyancy level. The diameter of plumes from large eruptions such as the CI is such that Coriolis forces affect the resultant plumes, leading to formation of a giant rotating tephra cloud (Baines & Sparks 2005; Baines *et al.* 2008). This enables these eruptions to cover continental-sized areas with tephra. Further away from the source, the evolution of flow is towards a limiting state of uniform layer depth and velocity (Baines 2013) and spreading is controlled by ambient turbulence (Bursik *et al.* 1992). The rate of spreading at such distances is greater than the terminal velocities of the individual particles, allowing the plume to spread over vast areas. The unimodal data (i.e. >1400 km from source) relates to this distal plume. At such distances, the tephra from the two phases is completely mixed in the atmosphere. The coarsest particles are deposited preferentially, but because they constitute such a small proportion of the grain size distribution in the plume, make little difference to the overall grain size distribution. Fine ash at the bottom edge of the plume produces aggregates and is deposited at rates faster than if the particles had settled individually. This explains the relatively unchanging mode at distances between 800 and 2300 km and also the decrease in maximum grain size. Although the grain size in the plume varies little, the decrease in thickness represents a decrease in tephra concentration in the plume at great distance.

The trends in grain size here can be compared with those of other eruptions. The bimodal grain size distribution seen in CI deposits is also seen in deep sea deposits from the Minoan eruption of Santorini (Sparks & Huang 1980). Analysis of deep sea tephra from the Toba (75 ka) eruption shows a similar pattern to the very distal tephra seen here. Grain size distributions (Rose & Chesner 1987) and modelling (Matthews *et al.* 2012) of distal deposits indicate that the Toba eruption was predominantly co-ignimbrite, producing unimodal deposits that show little change in mean or maximum diameter over great areas (Rose & Chesner 1987, Matthews *et al.* 2012). These observations show that the trends in grain size presented here are common to many large explosive eruptions, and therefore can be used to infer dynamics of other large explosive eruptions.

Conclusions

This work comprises high-resolution, systematic analysis of tephra physical characteristics with distance from source for the CI eruption (39 kyr). The physical characteristics of the CI tephra layer yield important information regarding tephra transport within the atmosphere, particularly at great distances from the source, which is difficult to study for smaller eruptions owing to preservation limitations.

The information, particularly thickness trends, preserved in deep sea deposits is shown to be equivalent, if not better, than that from subaerial locations. Thickness is shown to follow an exponential decrease in the medial to very distal reaches (400–1500 km). Both Plinian and co-ignimbrite contributions can be identified to downwind distances of 1400 km, with the Plinian fraction comprising 40 ± 5% of the deposit between 130 and 900 km. The Plinian component forms the coarse mode and the co-ignimbrite component the fine mode in bimodal deposits. Both modes follow exponential trends; the coarse mode decreases rapidly while the fine mode increases slightly with distance from source. At greater distances there is very little change in grain size distribution over hundreds of kilometres. This is inferred to represent the distal plume where spreading is controlled by ambient turbulence but is still at speeds greater than the terminal velocity of the particles allowing transportation of tephra over continental sized areas.

European Research Council advanced grant (VOLDIES) supported S.E. and R.S.J.S. S.C. was supported by National Science Foundation grant EAR-0545112. Subaerial samples were provided by U. Mueller (Tenaghi Phillipon), W. Ferrand (Franchi Cave), D. Pyle (Russia) and D. Veres (Romanian Loess). The assistance of C. Lane and V. Smith for advice regarding core sampling, and that of S. Wulf, V. Margari, M. Paterne, G. Rothwell, S. MacLachlan and N. Anest when sampling cores, was greatly appreciated. Helpful comments from two anonymous reviewers are also gratefully acknowledged.

References

Baines, P. & Sparks, R. S. J. 2005. Dynamics of giant volcanic tephra clouds from supervolcanic eruptions. *Geophysical Research Letters*, **32**, L24808, http://dx.doi.org/10.1029/2005GL024597

Baines, P., Jones, M. T. & Sparks, R. S. J. 2008. The variation of large-magnitude volcanic tephra cloud formation. *Journal of Geophysical Research*, **113**, D21204, http://dx.doi.org/10.1029/2007jd009568

Baines, P. G. 2013. The dynamics of intrusions into a density-stratified crossflow. *Physics of Fluids*, **25**, 076601, http://dx.doi.org/10.1063/1.4811850

Barberi, F., Innocenti, F., Lirer, L., Munno, R., Pescatore, T. S. & Santacroce, R. 1978. The Campanian Ignimbrite: a major prehistoric eruption in the Neapolitan area (Italy). *Bulletin of Volcanology*, **41**, 10–22, http://dx.doi.org/10.1007/BF02597680

Biass, S. & Bonadonna, C. 2010. A quantitative assessment of eruptive parameters derived from tephra deposits: the example of two large eruptions of Cotopaxi volcano, Ecuador. *Bulletin of Volcanology*, **73**, 73–90, http://dx.doi.org/10.1007/s00445-010-0404-5

Bonadonna, C. & Houghton, B. F. 2005. Total grain size distribution and volume of tephra fall deposits. *Bulletin of Volcanology*, **67**, 441–456, http://dx.doi.org/10.1007/s00445-004-0386-2

Brazier, S., Sparks, R. S. J., Carey, S. N., Sigurdsson, H. & Westgate, J. A. 1983. Bimodal grain size distribution and secondary thickening in air-fall ash layers. *Nature*, **301**, 115–119, http://dx.doi.org/10.1038/301115a0

Bursik, M. I. & Woods, A. W. 1996. The dynamics and thermodynamics of large tephra flows. *Bulletin of Volcanology*, **58**, 175–193, http://dx.doi.org/10.1007/s004450050134

Bursik, M. I., Sparks, R. S. J., Gilbert, J. S. & Carey, S. N. 1992. Sedimentation of tephra by volcanic plumes: 1. Theory and its comparison with a study of the Fogo A Plinian deposit, Sao Miguel (Azores). *Bulletin of Volcanology*, **54**, 329–344, http://dx.doi.org/10.1007/BF00301486

Carey, S. N. 1997. Influence of convective sedimentation on the formation of widespread tephra fall layers in the deep sea. *Geology*, **25**, 839–842, http://dx.doi.org/10.1130/0091-7613(1997)025<0839:IOCSOT>2.3.CO;2

Carey, S. N. & Sigurdsson, H. 1982. Influence of particle aggregation on deposition of distal tephra from the May 18 1980 eruption of Mount St. Helens volcano. *Journal of Volcanology and Geothermal Research*, **87**, 7061–7072, http://dx.doi.org/10.1029/JB087iB08p07061

Carey, S. N. & Sparks, R. S. J. 1986. Quantitative models of the fallout and dispersal of tephra from volcanic eruption columns. *Bulletin of Volcanology*, **48**, 109–125.

Cashman, K. V. & Fiske, R. S. 1991. Fallout of pyroclastic debris from submarine volcanic eruptions. *Science*, **253**, 275–280.

Chiesa, S., Coltorti, M., Cremaschi, M., Ferraris, M., Floris, B. & Prosperi, L. 1990. Loess sedimentation and quaternary deposits in the Marche province. *In*: Cremaschi, M. (ed.) *The Loess in Northern and Central Italy. Quaderni di Geodinamica Alpina e Quaternaria*, **1**, 103–130.

Constantin, D., Timar-Gabor, A., Veres, D., Begy, R. & Cosma, C. 2012. SAR-OSL dating of different grain-sized quartz extracted from a sedimentary section in southern Romania interbedding the Campanian Ignimbrite/Y5 ash layer. *Quaternary Geochronology*, **10**, 81–86, http://dx.doi.org/10.1016/j.quaint.2012.02.042

Cornell, W., Carey, S. & Sigurdsson, H. 1983. Computer simulation of transport and deposition of the Campanian Y-5 tephra. *Journal of Volcanology and Geothermal Research*, **17**, 89–109, http://dx.doi.org/10.1016/0377-0273(83)90063-X

Costa, A., Gottsmann, J., Melnik, O. & Sparks, R. S. J. 2011. A stress-controlled mechanism for the intensity of colossal explosive eruptions. *Earth and Planetary Science Letters*, **310**, 161–166.

Costa, A., Folch, A., Macedonio, G., Giaccio, B., Isaia, R. & Smith, V. C. 2012. Quantifying volcanic ash dispersal and impact from Campanian super-eruption. *Geophysical Research Letters*, **39**, L10310, http://dx.doi.org/10.1029/2012GL051605

Cramp, A., Vitaliano, C. J. & Collins, M. B. 1989. Identification and dispersion of the Campanian tephra layer (Y-5) in the sediments of the Eastern Mediterranean. *Geo-Marine Letters*, **9**, 19–25, http://dx.doi.org/10.1007/BF02262814

De Vivo, B., Rolando, G., Gans, P. B., Calvert, A., Bohrson, W. A., Spera, F. J. & Belkin, H. E. 2001. New constraints on the pyroclastic eruptive history of the Campanian volcanic Plain (Italy). *Mineralogy and Petrology*, **73**, 47–65, http://dx.doi.org/10.1007/s007100170010

Evans, J. R., Huntoon, J. E., Rose, W. I., Varley, N. R. & Stevenson, J. A. 2009. Particle sizes of andesitic ash fallout from vertical eruptions and co-pyroclastic flow clouds, Volcán de Colima, Mexico. *Geology*, **37**, 935–938, http://dx.doi.org/10.1130/G30208A.1

Eychenne, J., Le Pennec, J. L., Troncoso, L., Gouhier, M. & Nedelec, J. M. 2012. Causes and consequences of bimodal grain-size distribution of tephra fall deposited during the August 2006 Tungurahua eruption (Ecuador). *Bulletin of Volcanology*, **74**, 187–205, http://dx.doi.org/10.1007/s00445-011-0517-5

Fedele, F. G., Giaccio, B., Isaia, R. & Orsi, G. 2002. Ecosystem impact of the Campanian Ignimbrite eruption in Late Pleistocene Europe. *Quaternary Research*, **57**, 420–424, http://dx.doi.org/10.1006/qres.2002.2331

Fitzsimmons, K. E., Hambach, U., Veres, D. & Iovita, R. 2013. The Campanian Ignimbrite eruption: new data on volcanic ash dispersal and its potential impact on human evolution. *PLoS ONE*, **8**, e65839, http://dx.doi.org/10.1371/journal.pone.0065839

Frezzotti, M. & Narcisi, B. 1996. Late Quaternary tephra-derived paleosols in central Italy's carbonate Apennine Range: stratigraphical and paleoclimatological implications. *Quaternary International*, **34–36**, 147–153, http://dx.doi.org/10.1016/1040-6182(95)00079-8

Gallo, F., Vernia, L. & Vinci, A. 1980. *Caratteristiche petrografiche e geochimiche di livelli pyroclastic vetrosi in sedimenti recenti del Mar Ionio orientale*, Ateneo Parmense, (ed.) STEP, Parma, 175–182 (special volume dedicated to S. Venzo).

Giaccio, B., Isaia, I. et al. 2008. The Campanian Ignimbrite and Codola tephra layers: two temporal/stratigraphic markers for the Early Upper Palaeolithic in southern Italy and eastern Europe. *Journal of Volcanology and Geothermal Research*, **177**, 208–226, http://dx.doi.org/10.1016/j.jvolgeores.2007.10.007

Harkovska, A. V., Zlatkova, M. J., Chochov, S. D. & Vergilova, Z. I. 1990. Campanian area tephra on the territory of Bulgaria. *In*: *IAVCEI International Volcanological Congress*, 3–8 September, Mainz, Abstracts.

Hildreth, W. & Drake, R. E. 1992. Volcan Quizapu, Chilean Andes. *Bulletin of Volcanology*, **54**, 93–125, http://dx.doi.org/10.1007/BF00278002

Keller, J., Ryan, W. B. F., Ninkovich, D. & Altherr, R. 1978. Explosive volcanic activity in the

Mediterranean over the past 200, 000 years as recorded in deep sea sediments. *Geological Society of America Bulletin*, **89**, 591–604, http://dx.doi.org/10.1130/0016-7606(1978)89<591:EVAITM>2.0.CO;2

Lewis, L., Ditchfield, P., Pal, J. N. & Petraglia, M. 2012. Grain size distribution analysis of sediments containing Younger Toba tephra from Ghoghara, Middle Son valley, India. *Quaternary International*, **258**, 180–190, http://dx.doi.org/10-1016/j.quaint.2011.12.002

Manville, V. & Wilson, C. J. 2004. Vertical density currents: a review of their potential role in the deposition and interpretation of deep sea tephra layers. *Journal of the Geological Society*, **161**, 947–958, http://dx.doi.org/10.1144/0016-764903-067

Margari, V., Pyle, D. M., Bryant, C. & Gibbard, P. L. 2007. Mediterranean tephra stratigraphy revisited: results from a long terrestrial sequence on Levos Island, Greece. *Journal of Volcanology and Geothermal Research*, **163**, 34–54, http://dx.doi.org/10.1016/j.jvolgeores.2007.02.002

Matthews, N. E., Smith, V. C., Costa, A., Durant, A. J., Pyle, D. M. & Pearce, N. J. G. 2012. Ultra-distal tephra deposits from super-eruptions: examples from Toba, Indonesia and Taupo Volcanic Zone, New Zealand. *Quaternary International*, **258**, 54–79, http://dx.doi.org/j.quaint.2011.07.010

McCoy, F. W. 1981. Areal Distribution, Redistribution and mixing of deep sea tephra within the Mediterranean Sea. *In*: Self, S. & Sparks, R. S. J. (eds) *Tephra Studies*. Reidel, Dordrecht.

Melekestsev, I. V., Kirianov, V. Y. & Praslov, N. D. 1984. Catastrophic eruption in the Phlegrean Fields region (Italy) – possible source for a volcanic tephra in late Pleistocene sediments of the European part of the USSR. *Vulcanologija i Seismologija*, **3**, 35–44.

Morley, M. W. & Woodward, J. C. 2011. The Campanian Ignimbrite (Y5) tephra at Crvena Stijena Rockshelter, Montenegro. *Quaternary Research*, **75**, 683–696, http://dx.doi.org/10.1016/j.yqres.2011.02.005

Müller, U. C., Pross, J. *et al.* 2011. The role of climate in the spread of modern humans into Europe. *Quaternary Science Reviews*, **30**, 273–279, http://dx.doi.org/10.1016/j.quascirev.2010.11.016

Munno, R. & Petrosino, P. 2007. The late Quaternary tephrostratigraphical record of the San Gregorio Magno basin (southern Italy). *Journal of Quaternary Science*, **22**, 247–266, http://dx.doi.org/10.1002/jqs.1025

Narcisi, B. 1996. Tephrochronology of a late quaternary lacustrine record from the Monticchio maar (Vulture volcano, southern Italy). *Quaternary Science Reviews*, **15**, 155–165, http://dx.doi.org/10.1016/0277-3791(95)00045-3

Ninkovich, D. & Heezen, B. C. 1967. Physical and chemical properties of volcanic glass shards from Pozzolana tephra, Thera Island and from upper and lower tephra layers in Eastern Mediterranean deep sea cores. *Nature*, **213**, 582–584.

Paterne, M., Guichard, F., Labeyrie, J., Gillot, P. Y. & Duplessy, J. C. 1986. Tyrrhenian Sea tephrochronology of the oxygen isotope record for the past 60,000 yrs. *Marine Geology*, **72**, 259–285, http://dx.doi.org/10.1016/0025-3227(86)90123-4

Paterne, M., Guichard, F. & Labeyrie, J. 1988. Explosive activity of the South ~Italian volcanoes during the past 80000 years as determined by marine tephrochronology. *Journal of Volcanology and Geothermal Research*, **34**, 153–172, http://dx.doi.org/10.1016/0377-0273(88)90030-3

Paterne, M., Labeyrie, J., Guichard, F., Mazaud, A. & Maitre, F. 1990. Fluctuations of the Campanian explosive volcanic activity South Italy during the past 190,000 years, as determined by marine tephrochronology. *Earth and Planetary Science Letters*, **98**, 166–174, http://dx.doi.org/10.1016/0012-821X(90)90057-5

Paterne, M., Guichard, F., Duplessy, J. C., Siani, G., Sulpizio, R. & Labeyrie, J. 2008. A 90 000–200 000 yrs marine tephra record of Italian volcanic activity in the Central Mediterranean Sea. *Journal of Volcanology and Geothermal Research*, **177**, 187–19, http://dx.doi.org/10.1016/j.jvolgeores.2007.11.028

Perrotta, A. & Scarpati, C. 2003. Volume partition between the plinian and co-ignimbrite air fall deposits of the Campanian Ignimbrite eruption. *Mineralogy and Petrology*, **79**, 67–78, http://dx.doi.org/10.1007/s00710-003-0002-8

Pyle, D. M. 1989. The thickness, volume and grain size of tephra fall deposits. *Bulletin of Volcanology*, **51**, 1–15.

Pyle, D. M., Ricketts, G. D., Margari, V., van Andel, T. H., Sinitsyn, A. A., Praslov, N. & Lisitsyn, S. 2006. Wide dispersal and deposition of distal tephra during the Pleistocene "CI/Y5" eruption, Italy. *Quaternary Science Review*, **25**, 2713–2728, http://dx.doi.org/10.1007/BF01086757

Reide, F., Bazely, O., Newton, A. J. & Lane, C. S. 2011. A Laacher See-eruption supplement to Tephrabase: investigating distal tephra fallout dynamics. *Quaternary International*, **246**, 134–144, http://dx.doi.org/10/1016/j.quaint.2011.06.029

Rolandi, G., Bellucci, F., Heitzler, M. T., Belkin, H. E. & De Vivo, B. 2003. Tectonic controls on the genesis of the ignimbrites from the Campanian volcanic zone, southern Italy. *In*: De Vivo, B. & Scandone, R. (eds) Ignimbrites of the Campanian, Plain Special Issue. *Mineralogy and Petrology*, **79**, 3–31, http://dx.doi.org/10.1007/s00710-003-0014-4

Rose, W. I. & Chesner, C. A. 1987. Dispersal of tephra in the great Toba eruption, 75 ka. *Geology*, **15**, 913–917.

Rose, W. I. & Durant, A. J. 2009. Fine ash content of explosive eruptions. *Journal of Volcanology and Geothermal Research*, **186**, 32–33, http://dx.doi.org/10.1016/j.jvolgeores.2009.01.010

Rosi, M., Vezzoli, L., Castelmenzano, A. & Grieco, G. 1999. Plinian pumice fall deposit of the Cam-panian Ignimbrite eruption (Phlegraean fields, Italy). *Journal of Volcanology and Geothermal Research*, **91**, 179–198, http://dx.doi.org/10.1016/S0377–0273(99)00035-9

Rust, A. C. & Cashman, K. V. 2011. Permeability controls on expansion and size distributions of pyroclasts. *Journal of Geophysical Research*, **116**, B11202, http://dx.doi.org/10.1029/2011JB008494

Self, S., Gertisser, R., Thordarson, T., Rampino, M. R. & Wolff, J. A. 2004. Magma volume, volatile emissions, and stratospheric aerosols from the 1815 eruption of Tambora. *Geophysical Research Letters*, **31**, L20608, http://dx.doi.org/10.1029/2004GL020925

Sigurdsson, H. & Carey, S. 1989. Plinian and co-ignimbrite tephra fall from the 1815 eruption of Tambora volcano. *Bulletin of Volcanology*, **51**, 243–270, http://dx.doi.org/10.1007/BF01073515

Smith, V. C., Isaia, R. & Pearce, N. J. G. 2011. Tephrostratigraphy and glass compositions of post-15 kyr Campi Flegrei eruptions: implications for eruption history and chronostratigraphic markers. *Quaternary Science Reviews*, **30**, 3638–3660, http://dx.doi.org/10.1016/j.quascirev.2011.07.012

St Seymour, K. S. & Christanis, K. 1995. Correlation of a tephra layer in western Greece with a late Pleistocene eruption in the Campanian province of Italy. *Quaternary Research*, **43**, 46–54, http://dx.doi.org/10.1006/qres.1995.1005

Sparks, R. S. J. & Huang, T. C. 1980. The volcanological significance of deep sea tephra layers associated with ignimbrites. *Geological Magazine*, **117**, 425–436.

Sparks, R. S. J., Bursik, M. I. *et al.* 1997. *Volcanic Plumes*. John Wiley & Sons, Chichester.

Sulpizio, R., Zanchetta, G., D'Orazio, M., Vogel, H. & Wagner, B. 2010. Tephrostratigraphy and tephrochronology of lakes Ohrid and Prespa, Balkans. *Biogeosciences Discuss*, **7**, 3931–3967, http://dx.doi.org/10.5194/bgd-7-3931-2010

Thorarinsson, S. 1954. The eruptions of Hekla 1947–1948. *In*: *The Tephra fall from Hekla*. Vis Islendinga, Reykjavik, **2**, 68.

Thunell, R., Federman, A., Sparks, R. S. J. & Williams, D. 1979. The age, origin and volcanological significance of the Y-5 tephra layer in the Mediterranean. *Quaternary Research*, **12**, 241–252, http://dx.doi.org/10.1016/0033-5894(79)90060-7

Tomlinson, E., Smith, V. C. *et al.* 2012. Geochemistry of the Phlegraean Fields (Italy) proximal sources for major Mediterranean tephras: implications for the dispersal of Plinian & co-ignimbritic components of explosive eruptions. *Geochimica et Cosmochimica Acta*, **93**, 102–128, http://dx.doi.org/10.1016/j.gca.2012.05.043

Ton-That, T., Singer, B. & Paterne, M. 2001. ^{40}Ar/^{39}Ar dating of latest Pleistocene (41 ka) marine tephra in the Mediterranean Sea: implications for global climate records. *Earth and Planetary Science Letters*, **184**, 645–658, http://dx.doi.org/10.1016/S0012-821X(00)00358-7

Turney, C. S. M., Blockley, S. P. E. *et al.* 2008. Geochemical characterization of Quaternary tephras from the Campanian Province, Italy. *Quaternary International*, **178**, 288–305, http://dx.doi.org/10.1016/j.quaint.2007.02.007

Veres, D., Lane, C. S. *et al.* 2013. The Campanian Ignimbrite/Y5 tephra layer – a regional stratigraphic marker for isotope stage 3 deposits in the Lower Danube region, Romania. *Quarternary International*, **293**, 22–33, http://dx.doi.org/10.1016/j.quaint.2012.02.042

Vezzoli, L. 1991. Tephra layers in Bannock Basin (eastern Mediterranean). *Marine Geology*, **100**, 21–34, http://dx.doi.org/10.1016/0025-3227(91)90221-O

Vinci, A. 1985. Distribution and chemical composition of tephra layers from Eastern Mediterranean abyssal sediments. *Marine Geology*, **64**, 143–155, http://dx.doi.org/10.1016/0025-3227(85)90165-3

Vitaliano, C. J., Taylor, S. R., Farrand, W. R. & Jacobsen, T. W. 1981. Tephra layer in Franchthi cave, Peloponnesos, Greece. *In*: Self, S. & Sparks, R. S. J. (eds) *Tephra Studies*. Reidel, Dordrecht, 373–379.

Wagner, B., Sulpizio, R., Zanchetta, G., Wulf, S., Wessels, M., Daut, G. & Novaczyk, N. 2008. The last 40 ka tephrostratigraphic record of Lake Ohrid, Albania and Macedonia: a very distal archive for ash dispersal from Italian volcanoes. *Journal of Volcanology and Geothermal Research*, **177**, 71–80, http://dx.doi.org/10.1016/j.jvolgeores.2007.08.018

Walker, G. P. L. 1981. Generation and dispersal of fine tephra and dust by volcanic eruptions. *Journal of Volcanology and Geothermal Research*, **11**, 81–92, http://dx.doi.org/10.1016/0377-0273(81)90077-9

Watt, S. F. L., Pyle, D. M., Mather, T. A., Martin, R. S. & Matthews, N. E. 2009. Fallout and distribution of volcanic ash over Argentina following the May 2008 explosive eruption of Chaitén, Chile. *Journal of Geophysical Research*, **114**, B04207, http://dx.doi.org/10.1029/2008JB006219

Weisner, M. G., Wang, Y. & Zheng, L. 1995. Fallout of volcanic ash to the deep South China Sea induced by the 1991 eruption of Mount Pinatubo (Philippines). *Geology*, **23**, 885–888, http://dx.doi.org/10.1130/0091-7613(1995)023<0885:FOVATT>2.3.co;2

Wohletz, K. H., Sheridan, M. F. & Brown, W. K. 1989. Particle size distributions and the sequential fragmentation/transport theory applied to volcanic tephra. *Journal of Volcanology and Geothermal Research*, **94**, 15703–15721.

Wulf, S. W., Kraml, M., Brauer, A., Keller, J. & Negendanl, J. F. W. 2004. Tephrochronology of the 100 ka lacustrine sediment record of Lago Grande di Monticchio (southern Italy). *Quarternary International*, **122**, 7–30, http://dx.doi.org/10.1016/j.quaint.2004.01.028

Identification of a MIS 6 age (*c.* 180 ka) Icelandic tephra within NE Atlantic sediments: a new potential chronostratigraphic marker

F. D. HIBBERT[1,3]*, S. WASTEGÅRD[2], R. GWYNN[1,4] & W. E. N. AUSTIN[1]

[1]*School of Geography and Geosciences, University of St Andrews, Irvine Building, North Street, St Andrews, Fife KY16 9AL, UK*

[2]*Department of Physical Geography and Quaternary Geology, Stockholm University, Stockholm, Sweden*

[3]*Present address: Ocean and Earth Sciences, University of Southampton, National Oceanography Centre, Southampton, European Way, Southampton, UK*

[4]*Present address: Environmental Change Research Centre, Department of Geography, University College London, Gower Street, London WC1E 6BT, UK*

**Corresponding author (e-mail: F.Hibbert@noc.soton.ac.uk)*

Abstract: The incidence of volcanic ash (tephra) within marine sediments serves as a useful stratigraphic marker and tool for correlation. In addition, where an independent age estimate exists, tephra layers can provide a means of dating the sediments themselves. Here we present a geochemically characterized, size sorted tephra layer within Marine Isotope Stage (MIS) 6, most likely resulting from primary air-fall from an Icelandic volcanic source. This tephra layer is tentatively correlated to the Kerlingarfjöll volcanic system using major element geochemistry. The ash layer has an interpolated age of 181 ± 6 ka based on the age model for MD04-2822. We briefly review the occurrence of silicic tephra in the North Atlantic region from MIS 7 to MIS 5e inclusive and find potential correlatives to the MD04-2822 MIS 6 ash layer in the Norway Basin and Irminger Sea.

Geochemically characterized tephra horizons within marine sediment sequences provide a valuable means of correlation and, where independent age control exists (e.g. derived from annually counted ice cores or radiometrically dated proximal deposits), a means of age control for the sediments themselves. In addition, where numerous incidences of a particular tephra layer are documented (e.g. the widely documented Vedde Ash and North Atlantic Ash Zone (NAAZ) II, etc.), inferences about past atmosphere and/or ocean circulation can be made (e.g. Ruddiman & McIntyre 1973, review of Davies *et al.* 2012).

While numerous well-characterized tephra horizons have been identified within the North Atlantic region during the Holocene and last glacial cycle (e.g. Turney *et al.* 1997; Austin *et al.* 2004; Abbott *et al.* 2012), very few are documented during the penultimate glacial interval (Marine Isotope Stage (MIS) 6). We document the incidence and geochemistry of a size-sorted, rhyolitic tephra layer (with Kerlingarfjöll affinity) of early MIS 6 age from a North Atlantic marine sediment core (MD04-2822, Rockall Trough) and review the tephra stratigraphy of the North Atlantic region for this interval.

The identification and characterization of tephra layers within MIS 6 have the potential to significantly aid correlation between marine sediment cores (and the construction of age models if independent age control can be achieved). This is especially true where benthic records are 'patchy', incomplete or at low resolution. In addition, this interval predates the available Greenland ice cores and radiocarbon dating methods and, as such, most age control for marine sediment cores through this interval derives from correlation to benthic $\delta^{18}O$ stacks with potentially large age uncertainties deriving from both the stack chronology and correlation process (see review of Austin & Hibbert 2012).

Material and oceanographic setting

The giant piston core MD04-2822 was collected in 2004 by the RV *Marion Dufresene* from the Rockall Trough (56°50.54′N, 11°22.96′W, *c.* 37.7 m recovery) at a present water depth of 2344 m (Fig. 1). This high-resolution record provides the first insights into the history of the last British–Irish Ice Sheet prior to the last glacial period, confirming the presence of a British–Irish

From: Austin, W. E. N., Abbott, P. M., Davies, S. M., Pearce, N. J. G. & Wastegård, S. (eds) 2014. *Marine Tephrochronology*. Geological Society, London, Special Publications, **398**, 65–80.
First published online February 20, 2014, updated August 29, 2014, http://dx.doi.org/10.1144/SP398.10

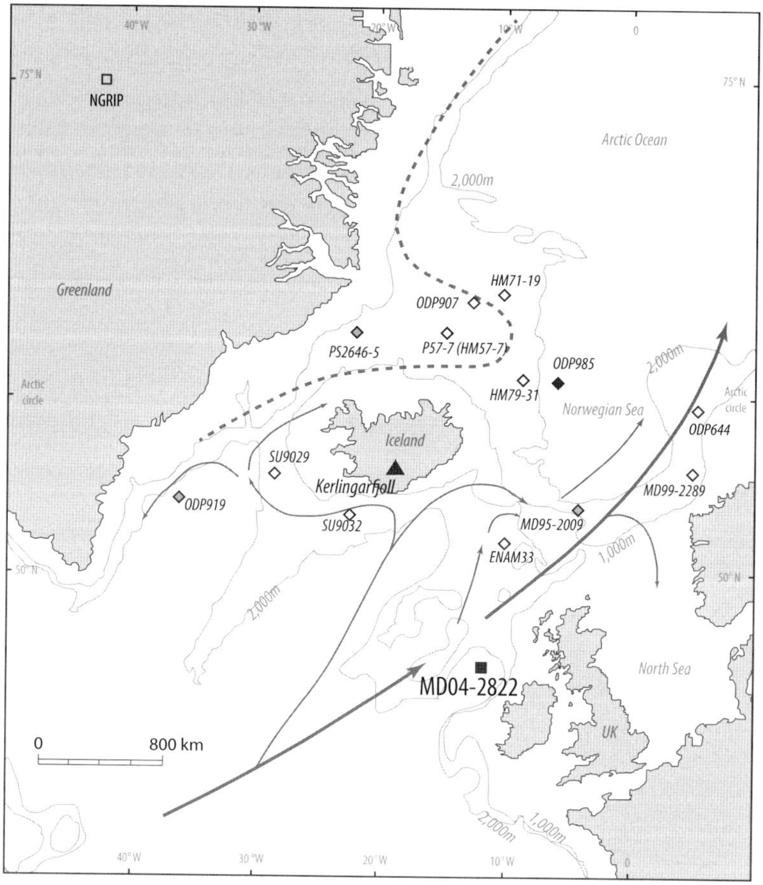

Fig. 1. Location of core MD04-2822 and simplified modern oceanography (red lines – North Atlantic Current; blue dashed line gives the approximate position of the polar front; after Hansen and Østerhus 2000) and cores mentioned in the text.

Ice Sheet on the Hebridean shelf during the penultimate glacial (MIS 6) (Hibbert *et al.* 2010).

Present-day surface conditions are primarily influenced by the North Atlantic Current; however, during glacial episodes, the site was bathed in polar waters as the polar front was positioned far to the south of the core site (e.g. Shackleton *et al.* 2000; Fig. 1).

Methods

Sample preparation

The relative abundance of the planktonic, polar species of foraminifera *N. pachyderma* (sinistral), ice-rafted debris (IRD) and total tephra shard concentrations were determined from the >150 μm size fraction using a light microscope. Counts were a minimum of 300 for both total planktonic foraminifera and tephra. This was converted to IRD and tephra concentration (shards >150 μm g^{-1} of dry sediment processed). Sampling was initially undertaken at 10 cm resolution with subsequent sampling at 1 cm over an interval of increased tephra abundance (30 cm). Shard morphology was recorded using a binocular microscope and the length of the longest axis of individual shards determined using a graticule eye piece.

Shards for geochemical analysis were also selected from the >150 μm size range. These were mounted onto double-sided tape and embedded in clear Araldite™, allowed to harden and polished. The sample blocks were then evaporated-carbon vacuum-coated. Major element analyses were carried out using an electron microprobe at the Facility for Earth and Environmental Analysis (FEEA; University of St Andrews) and the Tephra Analytical Unit (TAU; University of

Edinburgh), and expressed as oxide percentages (wt%; Table 2).

Preliminary analyses were undertaken at FEEA, University of St Andrews with follow-up analyses at TAU, University of Edinburgh. The TAU facilities also enabled phosphorous oxides to be measured for the identified tephra horizon. Analyses at FEEA were undertaken using a Jeol JXA8600 electron microprobe operated using an accelerating voltage of 15 kV, a beam current of 10 nA and a spot size of 5 μm. Primary calibrations were established using a combination of synthetic oxides, pure metals and well-characterized minerals. Secondary rhyolitic and basaltic glass standards (TB1G, Lipari, BCR2G, S-OBS, ATHO and A-ALK) were routinely analysed and used to assess sample stability. A total of 11 analyses on eight shards were undertaken and analyses with totals greater than 95% were used in the characterization of the tephra shards.

Analyses were conducted at TAU, School of Geosciences, University of Edinburgh using a Cameca SX100 electron probe microanalyser with an accelerating voltage of 15 kV and beam current of 2 nA for the major elements and 80 nA for the minor elements and a beam diameter of 5 μm. Standard calibration blocks were used for calibration and glass standards were used to assess the accuracy during analyses. Four shards were analysed and only analyses above 93% are reported. Two secondary standards (BHVO2g and Lipari obsidian) were analysed at regular intervals to monitor instrumental drift and assess the accuracy and precision of the analyses.

Age model

The age model for the core is primarily based upon tuning: (a) the benthic $\delta^{18}O$ (*C. wuellerstorfi*) record to the LR04 stack (Lisiecki & Raymo 2005); (b) percentage abundance of *N. pachyderma* (sinistral) to the NGRIP $\delta^{18}O$ record (on the GICC05 modelext. timescale; Wolff *et al.* 2010 and references therein) and; (c) percentage *N. pachyderma* (sinistral) to the EPICA Dome C methane record (EDC3 timescale; Loulergue *et al.* 2008). A calibrated radiocarbon date was used for the core top age control (Marine09 calibration curve (Reimer *et al.* 2009) using OxCal version 4.2 (Bronk-Ramsey 2009; no ΔR correction applied). Additional age control was achieved by the inclusion of the II-Rhy-1 tephra isochron component of North Atlantic Ash Zone II (see Austin & Abbott 2010), utilizing its ice core age 55.4 $\pm$ 2 ka b2K in the GICC05 timescale (Svensson *et al.* 2008).

The age model presented here (tie-points are given in Table 1) is a refinement of that in Hibbert *et al.* (2010) prompted by the inclusion of additional benthic isotope analyses. This gives a revised age estimate for the lowermost portions of the core, with the entire MD04-2822 record spanning *c.* 8–190 ka (Fig. 2).

Age uncertainty estimates were propagated for each tie-point using a mean squared estimate from which we have been able to estimate the minimum uncertainty associated with the inferred age of the identified MIS 6 tephra horizon. The uncertainty estimate includes age uncertainties associated with the ice core chronology itself (for the MIS 6 portion this is the EDC3 chronology; Loulergue *et al.* 2008), tuning (e.g. the potential error in determining the position of the tie-point within the MD04-2822 record, sample spacing) and bioturbation of the sediments.

Results and discussion

A peak in colourless tephra shard concentrations (approximately 100 shards per gram of dry sediment) was identified at a depth of 36.27–36.28 m core depth (mcd), with elevated tephra concentrations occurring across 0.3 m. The peak in tephra concentrations has an interpolated age of 181 ($\pm$6) ka. This occurs during an interval of increasing benthic $\delta^{18}O$ and clement surface conditions (percentage *N. pachyderma* (sinistral) may be used as a semi-quantitative estimate of sea-surface temperatures) with very low IRD inputs. The high concentration of colourless shards occurs against a background of very low and intermittent inputs of volcanic material (occasional basaltic grains; typically <0.3% of total IRD concentrations; Fig. 3).

The elevated tephra horizon has an asymmetric stratigraphic distribution typical of bioturbated marine sediments. This distribution probably results from the upward migration of tephra deposited on the seafloor within the active mixed layer; as sediments continue to accumulate, the mixed layer migrates upwards and transports ash upwards. The stratigraphic profile below the main concentration peak reflects the mixing of tephra downwards within the original mixed layer (Ruddiman & Glover 1972, Ruddiman *et al.* 1980; Jumars & Wheatcroft 1989; Todd *et al.* 2014). Sedimentation rates during this interval are estimated to be 38 cm ka^{-1}, giving an approximate temporal resolution of 26 years cm^{-1}. This is at the upper end of that expected for deep-ocean sediments (in a highly productive open ocean setting, sedimentation rates can reach 20 cm ka^{-1}).

The shards have a 'winged', bubble-walled morphology (cf. II-RHY-1 tephra component of NAAZII; Austin *et al.* 2004) with shard sizes typically 120–180 μm and a maximum of 408 μm (one shard at 3636–3637 cm). The peak in tephra

Table 1. *Chronostratigraphic tie-points for sediment core MD04-2822*

Origin	Feature	MD04-2822 depth (cm)	Age (ka)	Total error associated with tie-point (ka)
Calibrated core top radiocarbon date	n/a	0.5	0.89	0.09
Percentage *N.* pachyderma (sinistral) tuned to NGRIP $\delta^{18}O$ (GICC05modelext)	start of Holocene	246.5	11.70	0.13
Percentage *N. pachyderma* (sinistral) tuned to NGRIP $\delta^{18}O$ (GICC05modelext)	GS1	287.5	12.90	0.15
Percentage *N. pachyderma* (sinistral) tuned to NGRIP $\delta^{18}O$ (GICC05modelext)	GI-1e	436.8	14.69	0.19
Percentage *N. pachyderma* (sinistral) tuned to NGRIP $\delta^{18}O$ (GICC05modelext)	GI-2	1319.6	23.34	0.60
Percentage *N. pachyderma* (sinistral) tuned to NGRIP $\delta^{18}O$ (GICC05modelext)	GI-3	1756.0	27.78	0.83
Percentage *N. pachyderma* (sinistral) tuned to NGRIP $\delta^{18}O$ (GICC05modelext)	GI-4	1805.4	28.90	0.90
Percentage *N. pachyderma* (sinistral) tuned to NGRIP $\delta^{18}O$ (GICC05modelext)	GI-5	1895.7	32.50	1.14
Percentage *N. pachyderma* (sinistral) tuned to NGRIP $\delta^{18}O$ (GICC05modelext)	GI-6	1930.5	33.74	1.22
Percentage *N. pachyderma* (sinistral) tuned to NGRIP $\delta^{18}O$ (GICC05modelext)	GI-7	1962.1	35.48	1.34
Percentage *N. pachyderma* (sinistral) tuned to NGRIP $\delta^{18}O$ (GICC05modelext)	GI-8	2006.6	38.22	1.45
Percentage *N. pachyderma* (sinistral) tuned to NGRIP $\delta^{18}O$ (GICC05modelext)	GI-9	2014.5	40.16	1.60
Percentage *N. pachyderma* (sinistral) tuned to NGRIP $\delta^{18}O$ (GICC05modelext)	GI-10	2044.1	41.46	1.65
Percentage *N. pachyderma* (sinistral) tuned to NGRIP $\delta^{18}O$ (GICC05modelext)	GI-11	2065.6	43.34	1.77
Percentage *N. pachyderma* (sinistral) tuned to NGRIP $\delta^{18}O$ (GICC05modelext)	GI-12	2095.2	46.86	1.95
Percentage *N. pachyderma* (sinistral) tuned to NGRIP $\delta^{18}O$ (GICC05modelext)	GI-13	2116.7	49.28	2.08
Percentage *N. pachyderma* (sinistral) tuned to NGRIP $\delta^{18}O$ (GICC05modelext)	GI-14	2151.6	54.22	1.16
NAAZII	n/a	2163.5	55.38	1.18
Percentage *N. pachyderma* (sinistral) tuned to NGRIP $\delta^{18}O$ (GICC05modelext)	GI-15	2168.0	55.80	1.20
Percentage *N. pachyderma* (sinistral) tuned to NGRIP $\delta^{18}O$ (GICC05modelext)	GI-16	2215.3	58.28	1.28
Percentage *N. pachyderma* (sinistral) tuned to NGRIP $\delta^{18}O$ (GICC05modelext)	GI-17	2223.6	59.45	1.29
Percentage *N. pachyderma* (sinistral) tuned to NGRIP $\delta^{18}O$ (GICC05modelext)	GS-18	2254.5	63.87	1.40
Percentage *N. pachyderma* (sinistral) tuned to NGRIP $\delta^{18}O$ (GICC05modelext)	GI-18	2265.8	64.13	1.40
Benthic isotope tuned to LR04		2352.0	70.15	4.00
Percentage *N. pachyderma* (sinistral) tuned to NGRIP $\delta^{18}O$ (GICC05modelext)	GI-19	2357.3	72.33	1.72
Percentage *N. pachyderma* (sinistral) tuned to NGRIP $\delta^{18}O$ (GICC05modelext)	GI-20	2376.8	76.45	1.92
Percentage *N. pachyderma* (sinistral) tuned to NGRIP $\delta^{18}O$ (GICC05modelext)	GI-21	2399.7	84.76	2.00
Percentage *N. pachyderma* (sinistral) tuned to NGRIP $\delta^{18}O$ (GICC05modelext)	GS-22	2407.5	87.71	2.05
Percentage *N. pachyderma* (sinistral) tuned to NGRIP $\delta^{18}O$ (GICC05modelext)	GI-22	2421.4	90.06	2.40
Percentage *N. pachyderma* (sinistral) tuned to NGRIP $\delta^{18}O$ (GICC05modelext)	GI-23	2455.4	104.05	2.53
Percentage *N. pachyderma* (sinistral) tuned to NGRIP $\delta^{18}O$ (GICC05modelext)	GI-24	2460.1	108.29	0.32
Benthic isotope tuned to LR04		2483.3	115.54	4.00
Benthic isotope tuned to LR04		2581.3	127.32	4.00
Benthic isotope tuned to LR04		2691.7	135.07	4.01
Benthic isotope tuned to LR04		2750.8	140.00	4.00
Percentage *N. pachyderma* (sinistral) tuned to Antarctic CH_4 (EDC3 timescale)		2985.8	147.63	6.00
Percentage *N. pachyderma* (sinistral) tuned to Antarctic CH_4 (EDC3 timescale)		3056.3	154.40	6.01
Percentage *N. pachyderma* (sinistral) tuned to Antarctic CH_4 (EDC3 timescale)		3156.6	166.13	6.00
Percentage *N. pachyderma* (sinistral) tuned to Antarctic CH_4 (EDC3 timescale)		3536.1	176.01	6.00
Benthic isotope tuned to LR04		3693.2	184.40	4.00
Benthic isotope tuned to LR04		3730.2	186.60	4.01

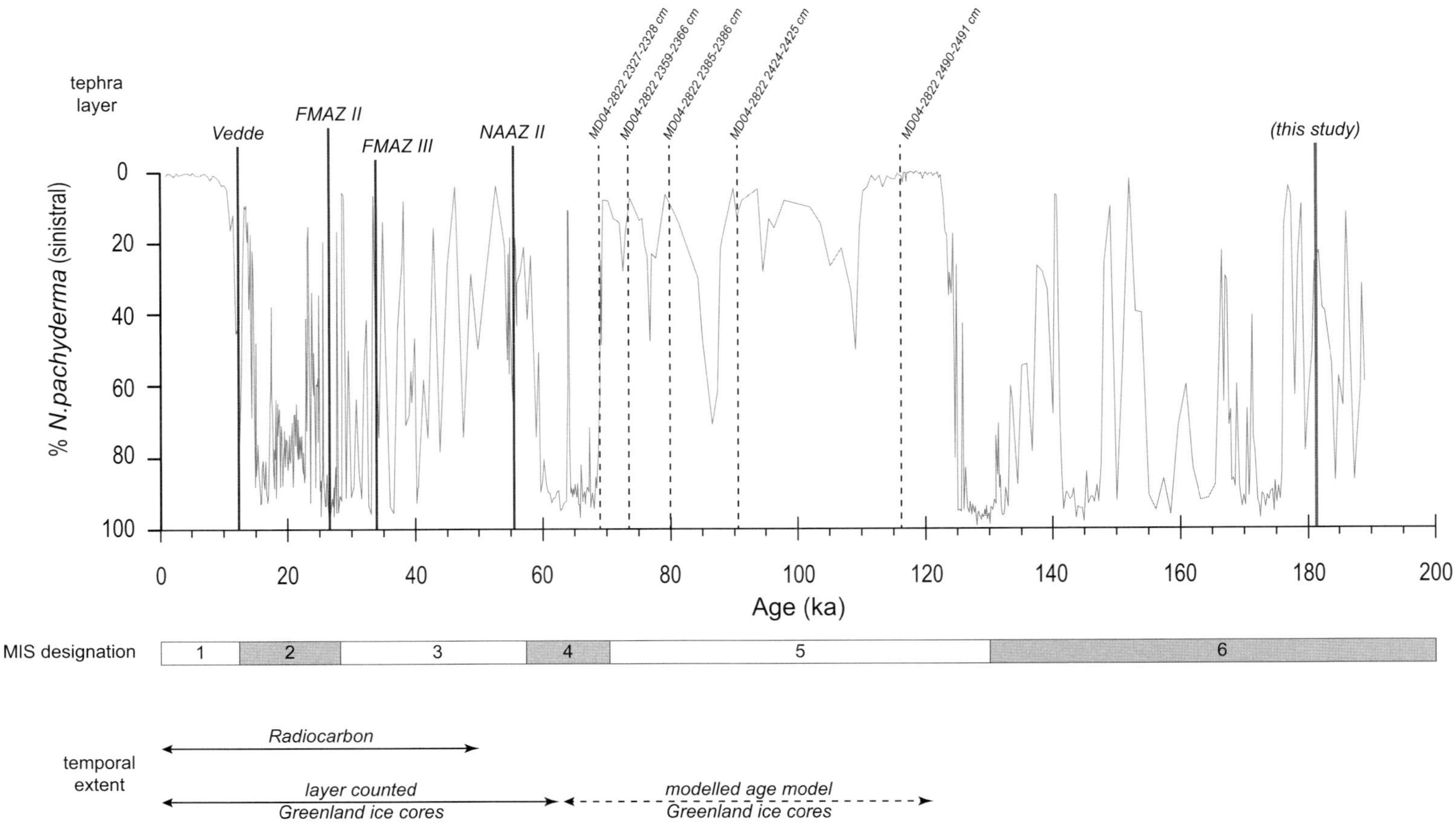

Fig. 2. MD04-2822 percentage *N. pachyderma* (sinistral) record with the stratigraphic positions of identified tephra horizons given by the vertical lines (Peters *et al.* 2010; Hibbert *et al.* 2010; Austin & Hibbert 2012; Abbott *et al.* 2011, 2013). FMAZIII has recently been identified and occurs at a depth of 2008–2010 cm in MD04-2822. The stratigraphic position of the newly identified MIS 6 tephra horizon is given by the red line. MIS designations are after Lisiecki & Raymo (2005); indicative temporal range of the radiocarbon method and Greenland ice core chronologies are given by the arrows.

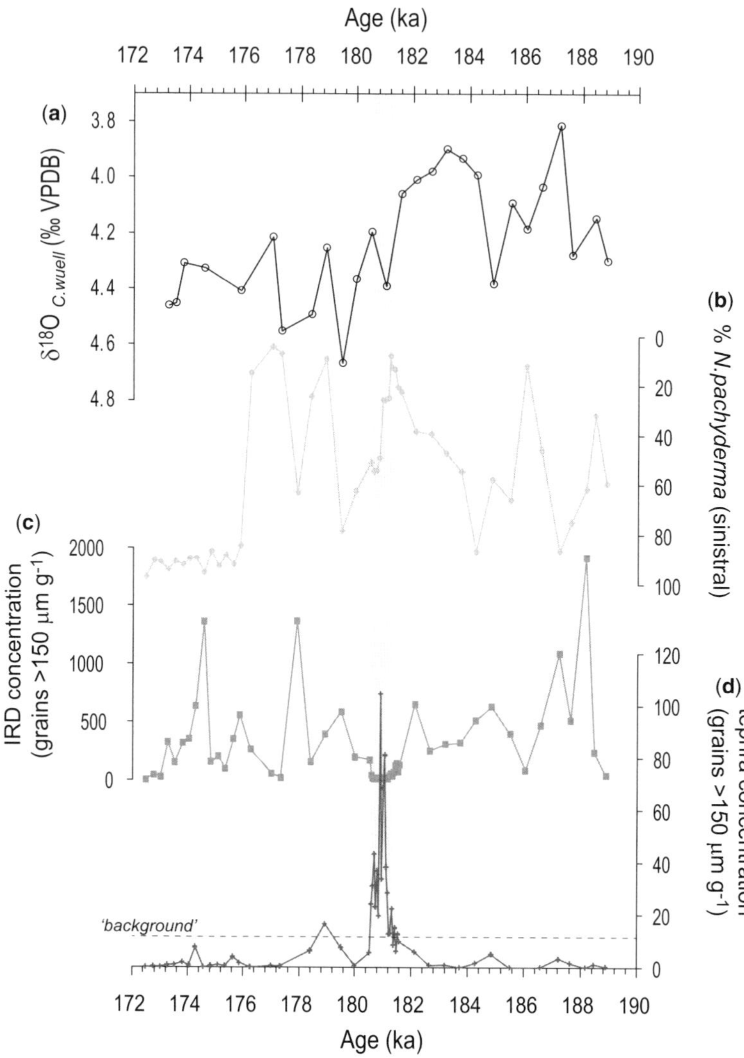

Fig. 3. MD04-2822 (**a**) benthic $\delta^{18}O_{C.wuellerstorfi}$; (**b**) percentage *N. pachyderma* (sinistral); (**c**) ice-rafted debris concentration; and (**d**) tephra concentration on the revised MD04-2822 age model.

concentration (3627–3628 cm) has a bi-modal size distribution about a mode of 127.5 μm and a maximum shard size of 348.5 μm (maximum shard size encountered through the horizon was 408 μm at 3636–3637 cm). The size distribution is similar for all intervals (Fig. 4), although in the 3630–3631 cm interval there is a shift to larger shard sizes which may be the result of sorting through the water column.

The size distribution and well-sorted nature of the horizon would suggest that the tephra was primarily deposited via primary air-fall. Atmospheric transport has been found to be responsible for the large-scale transport of shards produced in explosive eruptions (although grain size decreases exponentially away from the source; Ruddiman & Glover 1972; Ninkovich *et al.* 1978; Lacasse *et al.* 1996*a*, *b*). The MD04-2822 size distribution is indicative of primary air-fall or, alternatively, air-fall onto sea ice (and/or icebergs), which was subsequently transported to the site. The peak in tephra concentration occurs during an interval of generally clement, although cooling, sea-surface temperatures (increasing percentage abundance of *N. pachyderma* sinistral). This hydrographic control could give rise to the observed double peak in concentration, the former resulting from primary air-fall and the latter possible resulting from secondary inputs of sea ice rafted material (sea ice surviving and melting at the site). The lack of IRD during this interval would suggest that transport was not via icebergs.

The peak in shard concentration occurs at a depth of 3627–3628 cm; however, there is an earlier peak in colourless shards at a depth of 3630–3631 cm. This earlier peak marks the end of the first sharp increase in tephra concentration and, as such, it was from this depth interval that shards were selected for geochemical analysis.

Major oxide concentrations (expressed as weight percentage and normalized) from depths 36.30–36.31 mcd are given in Table 2. As mentioned previously, only analytical totals greater than 93 wt% were used in the characterization of the tephra. The shards analysed are alkali rhyolite in composition (see Table 2). The elevated SiO_2 for four analyses undertaken at FEEA suggests sodium mobilization during analysis and, as such, these values are excluded. Difficulties to achieve analyses >95 wt% may reflect the high vesicularity of the shards or a high magmatic water content (cf. Owen *et al.* 2013).

Comparison with volcanic source regions

The major element geochemistry of the shards, when compared with published major element compositions for various Icelandic volcanic systems (Haflidason *et al.* 2000; Jónasson 2007; Martin & Sigmarsson 2007) and averaged Holocene silicic tephra compositions (Larsen *et al.* 1999, 2001, 2002; Larsen & Eiríksson 2008), shows an affinity towards the Kerlingarfjöll volcanic system in central Iceland (Fig. 5). However, the comparison of glass and bulk XRF analyses is not straightforward, and major element geochemistry alone is insufficient to enable unambiguous identification of the volcanic system source (Davies *et al.* 2012); further trace element data would be needed to confirm this preliminary source determination.

Comparison with silicic tephra layers in marine sediment cores

A limited number of silicic tephra horizons have been identified in North Atlantic sediment sequences. North Atlantic Ash Zones (NAAZ) I, II and III are widely identified in the North Atlantic (e.g. Ruddiman & Glover 1972; Lacasse *et al.* 1996*a*, *b*), making them valuable stratigraphic marker horizons. In addition, an independent age determination for the rhyolitic component of NAAZ II (54 500 $\pm$ 1000 cal years BP, McIntosh in Southon 2004) makes this a particularly valuable isochron. NAAZs I and II have been identified in MD04-2822 (Fig. 2); NAAZ III has been excluded as a potential correlative on the basis of both stratigraphy and geochemistry.

The number of documented, and especially geochemically characterized, siliceous tephra horizons of late MIS 7 to MIS 5e age in the North Atlantic is small (e.g. Lacasse *et al.* 1996*a*, *b*; Haflidason *et al.* 2000; Brendryen *et al.* 2010; Table 3). We compared our data with both the widely distributed older (MIS 7) and younger (MIS 5e) silicic tephra layers in order to exclude the possibility of errors in our age model. We found no match, either in terms of geochemistry or stratigraphy, between our MIS 6-age tephra and the documented MIS 7 tephras (Fig. 6; summary of ash layers is given in Table 3). Similarly the tephra investigated here did not match the mid MIS 5e tephra (5e-Midt/RHY; see summary of the geographic distribution and correlation in Brendryen *et al.* 2010). The MIS 5 portion of MD04-2822 was carefully screened for tephra and a rhyolitic horizon was identified at 2490 cm, but this has a geochemical signature distinct from 5e-Midt/RYH (Abbott *et al.* 2013).

Two marine sediment cores (Norway Basin and Denmark Strait) contain geochemically characterized, MIS 6-age tephra horizons (age determination via interpolation using $\delta^{18}O$ stratigraphy; Voelker 2000; Lacasse & Garbe-Schönberg 2001). When we compare the MD04-2822 data with the major element geochemistry of these silicious tephra horizons, all except one can be excluded. ODP Site 985 from the Norway Basin contains a 2 cm-thick alkali rhyolite layer (average of 18 analyses), derived from an Icelandic off-rift source with an interpolated age of 184 ka (Lacasse & Garbe-Schönberg 2001; Lacasse & van den Bogaard 2002). This tephra layer most closely resembles the major element geochemistry of the MD04-2822 MIS 6 layer (Fig. 6). The interpolated age for this tephra within ODP985 (B 2H-1; 111-113) is also within error of our estimate. Additionally, a core from just north of the Iceland–Scotland Ridge (MD95-2009; Rasmussen *et al.* 1999) also has episodic, low-concentration (>100 shards g^{-1}) inputs of tephra (including rhyolitic) during MIS 6; however, no geochemical analyses are available. Similarly, St John *et al.* (2004) also document episodic input of colourless ash throughout MIS 6 to ODP919A in the Irminger Sea, but again, no geochemistry is available. One of these inputs of colourless shards to ODP Site 919A occurs during an interval of low IRD inputs and increasing $\delta^{18}O$ and, if its geochemistry was constrained by future analysis, could provide an additional correlative within the wider North Atlantic basin.

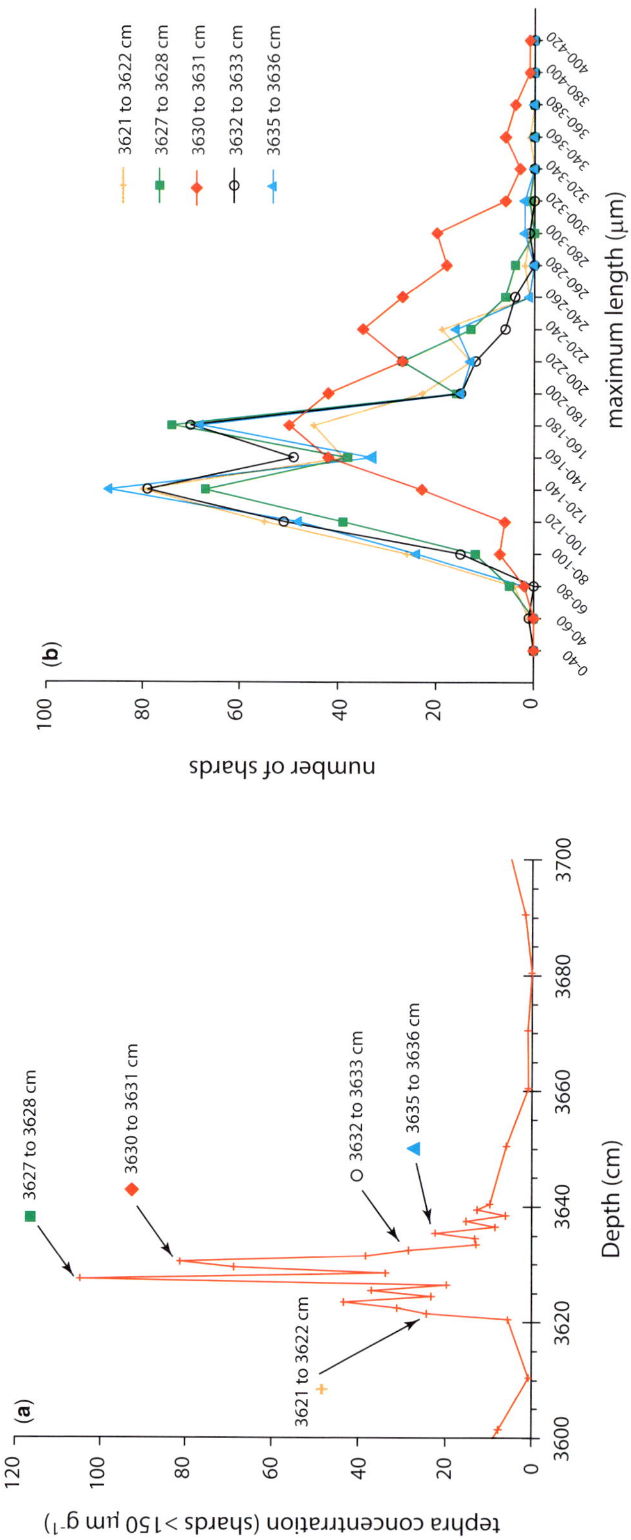

Fig. 4. **(a)** Tephra shard (>150 μm) concentration and **(b)** size distribution for selected depth intervals.

Table 2. *Major element compositions of MD04-2822 MIS 6 tephra layer (FEEA and TAU)*

Label	SiO_2	TiO_2	Al_2O_3	FeO	MnO	MgO	CaO	Na_2O	K_2O	P_2O_3	Total
FEEA (St Andrews)											
MIS6A2	71.80	0.18	12.31	2.28	0	0.09	0.63	4.21	4.11	–	95.61
MIS6D2	71.53	0.12	12.23	2.12	0.02	0.02	0.65	4.12	4.22	–	95.03
MIS6E2	72.35	0.21	12.69	2.41	0.09	0.02	0.63	4.26	4.26	–	96.92
MIS6F1	71.37	0.21	12.59	2.18	0.05	0.05	0.63	4.13	4.16	–	95.37
MIS6H2	71.84	0.09	12.49	2.34	0.09	0.13	0.60	5.20	4.22	–	97.00
MIS6I2	71.43	0.18	12.41	2.16	0	0.02	0.68	4.16	4.22	–	95.26
MIS6J1	70.55	0.22	12.25	2.29	0	0.10	0.64	5.42	4.27	–	95.74
Mean	71.55	0.17	12.42	2.25	0.04	0.06	0.64	4.50	4.21	–	
1σ	0.55	0.05	0.18	0.10	0.04	0.05	0.02	0.56	0.06	–	
Outlier											
MIS6E1	70.78	0.20	13.00	2.05	0.03	0.7154	0.70	5.90	4.35	–	97.72
Shards assumed to be affected by sodium migration during analysis											
MIS6J2	74.07	0.19	12.90	2.13	0.05	0.09	0.60	3.03	4.29	–	97.35
MIS6J3	73.51	0.21	12.78	2.46	0.02	0.13	0.74	2.05	4.27	–	96.17
MIS6K1	73.43	0.20	12.58	2.26	0.06	0.03	0.65	2.05	4.30	–	95.56
MIS6K2	73.01	0.15	12.80	2.49	0.04	0.16	0.71	2.37	3.95	–	95.68
TAU (Edinburgh)											
MIS6_5	69.66	0.16	12.20	2.04	0.06	0.07	0.85	4.22	4.08	0.02	93.36
MIS6_1	71.33	0.15	12.15	2.22	0.06	0.09	0.68	4.63	3.83	0.01	95.15
MIS6_2	69.55	0.17	12.38	2.19	0.08	0.09	0.78	4.48	3.98	0.02	93.72
MIS6_3b	69.50	0.17	12.20	2.24	0.10	0.06	0.76	4.73	4.07	0.01	93.84
Mean	70.01	0.16	12.23	2.17	0.08	0.08	0.77	4.52	3.99	0.02	
1σ	0.88	0.01	0.10	0.09	0.02	0.02	0.07	0.22	0.12	0.01	
All analyses (except those assumed to be affected by sodium migration)											
Mean	70.97	0.17	12.41	2.21	0.05	0.12	0.69	4.62	4.15	0.02	
1σ	0.97	0.04	0.25	0.11	0.04	0.19	0.07	0.59	0.14	0.01	

Comparison with proximal Icelandic terrestrial deposits

In addition to marine sediment sequences, we compared our MIS 6 tephra with Icelandic terrestrial silicic deposits (e.g. review of McGarvie 2009). Some of these proximal deposits have been geochemically characterized and Ar–Ar dated (e.g. McGarvie *et al.* 2006, 2007), although relatively few have been identified for the early MIS 5e to MIS 7 interval. Flude *et al.* (2010) have Ar–Ar dated and characterized (using XRF) 19 of the 21 different rhyolite eruptions (Grönvold 1972; Stevenson 2004) for the Kerlingarfjöll central volcano for the last 350 ka. Kerlingarfjöll forms Iceland's second largest outcrop of Quaternary rhyolites. Two units (Loðmundur and Snaekollur) have Ar–Ar ages (184 ± 10 and *c.* 150–190 ka) that overlap with our interpolated age and a similar geochemistry for the tephra horizon. However, these are tentative correlations given the absence of trace element data and the relatively wide error bars associated with the terrestrial deposits. Ljósufjöll volcano also produced silicic deposits during the time interval of interest (Flude *et al.* 2008); however, the major element geochemistry is very different from the MD04-2822 tephra horizon (Fig. 7).

Conclusions

The well-sorted size distribution, relative homogeneity in major element geochemistry and occurrence during an interval of limited IRD deposition strongly suggest that the tephra horizon characterized in MD04-2822 is a primary air-fall deposit. However, sea-ice rafting of the tephra (from air-fall onto sea-ice closer to the Icelandic source) cannot be excluded, but the timescales on which this occurs (a few months to years; Brendryen *et al.* 2010) would not compromise its potential use as an isochron. We have tentatively assigned a Kerlingarfjöll affinity to the tephra layer and the documentation of multiple rhyolitic eruptions from this region (Flude *et al.* 2010) during this interval might hold some promise if the correlation could be more securely attributed.

While relatively few MIS 6-age tephra horizons have been reported in the North Atlantic, the alkali rhyolite layer at ODP Site 985 does exhibit a similar major element geochemistry and

Table 3. *Summary of identified North Atlantic tephras documented within marine sediments from MIS 7 to MIS 5e (inclusive)*

Reference	Location	Core	Ash layer	Thickness (cm)	Shard type	Size (μm)	Sorting	Age (ka)	Chronology	Geochemistry	Deposition method
MIS 7											
Sjøholm (1987)	*Norwegian Sea*	*P57-5*						*Early MIS 7*			
Sejrup *et al.* (1989)	Iceland Plateau	P57-7	C	Not given	Not given		*Not given*	Early MIS 7	$\delta^{18}O$ stratigraphy (SPECMAP)	Yes*	Wind (absence of sea ice)
Sjøholm *et al.* (1991)	Iceland Plateau	P57-7	C	78	Colourless and black blocky (upper part, 410–412 cm dominated by silicic grains)		*Not given*		$\delta^{18}O$ stratigraphy (*N. pachyderma* sin.)	yes	'Wind deposited'
Lackschewitz (1991)†	Iceland Plateau	1852-2	Not analysed	6	Not given		*Not given*	MIS 7	*Not given*	No	
Birgisdottir (1991)†	Iceland Plateau	1243-1	Not analysed	*5*	Not given		*Not given*	MIS 7	*Not given*	No	
	Iceland Plateau	1244-2	Not analysed	*1*	Not given		*Not given*	MIS 7	*Not given*	No	
	Iceland Plateau	1245-1	Not analysed	*15*	Not given		*Not given*	MIS 7	*Not given*	No	
	Iceland Plateau	1246-4	Not analysed	*8*	Not given		*Not given*	MIS 7	*Not given*	No	
Baumann *et al.* (1993)	Iceland Plateau	1852	Unknown, not analysed	3	Not given		*Not given*	MIS 7	$\delta^{18}O$ stratigraphy	No	
Lacasse *et al.* (1996*a*, *b*)	Iceland Plateau	ODP 907A	C	<5	94% colourless shards (platy and bubble wall)		Well sorted	204 (170)	$\delta^{18}O$ stratigraphy (palaeomagntic)	Yes	'Likely air-borne'
			D	18	95–97% colourless shards (platy and bubble wall)		Well sorted	211 (177)	$\delta^{18}O$ stratigraphy (palaeomagntic)	Yes	'Likely air-borne'
Lacasse & Garbe-Schönberg (2001); Lacasse & van den Bogaard (2002)	Iceland Plateau	ODP 907A	C	*5*	Rhyolite	609 (mean max. diameter)	*Not given*	201	$\delta^{18}O$ stratigraphy (Fronval and Jansen 1996)	Yes	Fallout
	Iceland Plateau	ODP 907A	D	*16*	Comendite	524 (mean max. diameter)	*Not given*	207	$\delta^{18}O$ stratigraphy (Fronval and Jansen 1996)	Yes	Fallout
Wallrabe-Adams & Lackschewitz (2003)	Iceland Plateau	21852-2	Unknown	*Not given*	Colourless glass, platy and cuspate bubble wall shards	250–500	*Not given*	209‡	$\delta^{18}O$ stratigraphy (Lackschewitz 1991)	Yes	Fallout onto open water
	Iceland Plateau	21857-2	Unknown	*Not given*	Colourless glass, platy and cuspate bubble wall shards		*Not given*	MIS 7	n/a	Yes	Fallout onto open water
	Iceland Plateau	1244-2	Unknown	*Not given*	Colourless glass, platy and cuspate bubble wall shards		*Not given*	MIS 7	n/a	Yes	Fallout onto open water
	Iceland Plateau	1245-1	Unknown	*Not given*	Colourless glass, platy and cuspate bubble wall shards		*Not given*	MIS 7	n/a	Yes	Fallout onto open water
St John *et al.* (2004)	Irminger Basin	ODP919	Six major layers, no analysed	*Not given*	Colourless ash		*Not given*	*c.* 240–190	$\delta^{18}O$ stratigraphy	No	
MIS 6											
Lacasse & Garbe-Schönberg (2001); Lacasse & van den Bogaard (2002)	Norway Basin	ODP 985A	n/a	*3*	Alkali rhyolite	807 (mean max. diameter)	*Not given*	133	Risebrobakken (1999)	Yes	Fallout
	Norway Basin	ODP 985A	n/a	*1*	Rhyolite		*Not given*	157	Risebrobakken (1999)	Yes	
	Norway Basin	ODP 985B	n/a	*2*	Alkali rhyolite		*Not given*	184	Risebrobakken (1999)	Yes	

Rasmussen *et al.* (1999)	North of Iceland–Scotland Ridge	MD95-2009	Intermittent inputs during MIS 6, not analysed	*Not given*	Not given (rhyolitic and basaltic ash)		*Not given*	*c.* 150 (largest peak)	$\delta^{18}O$ stratigraphy	No	
Voelker (2000)	Denmark Strait	PS2646-5	n/a	*Not given*	Not given		*Not given*	MIS 6	$\delta^{18}O$ stratigraphy	Yes	
St John *et al.* (2004)	Irminger Basin	ODP919	Two major layers, not analysed	*Not given*	Colourless ash		*Not given*	155 and 175	$\delta^{18}O$ stratigraphy	No	
This study	Rockall Trough	MD04-2822	Unknown		Colourless, platy bubble wall shards	63–408	Well sorted	181 (± 6)	$\delta^{18}O$ stratigraphy	Yes	Primary air-fall
MIS 5e											
Sejrup *et al.* (1989)	Iceland Plateau	P57-7	B; 5e-Midt/RHY	*Not given*	Not given		*Not given*	MIS 5e	$\delta^{18}O$ stratigraphy (SPECMAP)	Yes*	
Sjøholm *et al.* (1991)	Iceland Plateau	P57-7	B; 5e-Midt/RHY	1	Mostly silicic shards		*Not given*		$\delta^{18}O$ stratigraphy	Yes	
Lacasse *et al.* (1996*a*, *b*)	Iceland Plateau	ODP 907A	B	1	Not given		*Not given*	122.26 (97)	$\delta^{18}O$ stratigraphy (palaeomagntic)	Yes	
Fronval *et al.* (1998)	Norwegian Sea	ODP644	5e-Midt/RHY	3	Not given		*Not given*	125	$\delta^{18}O$ stratigraphy	Yes	
	Iceland Plateau	HM57-7	5e-Midt/RHY	1–2	Not given		*Not given*			Yes	
	Iceland Plateau	HM71-19	5e-Midt/RHY	1	Not given		*Not given*			Yes	
	Iceland Plateau	HM79-31	5e-Midt/RHY	1	Not given		*Not given*			Yes	
Rasmussen *et al.* (1999)	Faroe Islands	MD95-2009	Not analysed		Not given		*Not given*	*c.* 123	$\delta^{18}O$ stratigraphy	No	
Haflidason *et al.* (2000)	Iceland Plateau	HM79-31	5e-Midt/RHY		Not given		*Not given*	125	$\delta^{18}O$ stratigraphy	Yes	
Wastegård & Rasmussen (2001)	North of Iceland–Scotland Ridge	MD95-2009	5e-Midt/RHY		Rhyolitic		*Not given*	124	$\delta^{18}O$ stratigraphy	Yes	
	South of Iceland–Scotland Ridge	ENAM33	5e-Top/RHY		Rhyolitic		*Not given*	*Not given (uncertain)*		Yes	
Lacasse & Garbe-Schönberg (2001)	Iceland Plateau	ODP 907A	B; 5e-Midt/RHY		Not given		*Not given*	125	$\delta^{18}O$ stratigraphy (Fronval and Jansen 1996)	Yes	
Wallrabe-Adams & Lackschewitz (2003)	Iceland Plateau	23243-1	5e-Midt/RHY		Colourless, platy bubble wall shards		*Not given*	125–130	$\delta^{18}O$ stratigraphy	Yes	
		23244-2	5e-Midt/RHY		Colourless, platy bubble wall shards		*Not given*	125–130	$\delta^{18}O$ stratigraphy	Yes	
		23245-1	5e-Midt/RHY		Colourless, platy bubble wall shards		*Not given*	125–130	$\delta^{18}O$ stratigraphy	Yes	
St John *et al.* (2004)	Irminger Basin	ODP919	Unknown, not analysed	*Not given*	Colourless ash		*Not given*	*c.* 125	$\delta^{18}O$ stratigraphy	No	
Brendryen *et al.* (2010)	Norwegian Sea	MD99-2289	5e-Eem/RHY I; 5e-Midt/RHY	1			*Not given*	121.8	$\delta^{18}O$ stratigraphy	Yes	
Abbott *et al.* (2013)	Rockall Trough	MD04-2822	MD04-2822 2490–2491cm	1	Rhyolitic	35–250	Well sorted	116.4 ± 4	$\delta^{18}O$ stratigraphy	Yes	Primary air-fall
Terrestrial											
Wastegård *et al.* (2005)	Faroe Islands	Klaksvík sequence	5e-Midt/RHY; Klaksvík		Rhyolitic		*Not given*	124	Correlation with marine sediments	Yes	Primary air-fall
		Klaksvík sequence	Klaksvík dacite		Dacitic		*Not given*	124	Correlation with marine sediments	Yes	Primary air-fall

*No values given in the original publication.
†Unpublished PhD thesis.
‡Interpolated age from the $\delta^{18}O$ age model of core 21852-2 (Lackschewitz 1991).

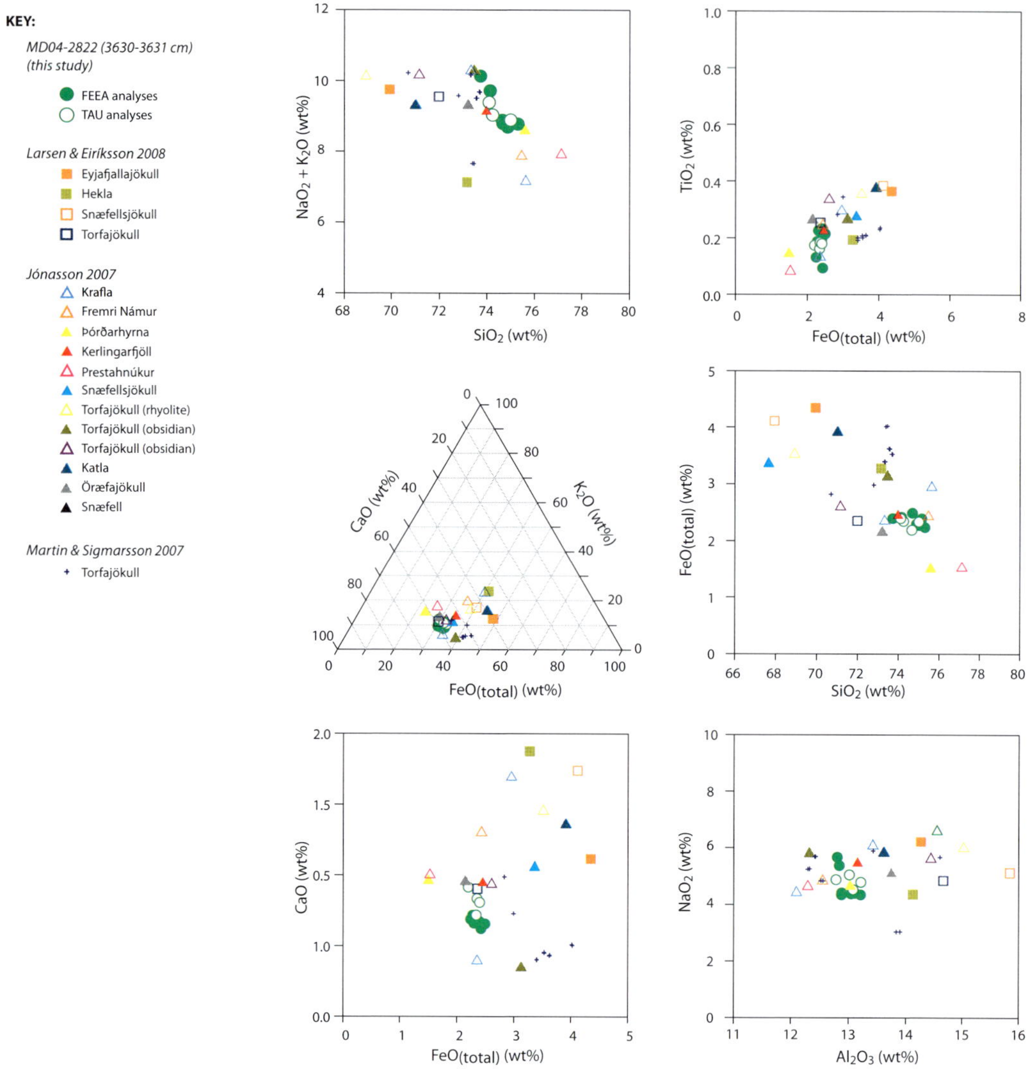

Fig. 5. Selected major element geochemistry plots for MD04-2822 MIS 6 tephra layer (FEEA and TAU data) and average data for Icelandic volcanic systems (Larsen & Eiríksson 2008; Jónasson 2007; Martin & Sigmarsson 2007). All data has been normalized to 100 wt% prior to plotting.

interpolated age. In addition, a tephra layer in a core in the Irminger Basin (St John *et al.* 2004) shares stratigraphic similarity's that would make it a good target for further analysis. If this tephra layer was confirmed in other North Atlantic sediment cores, it could provide information on the phasing of events during a period of increasing global ice volumes.

Compared with the last glacial period, relatively few tephra horizons have been identified within the North Atlantic during MIS6. The identification and characterization of this tephra horizon within the well-defined benthic isotope stratigraphy of MD04-2822 (Hibbert *et al.* 2010, this work) is potentially valuable for North Atlantic geochronology because this interval lies beyond the scope of two commonly utilized methods for age control: correlation of event-stratigraphies to the Greenland ice cores and radiocarbon. Notwithstanding the current lack of independent age control, the stratigraphic position of this tephra horizon during a well-characterized transition in the benthic oxygen isotope record makes this horizon potentially very useful for correlating between marine sediment

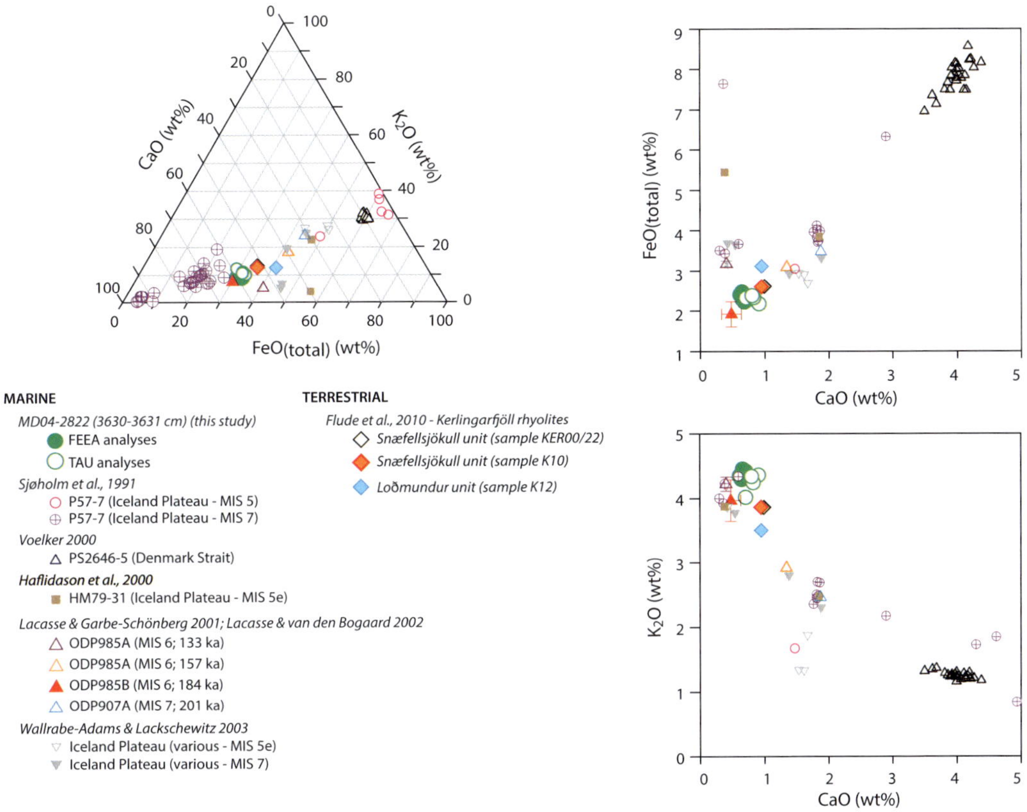

Fig. 6. Selected major element geochemistry plots for MD04-2822 MIS 6 tephra layer (FEEA and TAU data) and other marine (Sjøholm *et al.* 1991; Voelker 2000; Haflidason *et al.* 2000; Lacasse & Garbe-Schönberg 2001; Wallrabe-Adams & Lackschewitz 2003) and dated terrestrial deposits (Flude *et al.* 2010).

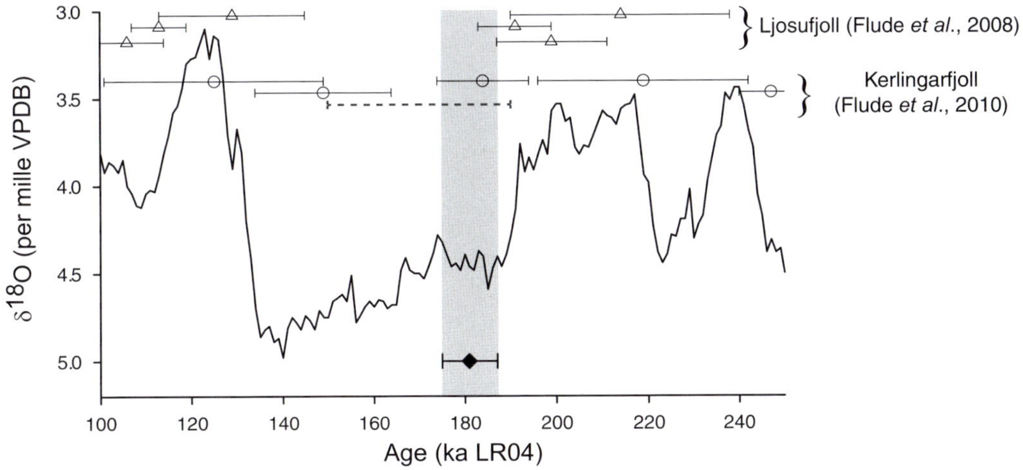

Fig. 7. LR04 benthic stack (Lisiecki & Raymo 2005) with Ar–Ar ages for terrestrial Icelandic rhyolitic deposits (Ljósufjoll, open triangles (Flude *et al.* 2008); Kerlingarfjöll, open circles (Flude *et al.* 2010)) and the interpolated age for the MD04-2822 tephra layer (solid diamond).

records. If an independent age was confirmed for the horizon, this would provide a valuable means of constraining those marine stratigaphies and, with further occurrences, it could provide one test of assumptions underpinning global benthic $\delta^{18}O$ stacks.

F.D.H. acknowledges the British Geological Survey and the University of St Andrews funding of her PhD. The Natural Environment Research Council analytical facility support via both the National Isotope Geochemistry Laboratory and NERC Radiocarbon Laboratory (SUERC) is also gratefully acknowledged. We thank the captain and crew of the RV *Marion Dufresne* for their assistance in recovering core MD04-2822. The assistance of D. Herd (Facility for Earth and Environmental Analysis, University of St Andrews) and C. Hayward (Tephra Analytical Unit, University of Edinburgh) in the use of the electron microprobe is gratefully acknowledged.

References

ABBOTT, P. M., DAVIES, S. M., AUSTIN, W. E. N., PEARCE, N. J. G. & HIBBERT, F. D. 2011. Identification of cryptotephra horizons in a North East Atlantic marine record spanning marine isotope stages 4 and 5a (~60 000–82 000 a b2k). *Quaternary International*, **246**, 177–189.

ABBOTT, P. M., DAVIES, S. M. *ET AL.* 2012. A detailed framework of Marine Isotope Stages 4 and 5 volcanic events recorded in two Greenland ice-cores. *Quaternary Science Reviews*, **36**, 59–77.

ABBOTT, P. M., AUSTIN, W. E. N., DAVIES, S. M., PEARCE, N. J. G. & HIBBERT, F. D. 2013. Cryptotephrochronology over the Eemian and the last interglacial–glacial transition in the North East Atlantic. *Journal of Quaternary Science*, **28**, 501–514.

AUSTIN, W. E. N. & ABBOTT, P. M. 2010. Comment: where last glacial climate event simultaneous between Greenland and France? A quantitative comparison using non-tuned chronologies. *Journal of Quaternary Science*, **25**, 1045–1046.

AUSTIN, W. E. N. & HIBBERT, F. D. 2012. Tracing time in the ocean: a brief review of chronological constraints in marine event-based stratigraphies 8–60 ka. *Quaternary Science Reviews*, **36**, 28–37.

AUSTIN, W. E. N., WILSON, L. J. & HUNT, J. B. 2004. The age and chronostratigraphic significance of North Atlantic Ash Zone II. *Journal of Quaternary Science*, **19**, 1 37–146.

BAUMANN, K.-H., LACKSCHEWITZ, K. S., ERLENKEUSER, H., HENRICH, R. & JIINGER, B. 1993. Late Quaternary calcium carbonate sedimentation and terrigenous input along the east Greenland continental margin. *Marine Geology*, **114**, 13–36.

BIRGISDOTTIR, L. 1991. *Die paläo-ozeanographische Entwicklung der Island-See in den letzten 550.000 Jahren. PhD thesis*, University of Kiel.

BRENDRYEN, J., HAFLIDASON, H. & SEJRUP, H.-P. 2010. Norwegian Sea tephrostratigraphy of marine isotope stages 4 and 5: prospects and problems for tephrochronology in the North Atlantic region. *Quaternary Science Reviews*, **29**, 847–864.

BRONK RAMSEY, C. 2009. Bayesian analysis of radiocarbon dates. *Radiocarbon*, **51**, 337–360.

DAVIES, S. M., ABBOTT, P. M., PEARCE, N. J., WASTEGÅRD, S. & BLOCKLEY, S. P. E. 2012. Integrating the INTIMATE records using tephrochronology. *Quaternary Science Reviews*, **36**, 11–27.

FLUDE, S., BURGESS, R. & MCGARVIE, D. W. 2008. Silicic volcanism at Ljósufjöll, Iceland: insights into evolution and eruptive history from Ar–Ar dating. *Journal of Volcanology and Geothermal Research*, **169**, 154–175.

FLUDE, S., MCGARVIE, D. W., BURGESS, R. & TINDLE, A. G. 2010. Rhyolites at Kerlingarfjöll, Iceland: the evolution and lifespan of silicic central volcanoes. *Bulletin of Volcanology*, **72**, 523–538.

FRONVAL, T. & JANSEN, E. 1996. Late Neogene paleoclimates and paleoceanography in the Iceland-Norwegian Sea: evidence from the Iceland and Voring Plateaus. *In*: THIEDE, J., MYHRE, A. M., FIRTH, J. V., JOHNSON, G. L. & RUDDIMAN, W. F. (eds) *Proceedings of the Ocean Drilling Programme, Scientific Results*, College Station, Texas, Ocean Drilling Programme, **1551**, 455–468.

FRONVAL, T., JANSEN, E., HAFLIDASON, H. & SEJRUP, H.-P. 1998. Variability in surface and deep water conditions in the Nordic Seas during the last interglacial period. *Quaternary Science Reviews*, **17**, 963–985.

GRÖNVOLD, K. 1972. *Structural and petrochemical studies in the Kerlingarfjöll region, central Iceland. PhD thesis*, University of Oxford.

HAFLIDASON, H., EIRÍKSSON, J. & VAN KREVELD, S. 2000. The tephrochronology of Iceland and the North Atlantic region during the Middle and Late Quaternary: a review. *Journal and Quaternary Science*, **15**, 3–22.

HANSEN, B. & ØSTERHUS, S. 2000. North Atlantic–Nordic Seas exchanges. *Progress in Oceanography*, **45**, 109– 208.

HIBBERT, F. D., AUSTIN, W. E. N., LENG, M. J. & GATLIFF, R. W. 2010. British Ice Sheet dynamics inferred from North Atlantic ice-rafted debris records spanning the last 175 000 years. *Journal of Quaternary Science*, **25**, 461–482, http://dx.doi.org/10.1002/jqs.1331

JÓNASSON, K. 2007. Silicic volcanism in Iceland: composition and distribution within the active volcanic zones. *Journal of Geodynamics*, **43**, 101–117.

JUMARS, P. A. & WHEATCROFT, R. A. 1989. Responses of benthos to changing food quality andquantity, with a focus on deposit feeding and bioturbation. *In*: BERGER, W. H., SMETACEK, V. S. & WEFER, G. (eds) *Productivity of the Ocean: Present and Past*, Wiley, New York, 235–253.

LACASSE, C. & GARBE-SCHÖNBERG, C. D. 2001. Explosive silicic volcanism in Iceland and the Jan Mayen area during the last 6 Ma: sources and timing of major eruptions. *Journal of Volcanology and Geothermal Research*, **107**, 113–147.

LACASSE, C. & VAN DEN BOGAARD, P. 2002. Enhanced airbourne dispersal of silicic tephras during the onset of Northern Hemisphere glaciations, from 6 to 0 Ma records of explosive volcanism and climate change in the subpolar North Atlantic. *Geology*, **30**, 623–626.

LACASSE, C., PATERNE, M., WERNER, R., WALLRABE-ADAMS, H.-J., SIGUDSSON, H., CAREY, S. & PINTE, G. 1996*a*. Geochemistry and origin of Pliocene and Pleistocene ash layers from the Iceland Plateau, Site 907. *In*: THIEDE, J., MYHRE, A. M., FIRTH, J. V., JOHSON, G. L. & RUDDIMAN, W. F. (eds) *Proceedings of the Ocean Drilling Program, Scientific Results*, **151**. Ocean Drilling Program, Texas A & M University, College Station, TX, 309–329.

LACASSE, C., SIGURDSSON, H., CAREY, S., PATERNE, M. & GUICHARD, F. 1996*b*. North Atlantic deep sea sedimentation of late Quaternary tephra from the Iceland hotspot. *Marine Geology*, **129**, 207–235.

LACKSCHEWITZ, K. S. 1991. *Sedimentationsprozesse am aktiven mittelozeanischen Kolbeinsey Rücken (nördlich von Island)*. PhD thesis. Geomar Rep., **9**, 1–133.

LARSEN, G. & EIRÍKSSON, J. 2008. Late Quaternary terrestrial tephrachronology of Iceland – frequency of explosive erruptions, type and volume of tephra deposits. *Journal of Quaternary Science*, **23**, 109–120.

LARSEN, G., DUGMORE, A. J. & NEWTON, A. J. 1999. Geochemistry of historical-age silicic tephras in Iceland. *The Holocene*, **9**, 463–471.

LARSEN, G., NEWTON, A. J., DUGMORE, A. J. & VILMUNDARDÓTTIR, E. G. 2001. Geochemistry, dispersal, volumes and chronology of Holocene silicic tephras from the Katla volcanic system, Iceland. *Journal of Quaternary Science*, **16**, 119–132.

LARSEN, G., EIRÍKSSON, J., KNUDSEN, K. L. & HEINEMEIER, J. 2002. Correlation of late Holocene terrestrial and marine tephra markers in North Iceland: implications for reservoir age changes and linking land-sea chronologies in the North Atlantic. *Polar Research*, **21**, 283–290.

LISIECKI, L. E. & RAYMO, M. E. 2005. A Pliocene–Pleistocene stack of 57 globally distributed benthic $\delta^{18}O$ records. *Paleoceanography*, **20**, PA1003.

LOULERGUE, L., SCHILT, A. ET AL. 2008. Orbital and millennial-scale features ofatmospheric CH_4 over the past 800 000 years. *Nature*, **453**, 383–386.

MARTIN, E. & SIGMARSSON, O. 2007. Crustal thermal state and origin of silicic magma in Iceland: the case of Torfajökull, Ljósufjöll and Snæfellsjökull volcanoes. *Contributions to Mineralogy and Petrology*, **153**, 593–605.

MCGARVIE, D. W. 2009. Rhyolitic volcano–ice interactions in Iceland. *Journal of Volcanology and Geothermal Research*, **185**, 367–389.

MCGARVIE, D. W., BURGESS, R., TINDLE, A. G., TUFFEN, H. & STEVENSON, J. A. 2006. Pleistocenerhyolitic volcanism at the Torfajökull central volcano, Iceland: eruption ages, glaciovolcanism, and geochemical evolution. *Jökull*, **56**, 57–75.

MCGARVIE, D. W., STEVENSON, J. A., BURGESS, R., TUFFEN, H. & TINDLE, A. G. 2007. Volcano–iceinteractions at Prestahnúkur, Iceland: rhyolite eruption during the last interglacial–glacial transition. *Annals of Glaciology*, **45**, 38–47.

NINKOVICH, D., SPARKS, R. S. J. & LEDBETTER, M. T. 1978. The exceptional magnitude and intensity of the Toba eruption, Sumatra: an example of the use of deep-sea tephra layers as a geological tool. *Bulletin of Volcanology*, **41**, 286–298.

OWEN, J., TUFFEN, H. & MCGARVIE, D. W. 2013. Explosive subglacial rhyolitic eruptions in Iceland are fuelled by high magmatic H_2O and closed-system degassing. *Geology*, **41**, 251–254.

PETERS, C., AUSTIN, W. E. N., WALDEN, J. & HIBBERT, F. D. 2010. Magnetic characterisation and correlation of a Younger Dryas tephra in North Atlantic marine sediments. *Journal of Quaternary Science*, **25**, 339– 347.

RASMUSSEN, T. L., BALBON, E., THOMSEN, E., LABEYRIE, L. & VAN WEERING, T. C. E. 1999. Climate records and changes in deep outflow from the Norwegian Sea ~150–55 ka. *Terra Nova*, **11**, 61–66.

REIMER, P. J., BAILLIE, M. G. L. ET AL. 2009. IntCal09 and Marine09 radiocarbon age calibration curve, 0–50 000 years cal BP. *Radiocarbon*, **51**, 1111–1150.

RISEBROBAKKEN, B. 1999. *Midt Pleistocene klimasvingninger i de Nordiske havene*. M.Sc. thesis, University of Bergen.

RUDDIMAN, W. F. & GLOVER, L. K. 1972. Vertical mixing of ice-rafted volcanic ash in North Atlantic sediments. *Geological Society American Bulletin*, **83**, 2817–2836.

RUDDIMAN, W. F. & MCINTYRE, A. 1973. Time-transgressive deglacial retreat of polar waters from the North Atlantic. *Quaternary Research*, **3**, 117–130.

RUDDIMAN, W. F., MOLFINO, B., ESMAY, A. & POKRAS, E. 1980. Evidence bearing on the mechanism of rapid deglaciation. *Climate Change*, **3**, 65–87.

SEJRUP, H. P., SJØHOLM, J., FURNES, H., BEYER, I., EIDE, L., JANSEN, E. & MANGERUD, J. 1989. Quaternary tephrachronology on the Iceland Plateau, north of Iceland. *Journal of Quaternary Science*, **4**, 109–114.

SHACKLETON, N. J., HALL, M. A. & VINCENT, E. 2000. Phase relationships between millennial scale events 64 000 to 24 000 years ago. *Paleoceanography*, **15**, 565–569.

SJØHOLM, J. 1987. *Kvartae askesoner og sedimenter pA Islandspladet*. Unpublished thesis, University of Bergen.

SJØHOLM, J., SEJRUP, H.-P. & FURNES, H. 1991. Quaternary volcanic ash zones on the Iceland Plateau, southern Norwegian Sea. *Journal of Quaternary Science*, **6**, 159–173.

SOUTHON, J. 2004. A radiocarbon perspective on Greenland ice-core chronologies: can we use ice cores for ^{14}C calibration? *Radiocarbon*, **46**, 1239– 1259.

STEVENSON, J. A. 2004. *Volcano-ice interaction at Öraefajökull and Kerlingarfjöll, Iceland*. PhD thesis, The Open University, Lancaster University.

ST JOHN, K., FLOWER, B. P. & KRISSEK, L. 2004. Evolution of iceberg melting, biological productivity and the record of Icelandic volcanism in the Irminger basin since 630 ka. *Marine Geology*, **212**, 133–152.

SVENSSON, A., ANDERSEN, K. K. ET AL. 2008. A 60 000 year Greenland stratigraphic ice core chronology. *Climate of the Past*, **4**, 47–57.

TODD, J. A., AUSTIN, W. E. N. & ABBOTT, P. M. 2014. Quantifying bioturbation of a simulated ash fall event. *In*: AUSTIN, W. E. N., ABBOTT, P. M., DAVIES, S. M., PEARCE, N. J. G. & WASTEGÅRD, S. (eds) *Marine Tephrochronology*. Geological Society, London, Special Publications, **398**. First published online February 5, 2014, http://dx.doi.org/10.1144/SP398.9

TURNEY, C. S. M., HARKNESS, D. D. & LOWE, J. J. 1997. The use of microtephra horizons to correlate late-glacial lake sediment successions in Scotland. *Journal of Quaternary Science*, **16**, 525–531.

VOELKER, A. H. L. 2000. *Tephra chemistry of sediment core PS2646–5*, http://dx.doi.org/10.1594/PANGAEA.57664. Supplement to: Voelker, Antje HL (1999): Zur Deutung der Dansgaard–Oeschger Ereignisse in ultra-hochauflösenden Sedimentprofilen aus dem Europäischen Nordmeer. (Dansgaard–Oeschger events in ultra-high resolution sediment records from the Nordic Seas.) Berichte-Reports, Institut für Geowissenschaften, Universität Kiel, **9**, http://dx.doi.org/10.2312/reports-ifg.1999.9

WALLRABE-ADAMS, H.-J. & LACKSCHEWITZ, K. S. 2003. Chemical composition, distribution and origin of silicic volcanic ash layers in the Greenland–Iceland–Norwegian Sea: explosive volcanism from 10 to 300 ka as recorded in deep-sea sediments. *Marine Geology*, **193**, 273–293.

WASTEGÅRD, S. & RASMUSSEN, T. L. 2001. New tephra horizons from Oxygen Isotope Stage 5 in the North Atlantic: correlation potential for terrestrial, marine and ice-core archives. *Quaternary Science Reviews*, **20**, 1587–1593.

WASTEGÅRD, S., BJÖRCK, S., GREVE, C. & RASMUSSEN, T. L. 2005. A tephra-based correlation between the Faroe Islands and the Norwegian Seas raises questions about chronological relationships during the last interglacial. *Terra Nova*, **17**, 7–12.

WOLFF, E. W., CHAPPELLAZ, J., BLUNIER, T., RASMUSSEN, S. O. & SVENSSON, A. 2010. Millennialscale variability during the last glacial: the ice core record. *Quaternary Science Reviews*, **29**, 2828–2838.

Faroe Marine Ash Zone IV: a new MIS 3 ash zone on the Faroe Islands margin

STEFAN WASTEGÅRD[1*] & TINE L. RASMUSSEN[2]

[1]*Department of Physical Geography and Quaternary Geology, Stockholm University, SE-10691 Stockholm, Sweden*

[2]*Department of Geology, University of Tromsø, Dramsveien 210, NO-9037, Tromsø, Norway*

**Corresponding author (e-mail: stefan.wastegard@geo.su.se)*

Abstract: A basaltic tephra layer from MIS 3 has been discovered by analysis of cores from the Faroe Islands margin. The tephra layer appears up to 20 cm thick in some records. After the first main fall-out event the tephra is believed to be mainly deposited and redistributed by bottom currents. Geochemical analyses suggest that the tephra is relatively undisturbed by allochtonous tephra grains and unmixed. The peak occurrences are in the lower part of GIS (Greenland Interstadial) 12 and we suggest naming this new tephra Faroe Marine Ash Zone IV (FMAZ IV), following the nomenclature adopted for previous ash zones found on the Faroe Islands margin. Geochemical analyses of the tephra show affinities with the Grímsvötn volcanic system in the Eastern Volcanic Zone in south Iceland. The average age of FMAZ IV from four independent age models is 46 800 ± 1000 years BP. We suggest that the V5 ash zone, found on the Reykjanes Ridge is a correlative to the FMAZ IV.

Supplementary material: Single-grain analyses of tephra from FMAZ IV in cores from the northern Faroe margin and the central Faroe-Shetland Channel are available at http://www.geolsoc.org.uk/SUP18714

The need for exact dating and correlation methods between marine and terrestrial archives has increased in the palaeoclimate community during recent decades as more investigations now aim at high temporal resolutions. The last glacial period of NW Europe and the North Atlantic region was characterized by a number of rapid and pronounced climatic fluctuations that are poorly understood (e.g. Rahmstorf 2002; Sanchez-Goñi *et al.* 2008; Ampel *et al.* 2010). Although close similarities are evident in the palaeoclimatic reconstructions obtained from terrestrial, marine and ice-core records from this period, uncertainties exist as to the degree of synchroneity (or asynchroneity) between them, largely owing to the limitations of the radiocarbon dating method (radiocarbon plateaus, the marine reservoir effects, etc.; e.g. Björck *et al.* 1998). In addition, the lack of suitable and accurate dating methods for the time period before *c.* 40 ka BP is a problem owing to the end of the radiocarbon window.

One method that holds much promise of effecting more precise regional correlations is tephrochronology. Marine and terrestrial records from the last ice age are often tuned to ice-core records, assuming synchroneity of climate events such as Dansgaard–Oeschger events, which may lead to erroneous conclusions about lead and lag relationships (cf. Blaauw *et al.* 2010). Tephrochronology has the potential to provide exact time markers, but the full potential has not yet been exploited. Tephrochronology involves the use of distinctive and widespread tephra layers as time-parallel markers that permit reliable correlation over long distances. Tephra horizons can be geochemically fingerprinted using major element composition (e.g. Larsen 1981) or trace and rare earth element compositions (e.g. Pearce *et al.* 2004) and used for precise correlations over large areas. Recent reviews (Haflidason *et al.* 2000; Wastegård & Davies 2009; Lawson *et al.* 2012) show that more than 150 tephra horizons from the last 400 ka have been identified in the North Atlantic region.

A majority of tephra deposits in the North Atlantic region originate from eruptions on Iceland, which is one of the most active volcanic areas in the world. Tephra layers from Jan Mayen have also been identified in the Greenland ice-cores (Abbott *et al.* 2012) and in marine cores from the Nordic Seas (Lacasse & Garbe-Schönberg 2001; Hunt 2004), but as yet the importance of these as chronological markers is uncertain. The recent finding of tephra from the Alaskan Aniakchak volcano (*c.* 3600 BP) in a marine core from the SE Greenland shelf (Jennings *et al.* 2014) opens new possibilities of correlations of Holocene marine records between the North Pacific and the North Atlantic.

From: Austin, W. E. N., Abbott, P. M., Davies, S. M., Pearce, N. J. G. & Wastegård, S. (eds) 2014. *Marine Tephrochronology*. Geological Society, London, Special Publications, **398**, 81–93.
First published online March 4, 2014, http://dx.doi.org/10.1144/SP398.3

Ash zones in North Atlantic marine sediments were first described by Bramlette & Bradley (1941) and later by Ruddiman & Glover (1972), who described the so-called North Atlantic Ash Zones (NAAZ) I (*c.* 12 ka BP), II (*c.* 55 ka BP) and III (*c.* 305 ka BP). Usually the ash zones consist of a mixture of basaltic and rhyolitic tephra from major Icelandic eruptions, but they were originally defined as intervals of increased concentration of rhyolitic shards. The ash was transported southward from Iceland by ice-rafting by icebergs as far south as 45°N or by airfall on sea-ice, which was then rafted to the depositional site (e.g. Austin *et al.* 2004). Until recently, NAAZ I and II provided the only fix-points between ice-core records from Greenland and marine cores from the North Atlantic (Bond *et al.* 1993; Grönvold *et al.* 1995; Haflidason *et al.* 1995; Austin *et al.* 2004).

During recent years, several investigations of tephra in marine and ice-core records have been completed in the North Atlantic region. Two new isochrones, the Fugloyarbanki tephra/FMAZ II (Faroe Marine Ash Zone) and FMAZ III have been established, which enable correlations between marine cores in the North Atlantic region and the Greenland ice-cores (Davies *et al.* 2010; Austin *et al.* 2012). The Fugloyarbanki tephra/FMAZ II was first described in marine cores from the Faroe Islands region, where it is 2–10 cm thick (Rasmussen *et al.* 2003). The geochemistry is similar in all cores and an origin in the Hekla–Vatnafjöll volcanic system has been suggested (Wastegård *et al.* 2006; Davies *et al.* 2008). A record of the FMAZ II in the Labrador Sea (Wastegård *et al.* 2006) suggests a wide distribution in the North Atlantic, probably aided by iceberg or sea-ice rafting.

The Faroe Marine Ash Zone III (FMAZ III) is an important marker horizon in several marine records within the Faroe Islands margin. It was found at the transition from Heinrich 4 to GIS (Greenland Interstadial) 8 and first described as the 33 ka tephra (Rasmussen *et al.* 2003), from its estimated radiocarbon age. In some cores the zone consists of a small lower peak of tephra composed of fresh porous basaltic particles considered as airborne, and a more prominent, thick upper peak composed mainly of larger and heavier glass particles probably transported to the core locations by ice-rafting and/or bottom currents. Geochemical analyses suggest that the source is the Grímsvötn system on south Iceland (Wastegård *et al.* 2006).

For this study we targeted a tephra layer found between the North Atlantic Ash Zone II (Ruddiman & Glover 1972; Haflidason *et al.* 2000; Wastegård *et al.* 2006) and the FMAZ III (Rasmussen *et al.* 2003; Wastegård *et al.* 2006; Davies *et al.* 2010). The tephra layer is basaltic and was discovered by analysis of piston cores LINK16 (see Rasmussen & Thomsen 2009) and LINK15 (Rasmussen *et al.* 2012) from the Faroe Islands margin (Fig. 1). The tephra layer appeared as very thick (*c.* 20 cm in core LINK16 and *c.* 10 cm in core LINK15), with most tephra shards in the 63–100 μm size fraction, and it was believed to be mainly deposited and redistributed by bottom currents. Preliminary geochemical analyses on shards from LINK15 suggested that the tephra was relatively undisturbed and unmixed (Wastegård 2008). Therefore, a number of new samples from a series of five marine cores from the Faroe margin were analysed. The peak occurrences are in the lower part of GIS 12 and we suggest naming this new tephra FMAZ IV, following the nomenclature adopted for previous ash zones found on the Faroe Islands margin. Geochemical analyses of the tephra from five different core records show affinities with the Grímsvötn volcanic system in the Eastern Volcanic Zone in south Iceland. The FMAZ IV tephra dates to *c.* 44 000 ^{14}C years BP according to the age models, which calibrates to 47 000 years BP.

Materials and methods

This study is based on tephra shards from five marine cores from the northern Faroe margin and the central Faroe–Shetland Channel recovered during three different cruises: ENAM93-20 and ENAM93-21 in 1993 during a cruise with *RV Pelagia*; LINK15 and LINK16 in 2000 during a cruise with *RV Dana*; and JM11-19PC in 2011 during a cruise with *RV Jan Mayen* (Fig. 1; Table 1). The cores span a water depth from 773 to 1602 m.

New counts of tephra grains >100 μm for cores LINK15 and LINK16 and counts of tephra >150 μm in selected intervals in core ENAM93-21 were performed according to available size fractions from each record (Figs 2 & 3). The counting methods of tephra shards follow the procedures described in Wastegård and Rasmussen (2001). The magnetic susceptibility record for ENAM93-21 was published in Rasmussen *et al.* (1996), while the record of ENAM93-20 is found in Rasmussen *et al.* (2003). The magnetic susceptibility records for LINK15 and LINK16 were published in Nielsen *et al.* (2007). The magnetic susceptibility for core JM11-19PC was measured onboard *RV Jan Mayen* in 2011 with a loop sensor following the same procedures as published in Rasmussen *et al.* (1996) and Nielsen *et al.* (2007). The tephra layers in cores ENAM93-20 and JM11-19PC were pinpointed based on the magnetic susceptibility record (see Rasmussen *et al.* 1996, 1999). The shards were not counted and only one level in each core was analysed geochemically (Appendix 1). Owing

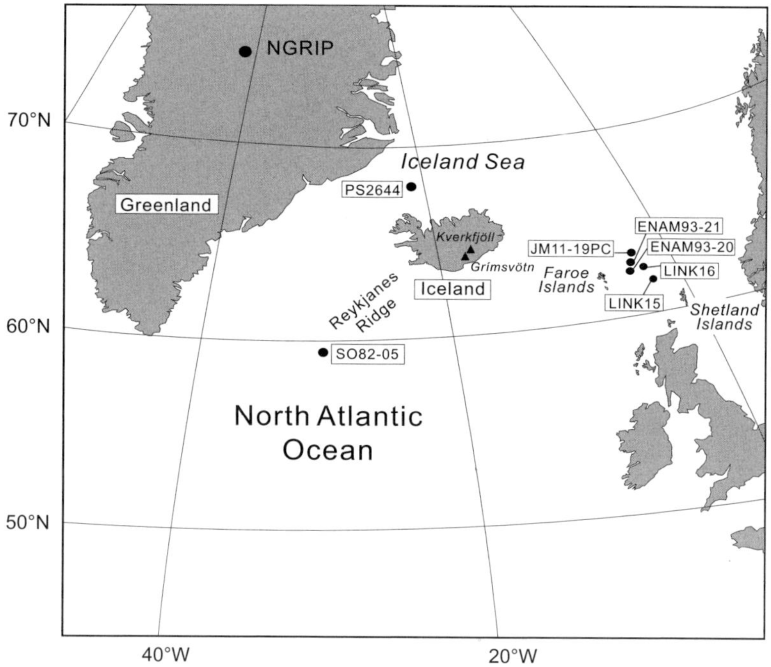

Fig. 1. Map of the Faroe Islands area showing location of cores investigated in this study. Other cores mentioned in the text are also indicated.

to a void of 20 cm in the record of JM11-19PC, the tephra appear to occur at 560 cm core depth (Fig. 3), but the true sediment depth is 540 cm.

We used previously published accelerator mass spectrometry (AMS) ^{14}C dates on cores ENAM93-21 and -20, and LINK16. New dates for LINK15 were obtained (Table 2; see also Olsen *et al.* 2014). The dates were calibrated using Calib6.01, Marine09 (Reimer *et al.* 2009). JM11-19PC was not dated, but correlated to the other records by means of the magnetic susceptibility record (see below).

For the geochemical analyses of tephra only the very light or highly porous basaltic tephra shards were used as these shards were most likely to have been airborne. Heavy, blocky fragments considered ice-rafted were avoided.

Quantitative geochemical analyses of tephra shards were performed on a wavelength dispersive spectroscopy (WDS)-equipped Cameca SX100 electron microprobe at the Tephra Analytical Unit at the Department of Geology and Geophysics, University of Edinburgh, UK. For analytical settings, see Appendix 1. Three different size fractions were investigated, 63–100 μm (LINK16, ENAM93-20 and ENAM93-21), >100 μm (JM11-19PC and LINK15) and >150 μm (ENAM93-20 and ENAM93-21; Table 3 and Appendix 1). For some samples only a few shards were analysed owing to small samples and preparation difficulties.

Table 1. *Locations of investigated cores and other cores mentioned in the text*

Core	Position		Water depth (m)
	Latitude (N)	Longitude (W)	
ENAM93-20	62°40′	3°59′	853
ENAM93-21	62°44′	3°53′	1020
LINK15	61°45′	2°41′	1602
LINK16	62°36′	3°31′	773
JM11-19PC	62°48′	3°50′	1180

Results

Geochemistry of the ash zone

The analyses show that tephra with tholeiitic basalt composition dominate in all cores. No differences in geochemistry between the analysed size fractions were observed. Biplots of, for example, K_2O vs TiO_2 and SiO_2 v. FeO_{tot}/TiO_2 suggest an origin in the Grímsvötn volcanic system in the Eastern Volcanic zone of Iceland (Fig. 4). Grímsvötn is one of the most active volcanic systems on Iceland (e.g. Thordarson & Larsen 2007; Óladóttir *et al.*

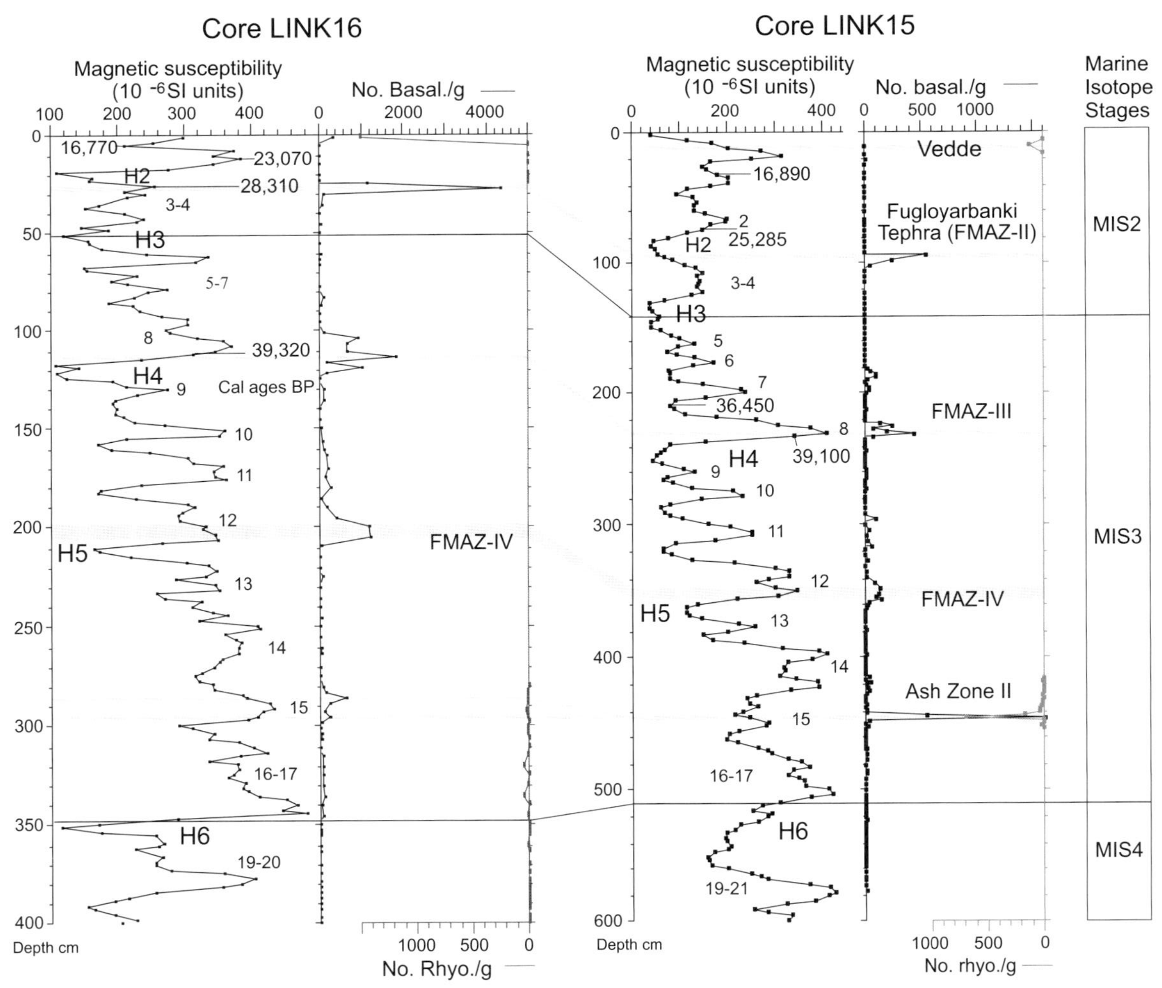

Fig. 2. Magnetic susceptibility and tephra counts >100 μm in numbers per gram dry weight sediment for cores LINK15 and LINK16 plotted v. depth (cm). Right column shows isotope stage boundaries. Grey horizontal bars mark locations of tephra layers. Calibrated ages and Greenland interstadial and Heinrich event (H) numbers are indicated.

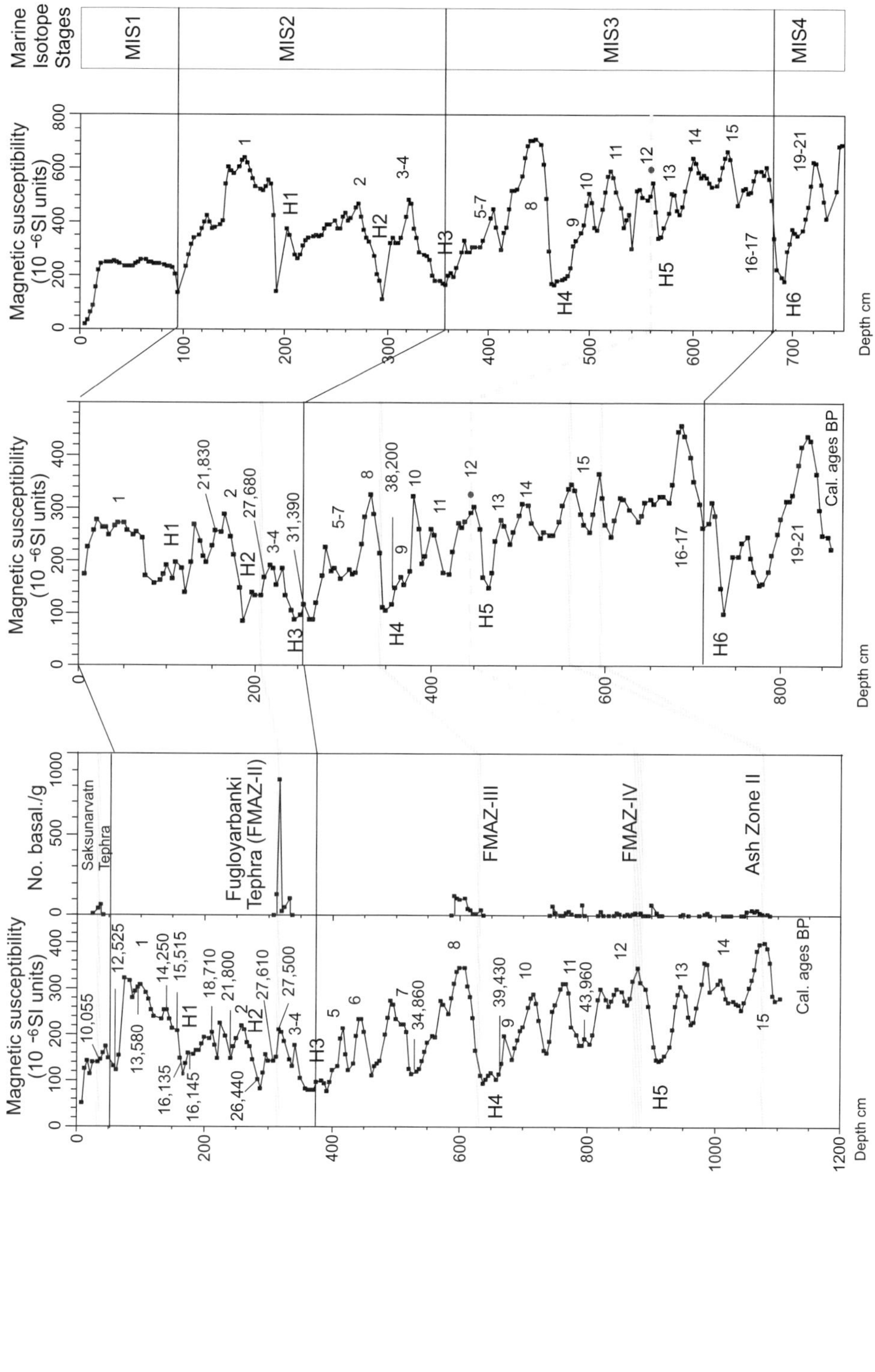

Fig. 3. Magnetic susceptibility and tephra counts >150 µm in numbers per gram dry weight sediment for core ENAM93-21 and magnetic susceptibility for cores ENAM93-20 and JM11-19PC. Right column shows isotope stage boundaries. Grey horizontal bars mark locations of tephra layers. Black dots in ENAM93-21 and JM11-19PC mark sample position for analysed tephra of FMAZ-IV. Calibrated ages and Greenland interstadial and Heinrich event (H) numbers are indicated.

Table 2. *New radiocarbon dates for LINK15 (see also Olsen* et al. *2014)*

Core	Depth cm	^{14}C age (corr.-400)*	Laboratory code	Material
LINK15	30	13 800 ± 80	AAR-12468	*N. pachyderma* s
LINK15	74	21 180 ± 160	AAR-12469	*N. pachyderma* s
LINK15	231	33 153 ± 191	UB-17235†	*N. pachyderma* s

*Marine reservoir correction of 400 years.
†Date from Olsen *et al.* (2014).

2011) and it is one of the main contributors to the tephra records in the Greenland ice-cores and marine North Atlantic cores (e.g. Brendryen *et al.* 2010; Abbott *et al.* 2012; Gudmundsdóttir *et al.* 2012). For example, the widespread Early Holocene Saksunarvatn Tephra (*c.* 10.3 ka BP) and FMAZ III both derive from Grímsvötn (Kvamme *et al.* 1989; Wastegård *et al.* 2006). A few outliers, not correlated with Grímsvötn, occur in most samples except in JM11-19PC, and a second population with affinities with the Katla volcanic system occurs at 355–356 cm in the LINK15 core (Fig. 4; Table 3). We believe that the FMAZ IV is identical in all cores and represents a major eruption in the Grímsvötn system. However, contribution of tephra from several eruptions close in time, as has been suggested for the Saksunarvatn Tephra (Kylander *et al.* 2012; Jennings *et al.* 2014), cannot be excluded. Only small differences in major element composition have been observed in cores where tephra from two different levels have been analysed, but we acknowledge that this might be due to the fact that only fresh porous shards considered as airborne tephra were analysed, which may have biased the results (Appendix 1).

Age and stratigraphy of FMAZ IV

The stratigraphy and geochemistry of several known tephra layers from ENAM93-21, ENAM93-20 and LINK16 have been published previously (Rasmussen *et al.* 2003; Wastegård *et al.* 2006). These tephra records comprised the Fugloyarbanki Tephra/FMAZ II and FMAZ-III. The FMAZ II tephra has been dated to *c.* 28 000 ± 360 cal years BP (Rasmussen *et al.* 2003, 2012), while FMAZ III has been dated to *c.* 38 000 ± 465 cal years BP (Olsen *et al.* 2014). Both tephras have recently been found in the NGRIP ice-core on Greenland (Davies *et al.* 2008, 2010). The Fugloyarbanki tephra/FMAZ II was found in a relatively cold period, above the warmest phase of the GIS 3, and was dated to *c.* 26 740 ± 390 years b2k (before year AD 2000; GICC05; Davies *et al.* 2008). A tephra correlated to the FMAZ III was described from the NGRIP ice-core (Davies *et al.* 2010), where it is a few millimetres thick and is dated to 38 122 ± 723 yr b2k. The tephra occurs directly at the thermal peak of GIS 8 and a marine-ice correlation, based on the position of this tephra, suggests a close coupling of the marine and atmospheric systems at this time (Davies *et al.* 2010).

In LINK15 the age of the FMAZ IV was calculated to 47 000 ± 1400 cal years BP (Rasmussen *et al.* 2012). Using the ^{40}Ar–^{39}Ar age of the Thorsmörk ignimbrite, which correlates to the rhyolitic component of NAAZ II (54 500 ± 1000 cal years BP; McIntosh, in Southon 2004) and assuming linear sedimentation rates between dated points (Figs 2 & 3), we calculate an age of FMAZ IV of 47 080 ± 1000 cal years BP for ENAM93-21 and 46 350 ± 850 cal years BP for ENAM93-20. In LINK16 we calculate an age of 46 720 ± 750 cal years BP. The average age of FMAZ IV from these four independent age models is 46 800 ± 1000 years BP.

The cores ENAM93-21, ENAM93-20 and LINK16 have previously been correlated to the Greenland ice-cores in great detail and the Dansgaard–Oeschger millennial scale events (DO) have been identified and numbered together with the Heinrich events H6-H1 (Rasmussen *et al.* 1996; Rasmussen & Thomsen 2004, 2009; Figs 2, 3 & 5). Interstadials are characterized by reduced relative abundance of the polar planktic foraminifera species *Neogloboquadrina pachyderma* s (Bond *et al.* 1993; Rasmussen *et al.* 1996; Fig. 5). Cores LINK15 and JM11-19PC show a very similar pattern of magnetic susceptibility to the former three records (Figs 2, 3 & 5). From the correlation the tephra layer in all cores is found in Greenland Interstadial 12 close to the highest peak in magnetic susceptibility values correlating with the warmest conditions (Figs 2, 3 & 5). The tephra is found above the cold H5 event of low magnetic susceptibility values.

Discussion and conclusions

The tephrostratigraphic record from the Faroe margin and the Faroe–Shetland Channel provides a detailed insight into Icelandic volcanism during the Late Quaternary. The records in this area have a high temporal resolution and are close enough

Table 3. *Major element composition of the FMAZ IV in cores from the northern Faroe margin and the central Faroe–Shetland Channel, shown as mean oxide concentrations (wt%)*

Tephra	Core/site	Depth	Fraction	*n*	SiO_2	TiO_2	Al_2O_3	FeO_{tot}	MnO	MgO	CaO	Na_2O	K_2O	P_2O_5	Cl	Total	Age (BP)	Source
FMAZ IV	JM11-19PC	540 cm	>100 μm	16	49.23	2.53	13.09	13.68	0.25	5.56	9.97	2.60	0.37	0.24		97.52	46.8 ka*	Grímsvötn
					0.61	0.08	0.24	0.26	0.02	0.37	0.27	0.11	0.04	0.01				
FMAZ IV	LINK15	347–48 cm	>100 μm	11	49.43	2.72	12.78	13.40	0.25	5.30	9.77	2.73	0.43	0.29	0.02	97.12	46.8 ka*	Grímsvötn
					0.92	0.30	0.22	0.49	0.10	0.55	0.66	0.13	0.06	0.04	0.01			
FMAZ IV	LINK15	355–56 cm	>100 μm	6	49.20	2.64	13.18	12.68	0.15	5.81	10.48	2.68	0.39	0.23	0.04	97.48	46.8 ka*	Grímsvötn
					0.52	0.23	0.17	0.48	0.03	0.35	0.40	0.06	0.05	0.10	0.01			
FMAZ IV-Katla	LINK15	355–56 cm	>100 μm	4	49.31	3.85	12.59	14.29	0.28	4.38	8.92	2.95	0.69	0.54	0.02	97.84	46.8 ka*	Katla
					0.25	0.29	0.19	0.69	0.09	0.10	0.20	0.09	0.05	0.05	0.01			
FMAZ IV	LINK16	205–06 cm	63–100 μm	3	49.68	2.52	13.33	13.58	0.26	5.45	9.80	2.71	0.36	0.23		97.93	46.8 ka*	Grímsvötn
					0.17	0.02	0.11	0.51	0.02	0.06	0.12	0.09	0.06	0.01				
FMAZ IV	ENAM93-20	455–56 cm	63–100 μm	6	49.52	2.49	13.37	13.50	0.24	5.66	10.01	2.46	0.35	0.23		97.83	46.8 ka*	Grímsvötn
					0.65	0.07	0.39	0.38	0.01	0.26	0.24	0.12	0.03	0.02				
FMAZ IV	ENAM93-20	455–56 cm	>150 μm	6	49.74	2.55	13.31	13.68	0.25	5.50	10.06	2.56	0.36	0.23		98.23	46.8 ka*	Grímsvötn
					0.85	0.05	0.32	0.72	0.03	0.75	0.32	0.16	0.03	0.01				
FMAZ IV	ENAM93-21	860–61 cm	63–100 μm	8	49.33	2.48	13.08	13.28	0.25	5.37	9.72	2.93	0.46	0.25		97.39	46.8 ka*	Grímsvötn
					0.94	0.10	0.32	0.60	0.02	0.61	0.74	0.26	0.11	0.02				
FMAZ IV	ENAM93-21	865–66 cm	>150 μm	5	49.92	2.61	12.79	13.72	0.25	5.30	9.69	2.56	0.40	0.25		97.49	46.8 ka*	Grímsvötn
					0.44	0.10	0.30	0.25	0.01	0.28	0.33	0.41	0.04	0.01				
VZ 5	Reykjanes Ridge		?	38	49.47	2.65	13.68	12.49	0.21	5.84	10.65	2.61	0.39	0.13	0.02	98.31	46.2–52.5 ka	Grímsvötn
						1.08	0.92	0.81	1.33	0.06	1.09	0.86	0.37	0.21	0.02	0.02		
AL	PS-2644	5.18 m	?	33	49.51	3.28	12.41	14.90	0.25	4.89	9.33	2.53	0.61	0.74	0.03	98.38†	*c.* 45.5 ka	Kverkfjöll
					0.45	0.12	0.20	0.43	0.04	0.37	0.33	0.08	0.07	0.04	0.01			
AL	PS-2644	5.22 m	?	8‡	50.19	2.61	12.61	13.50	0.21	5.53	10.09	2.54	0.47	0.57	0.03	99.35†	*c.* 45.5 ka	Grímsvötn
					0.62	0.06	0.20	0.46	0.04	0.45	0.39	0.11	0.08	0.13	0.01			
AL	PS-2644	5.22 m	?	17§	49.72	3.19	12.41	14.92	0.25	4.92	9.22	2.44	0.61	0.73	0.03	99.45†	*c.* 45.5 ka	Kverkfjöll
					0.26	0.11	0.07	0.31	0.02	0.07	0.14	0.05	0.02	0.03	0.02			

For analytical conditions, see Appendix 1. Analyses of other ash zones from GIS 12 are shown for comparison; VZ 5 from the Reykjanes Ridge (Lackschewitz & Wallrabe-Adams 1997) and PS-2644 from the Iceland Sea (Voelker 1999, 2000*a*, *b*; Voelker *et al.* 2000).

*Average of four independent age models (see text for explanation).

†Totals also include SO_2.

‡Eight of 39 analyses.

§17 of 39 analyses.

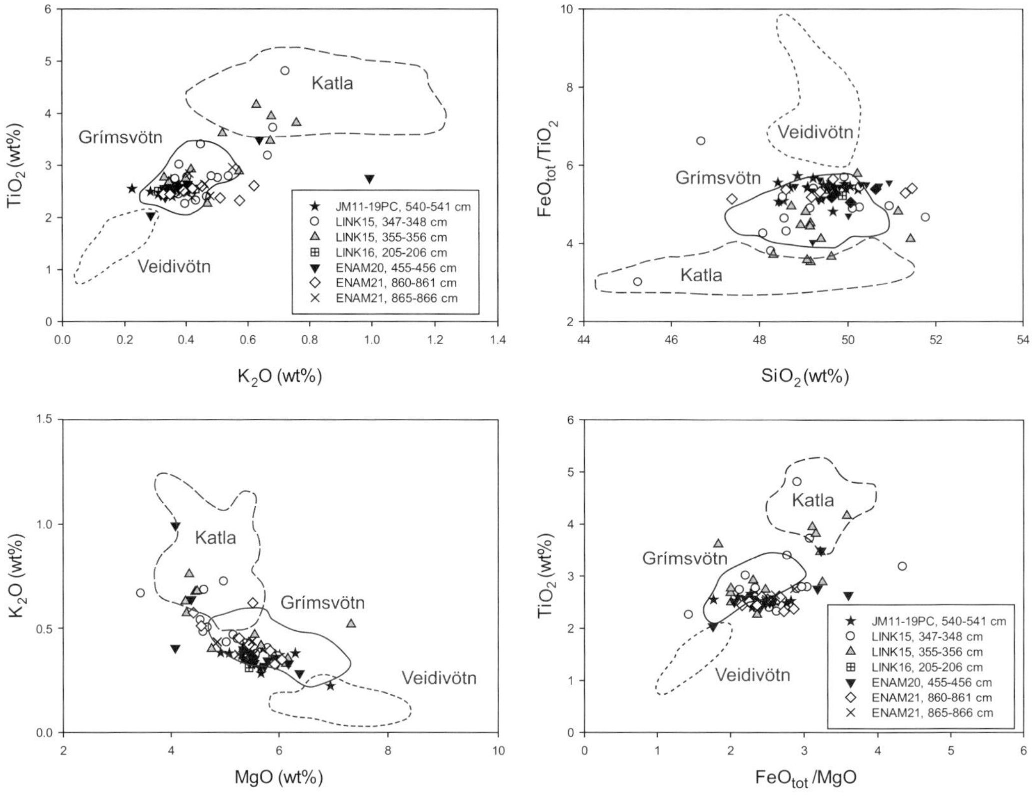

Fig. 4. Major oxides biplots of the FMAZ IV. Geochemical envelopes for source volcanoes are based on Davies *et al.* (2010, figure 3) and references therein. The results have not been normalized.

to Iceland to receive tephra derived from atmospheric transport, even from small and medium-sized basaltic eruptions (e.g. Wastegård *et al.* 2001; Wastegård 2002; Lind & Wastegård 2011). A notable result of recent studies of the last glacial period in the Greenland ice-cores is that more than half of the Icelandic tephra horizons have a basaltic composition (Abbott *et al.* 2012; Abbott & Davies 2012). This is surprising since basaltic volcanism normally is more effusive and not sufficiently explosive to eject tephra high into the atmosphere. This further emphasizes the Faroe region as important for marine-ice correlations in the North Atlantic regions based on basaltic tephra layers.

The FMAZ IV has been identified in several cores on the Faroe margin and the Faroe–Shetland Channel, but it has not been thoroughly investigated whether the distribution also includes other parts of the North Atlantic region, including the Greenland ice-cores. However, a data survey has resulted in a potential match between the FMAZ IV and an ash zone identified in marine cores from the Reykjanes Ridge SW of Iceland (Fig. 1; Lackschewitz & Wallrabe-Adams 1997). These authors described volcanic ash zones in cores spanning the last *c.* 70 000 years, including NAAZ I and II and with five ash zones in between. The age of the oldest of these, VZ 5 (Table 3), was estimated to *c.* 46.2–52.5 ka BP and the geochemistry is typical for Grímsvötn with TiO_2 around 2.65 wt% and MgO around 5.8 wt% (Table 3; Fig. 6). Standard deviations are relatively high (around 1% for many elements; Fig. 6), however, indicating contribution from other volcanic systems and/or several eruptions within the Grímsvötn volcanic system. Furthermore, ages reported are based on age–depth models based on few radiocarbon dates and the stratigraphical position of the ash zone in relation to Heinrich and DO events is not clear, although the tephra counts v. ice rafted debris (IRD) accumulation rates (figure 7 in Lackschewitz & Wallrabe-Adams 1997) indicate a position above an IRD peak, possibly Heinrich event H5. Most of the other ash zones described by Lackschewitz & Wallrabe-Adams (1997) are correlated with high amounts of IRD, indicating significant contribution

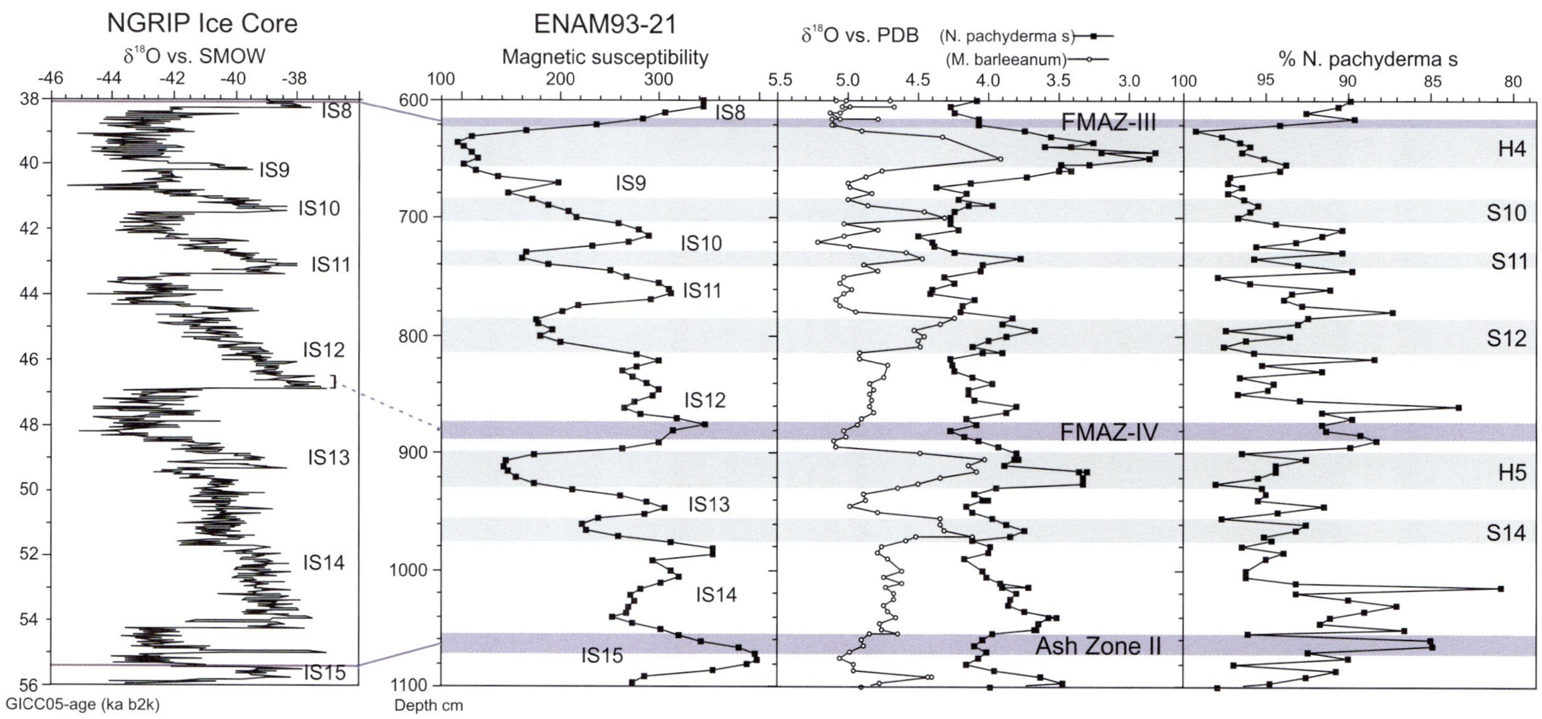

Fig. 5. NGRIP ice-core oxygen isotope data plotted v. ice-core years before 2000 years (b2k) (data from Svensson *et al.* 2008) and magnetic susceptibility and relative abundance of the polar planktic foraminifera species *Neogloboquadrina pachyderma* sinistral (s) from core ENAM93-21 plotted v. depth (cm; data from Rasmussen *et al.* 1996). Horizontal dark grey bars mark tephra layers. Greenland interstadial (GIS) and stadial (S) and Heinrich event (H) numbers are indicated. Light grey horizontal bars mark cold stadials and Heinrich events. The potential location of the FMAZ IV tephra in the ice-core is indicated by hatched line.

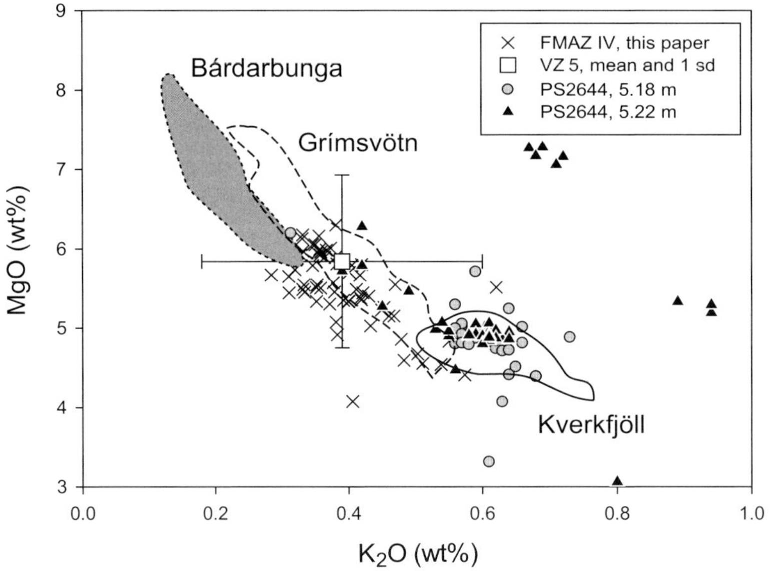

Fig. 6. Biplots of K_2O v. MgO for the main component of FMAZ IV from all cores compared with VZ 5 from the Reykjanes Ridge (Lackschewitz & Wallrabe-Adams 1997) and PS-2644 from the Iceland Sea (Voelker 1999, 2000*a*, *b*; Voelker *et al.* 2000). The results have not been normalized. Geochemical envelopes are based on electron microprobe analyses of tephra from Holocene eruptions of the basaltic tholeiitic volcanic systems Bárdarbunga, Grímsvötn and Kverkfjöll (Óladóttir *et al.* 2011).

of tephra from ice-rafting. VZ 5, however, is situated above an IRD peak, which indicates that this ash zone was deposited during a warm period and may represent an air-fall deposit. We tentatively correlate FMAZ IV with VZ 5, but further investigations are needed in order to confirm a link between the Faroe Margin and the Reykjanes Ridge.

A basaltic ash zone in the lower part of MIS 3 has also been reported in the deep-sea core PS2644 from the Iceland Sea (Fig. 1) by Voelker *et al.* (2000) and Voelker (1999, 2000*b* and pers. comm. 2007). A correlation between planktonic $\delta^{18}O$, $\delta^{13}C$ and IRD records of the sediment core PS2644 and the bidecadal $\delta^{18}O$ record of ice-core GISP2 places the ash zone approximately at the transition from H5 to GIS 12 (figure 2 in Voelker *et al.* 2000), that is, in a similar stratigraphical position to FMAZ IV on the Faroe Margin. Chemical analyses of tephra from two levels, 5.18 and 5.22 m (Table 3), show that the lower peak at 5.22 m has a mixed composition with several different components (Voelker 2000*a*, *b*). Only eight of 39 analyses have affinities with the Grímsvötn volcanic system, for example, MgO around 5.5wt% and TiO_2 around 2.6 wt% (Table 3, Fig. 6) and these probably represent a background of Grímsvötn-type material rather than a tephra isochron. Almost 50% of the analyses (17 of 39) at 5.22 m and all analyses of the upper peak (5.18 m) indicate an origin in the Kverkfjöll volcanic system (e.g. Óladóttir *et al.* 2011) rather than in the Grímsvötn system (e.g. MgO around 4.9 wt% and TiO_2 around 3.3 wt%). It might well be that the Grímsvötn component in the 5.22 m sample derives from the same eruption that resulted in the FMAZ IV on the Faroe Margin. However, since only eight of 72 analyses suggest an origin in the Grímsvötn system, a correlation of the GIS 12 ash zone from the Iceland Sea with FMAZ IV is not possible at the moment.

So far, no tephra layers between GIS 14 and GIS 8 (*c.* 54–38 ka BP; Abbott & Davies 2012) have been reported from the Greenland ice-cores, but investigations are underway. As outlined by Abbott & Davies (2012), tephras from explosive basaltic eruptions are more common in the Greenland ice-cores than in European continental records. Terrestrial records from Europe extending back to MIS 5 are rare and only a few sites have been investigated in detail for cryptotephras. One exception is the Les Echets sequence in France, but only material from Central European volcanoes has been found so far (Veres *et al.* 2008). As pointed out earlier, the area around the Faroe Islands may be ideal for linking the Greenland ice-cores with marine records and continental records using basaltic tephra layers. Both the FMAZ II/Fugloyarbanki tephra and the FMAZ III have been found in ice-core records from Greenland, which suggests that these eruptions were large enough to transport tephra in both easterly and westerly directions (Davies *et al.* 2008, 2010). This is important for our understanding of the dispersal of tephra from major basaltic eruptions. Both the FMAZ III and FMAZ IV occur at the warming limbs of DO events and we speculate that the processes that caused these warmings also may have triggered explosive basaltic eruptions on Iceland. There is strong potential for positive feedbacks between melting of ice and enhanced volcanism (e.g. Tuffen 2010), and the geological record shows that there are a number of explosive events that fall stratigraphically in close association with rapid climatic changes (e.g. Mayewski *et al.* 2004; Lowe *et al.* 2012; Storey *et al.* 2012), although the link between the casual forcing and the climate change is not always clear. The tephras from these events are particularly useful for comparing the strength, spatial distribution and timing of climate oscillations, and a link between volcanic activity, glaciation and millennial-scale rapid environmental change has often been discussed (e.g. Nowell *et al.* 2006; Huybers & Langmuir 2009).

This research is supported by the Swedish Research Council (VR; grant no. 2006-5868). C. Hayward, University of Edinburgh, is acknowledged for help with microprobe analyses. The marine core study is part of the 'Paleo-CIRCUS' project supported by the University of Tromsø and the Mohn Foundation. This project is also part of CAGE supported by the Research Council of Norway, Grant No. 223259/F5.

Appendix 1

Single-grain analyses of tephra from FMAZ IV in cores from the northern Faroe margin and the central Faroe–Shetland Channel Analyses were obtained on a CAMECA SX100 scanning electron microprobe by wavelength dispersive spectrometry with an accelerating voltage of 15 kV, a beam current of 2 nA for the major elements and 80 nA for the minor elements and a beam diameter of 5 μm. All oxides are expressed as wt%. Total iron is expressed as FeO_{tot}. Standard calibration blocks were used for calibration and glass standards were used to assess the accuracy during analyses. Two secondary glass standards, BHVO2 and Lipari obsidian (Hunt & Hill 1996; Kuehn *et al.* 2011), were analysed at regular intervals to control instrumental drift and the precision and accuracy of the glass analysis. Only analyses above 95% are included in the means; *n* represents the number of analyses.

References

Abbott, P. M. & Davies, S. M. 2012. Volcanism and the Greenland ice-cores: the tephra record. *Earth-Science Reviews*, **115**, 173–191.

Abbott, P. M., Davies, S. M. *et al.* 2012. A detailed framework of Marine Isotope stages 4 and 5 volcanic events recorded in two Greenland ice-cores. *Quaternary Science Reviews*, **36**, 59–77.

Ampel, L., Wohlfarth, B., Risberg, J., Veres, D., Leng, M. J. & Kaislahti Tillman, P. 2010. Diatom assemblage dynamics during abrupt climate change: the response of lacustrine diatoms to Dansgaard–Oeschger cycles during the last glacial period. *Journal of Paleolimnology*, **44**, 397–404.

Austin, W. E. N., Wilson, L. J. & Hunt, J. B. 2004. The age and chronostratigraphical significance of North Atlantic Ash zone II. *Journal of Quaternary Science*, **19**, 137–146.

Austin, W. E. N., Hibbert, F. D., Rasmussen, S. O., Peters, C., Abbott, P. M. & Bryant, C. 2012. The synchronization of palaeoclimatic events in the North Atlantic region during Greenland Stadial 3 (ca 27.5 to 23.3 kyr b2k). *Quaternary Science Reviews*, **36**, 154–163.

Björck, S., Walker, M. J. C., Cwynar, L. C., Johnsen, S., Knudsen, K. L., Lowe, J. J. & Wohlfarth, B. 1998. An event stratigraphy for the last termination in the North Atlantic region based on the Greenland ice-core record; a proposal by the INTIMATE group. *Journal of Quaternary Science*, **13**, 283–292.

Blaauw, M., Wohlfarth, B. *et al.* 2010. Were last glacial events simultaneous between Greenland and western Europe? *Journal of Quaternary Science*, **25**, 387–394.

Bond, G., Broecker, W., Johnsen, S., McManus, J., Labeyrie, L., Jouzel, J. & Bonani, G. 1993. Correlations between climate records from North-Atlantic sediments and Greenland ice. *Nature*, **365**, 143–147.

Bramlette, M. N. & Bradley, W. H. 1941. *Geology and Iology of North Atlantic Deep-sea Cores Between Newfoundland and Ireland: I. Lithology and Geologic Interpretation*. US Geological Survey, Reston, VA, Professional Papers, **196-A**, 1–34.

Brendryen, J., Haflidason, H. & Sejrup, H. P. 2010. Norwegian Sea tephrostratigraphy of marine isotope stages 4 and 5: prospects and problems for tephrochronology in the North Atlantic region. *Quaternary Science Reviews*, **29**, 847–864.

Davies, S. M., Wastegård, S., Rasmussen, T. L., Svensson, A., Johnsen, S. J., Steffensen, J. P. & Andersen, K. K. 2008. Identification of the Fugloyarbanki tephra in the NGRIP ice core: a key tie-point for marine and ice-core sequences during the last glacial period. *Journal of Quaternary Science*, **23**, 409–414.

Davies, S. M., Wastegård, S. *et al.* 2010. Tracing volcanic events in the NGRIP ice-core and synchronising North Atlantic marine records during the last glacial period. *Earth and Planetary Science Letters*, **294**, 69–79.

Grönvold, K., Óskarsson, K., Johnsen, S. J., Clausen, H. B., Hammer, C. U., Bond, G. & Bard, E. 1995. Ash layers from Iceland in the Greenland GRIP ice core correlated with oceanic and land sediments. *Earth and Planetary Science Letters*, **135**, 149–155.

Gudmundsdóttir, E. R., Larsen, G. & Eiríksson, J. 2012. Tephra stratigraphy on the North Icelandic shelf: extending tephrochronology into marine sediments off North Iceland. *Boreas*, **41**, 718–734.

HAFLIDASON, H., SEJRUP, H. P., KLITGAARD-KRISTENSEN, D. & JOHNSEN, S. 1995. Coupled response of the late glacial climatic shifts of Northwest Europe reflected in Greenland ice cores; evidence from the northern North Sea. *Geology*, **23**, 1059–1062.

HAFLIDASON, H., EIRÍKSSON, J. & VAN KREFELD, S. 2000. The tephrochronology of Iceland and the North Atlantic region during the Middle and Late Quaternary: a review. *Journal of Quaternary Science*, **15**, 3–22.

HUNT, J. B. 2004. Tephrostratigraphical evidence for the timing of Pleistocene explosive volcanism at Jan Mayen. *Journal of Quaternary Science*, **19**, 121–136.

HUNT, J. B. & HILL, P. G. 1996. An inter-laboratory comparison of the electron probe microanalysis of glass geochemistry. *Quaternary International*, **34–6**, 229–241.

HUYBERS, P. & LANGMUIR, C. 2009. Feedback between deglaciation, volcanism, and atmospheric CO_2. *Earth and Planetary Science Letters*, **286**, 479–491.

JENNINGS, A., THORDARSON, T., ZALZAL, K., STONER, J., HAYWARD, C., GEIRSDÓTTIR, Á. & MILLER, G. 2014. Holocene tephra from Iceland and Alaska in SE Greenland Shelf Sediments. *In*: AUSTIN, W. E. N., ABBOTT, P. M., DAVIES, S. M., PEARCE, N. J. G. & WASTEGÅRD, S. (eds) *Marine Tephrochronology*. Geological Society, London, Special Publications, **398**. First published online February 27, 2014, http://dx.doi.org/10.1144/SP398.6

KUEHN, S. C., FROESE, D. G., SHANE, P. A. R. & INTAV Intercomparison Participants 2011. The INTAV intercomparison of electron-beam microanalysis of glass by tephrochronology laboratories: results and recommendations. *Quaternary International*, **246**, 19–47.

KVAMME, T., MANGERUD, J., FURNES, H. & RUDDIMAN, W. 1989. Geochemistry of Pleistocene ash zones in cores from the North Atlantic. *Norsk Geologisk Tidskrift*, **69**, 251–272.

KYLANDER, M., LIND, E. M., WASTEGÅRD, S. & LÖWEMARK, L. 2012. Recommendations for using XRF core scanning as a tool in tephrochronology. *The Holocene*, **22**, 371–375.

LACASSE, C. & GARBE-SCHÖNBERG, C.-D. 2001. Explosive silicic volcanism in Iceland and the Jan Mayen area during the last 6 Ma: sources and timing of major eruptions. *Journal of Volcanology and Geothermal Research*, **107**, 113–147.

LACKSCHEWITZ, K. S. & WALLRABE-ADAMS, H.-J. 1997. Composition and origin of volcanic ash zones in late Quaternary sediments from the Reykjanes Ridge; evidence for ash fallout and ice-rafting. *Marine Geology*, **136**, 209–224.

LARSEN, G. 1981. Tephrochronology by microprobe glass analysis. *In*: SELF, S. & SPARKS, R. S. J. (eds) *Tephra Studies; Proceedings of the NATO Advanced Study Institute 'Tephra Studies as a Tool in Quaternary Research'*. Reidel, Dordrecht, 95–102.

LAWSON, I. T., SWINDLES, G. T., PLUNKETT, G. & GREENBERG, D. 2012. The spatial distribution of Holocene cryptotephras in north-west Europe since 7 ka: implications for understanding ash fall events from Icelandic eruptions. *Quaternary Science Reviews*, **41**, 57–66.

LIND, E. M. & WASTEGÅRD, S. 2011. Tephra horizons contemporary with short early Holocene climate fluctuations: new results from the Faroe Islands. *Quaternary International*, **246**, 157–167.

LOWE, J., BARTON, N. ET AL. 2012. Volcanic ash layers illuminate the resilience of Neanderthals and early modern humans to natural hazards. *Proceedings of the National Academy of Science of the USA*, **109**, 13532–13537.

MAYEWSKI, P. A., ROHLING, E. E. ET AL. 2004. Holocene climate variability. *Quaternary Research*, **62**, 243–255.

NIELSEN, T., RASMUSSEN, T. L., CERAMICOLA, S. & KUIJPERS, A. 2007. Quaternary sedimentation, margin architecture and ocean circulation variability around the Faroe Islands, North Atlantic. *Quaternary Science Reviews*, **26**, 1016–1036.

NOWELL, D. A. G., JONES, M. C. & PYLE, D. M. 2006. Episodic Quaternary voicanism in France and Germany. *Journal of Quaternary Science*, **21**, 645–675.

ÓLADÓTTIR, B. A., SIGMARSSON, O., LARSEN, G. & DEVIDAL, J. L. 2011. Provenance of basaltic tephra from Vatnajokull subglacial volcanoes, Iceland, as determined by major- and trace-element analyses. *The Holocene*, **21**, 1037–1048.

OLSEN, J., RASMUSSEN, T. L. & REIMER, P. 2014. North Atlantic marine radiocarbon reservoir ages through Heinrich event H4: a new method for marine age model construction. *In*: AUSTIN, W. E. N., ABBOTT, P. M., DAVIES, S. M., PEARCE, N. J. G. & WASTEGÅRD, S. (eds) *Marine Tephrochronology*. Geological Society, London, Special Publications, **398**. First published online February 6, 2014, http://dx.doi.org/10.1144/SP398.2

PEARCE, N. J. G., WESTGATE, J. A., PERKINS, W. T. & PREECE, S. J. 2004. The application of ICP-MS methods to tephrochronological problems. *Applied Geochemistry*, **19**, 289–322.

RAHMSTORF, S. 2002. Ocean circulation and climate during the past 120,000 years. *Nature*, **419**, 207–214.

RASMUSSEN, T. L. & THOMSEN, E. 2004. The role of the North Atlantic Drift in the millennial timescale glacial climate fluctuations. *Palaeogeography Palaeoclimatology Palaeoecology*, **210**, 101–116.

RASMUSSEN, T. L. & THOMSEN, E. 2009. Ventilation changes in intermediate water on millennial time scales in the SE Nordic seas, 65–14 kyr BP. *Geophysical Research Letters*, **36**, L01601, http://dx.doi.org/10.1029/2008GL036563

RASMUSSEN, T. L., THOMSEN, E., VAN WEERING, T. & LABEYRIE, L. 1996. Rapid changes in surface and deep water conditions at the Faeroe Margin during the last 58,000 years. *Paleoceanography*, **11**, 937–940.

RASMUSSEN, T. L., BALBON, E., THOMSEN, E., LABEYRIE, L. & VAN WEERING, T. C. E. 1999. Climate records and changes in deep outflow from the Norwegian Sea ~150–55 ka. *Terra Nova*, **11**, 61–66.

RASMUSSEN, T. L., WASTEGÅRD, S., KUIJPERS, A., VAN WEERING, T. C. E., HEINEMEIER, J. & THOMSEN, E. 2003. Stratigraphy and distribution of tephra layers in marine sediment cores from the Faeroe Islands, North Atlantic. *Marine Geology*, **199**, 263–277.

RASMUSSEN, T. L., OLSEN, J., WASTEGÅRD, S. & THOMSEN, E. 2012. Stratigraphy of three mid-late Weichselian tephra layers from the Faroe Islands margin. *EGU General Assembly*, Vienna, EGU2012-7087.

REIMER, P. J., BAILLIE, M. G. L. *ET AL.* 2009. IntCal09 and MARINE09 radiocarbon age calibration curves, 0-50,000 years cal BP. *Radiocarbon*, **51**, 1111-1150.

RUDDIMAN, W. & GLOVER, R. 1972. Vertical mixing of ice-rafted volcanic ash in North Atlantic sediments. *Geological Society of America Bulletin*, **83**, 2817-2836.

SANCHEZ-GOÑI, M. F., LANDAIS, A., FLETCHER, W. J., NAUGHTON, F., DESPRAT, S. & DUPRAT, J. 2008. Contrasting impacts of Dansgaard-Oeschger events over a western European latitudinal transect modulated by orbital parameters. *Quaternary Science Reviews*, **27**, 1136-1151.

SOUTHON, J. 2004. A radiocarbon perspective on Greenland ice-core chronologies: can we use ice cores for ^{14}C calibration? *Radiocarbon*, **46**, 1239-1259.

STOREY, M., ROBERTS, R. G. & SAIDIN, M. 2012. Astronomically calibrated $^{40}Ar/^{39}Ar$ age for the Toba supereruption and global synchronisation of late Quaternary records. *Proceedings of the National Academy of Science of the USA*, **109**, 18 684-18 688.

SVENSSON, A., ANDERSEN, K. K. *ET AL.* 2008. A 60 000 year Greenland stratigraphic ice core chronology. *Climate of the Past*, **4**, 47-57.

THORDARSON, T. & LARSEN, G. 2007. Volcanism in Iceland in historical time: volcano types, eruption styles and eruptive history. *Journal of Geodynamics*, **43**, 118-152.

TUFFEN, H. 2010. How will melting of ice affect volcanic hazards in the twenty-first century? *Philosophical transactions of the Royal Society A – Mathematical, Physical & Engineering Sciences*, **368**, 2535-2558.

VERES, D., DAVIES, S. M., WOHLFARTH, B., PREUSSER, F., WASTEGÅRD, S., AMPEL, L., HORMES, A., POSSNERT, G., RAYNAL, J.-P. & VERNET, G. 2008. Age, origin and significance of a new middle MIS 3 tephra horizon identified within a long-core sequence from Les Echets, France. *Boreas*, **37**, 434-443.

VOELKER, A. 1999. *Zur Deutung der Dansgaard-Oeschger Ereignisse in ultra-hochauflöenden Sedimentprofilen aus dem Europäischen Nordmeer.* PhD thesis, Christian-Albrechts-Universität zu Kiel, Kiel.

VOELKER, A. 2000*a*. *Age model of sediment core PS2644-5.* http://dx.doi.org/10.1594/PANGAEA.57797

VOELKER, A. 2000*b*. *Tephra cheminstry of sediment core PS2644-5.* http://doi.pangaea.de/10.1594/PANGAEA.57791

VOELKER, A. H. L., GROOTES, P. M., NADEAU, M. J. & SARNTHEIN, M. 2000. Radiocarbon levels in the Iceland Sea from 25-53 kyr and their link to the Earth's magnetic field intensity. *Radiocarbon*, **42**, 437-452.

WASTEGÅRD, S. 2002. Early to Middle Holocene silicic tephra horizons from the Katla volcanic system, Iceland: new results from the Faroe Islands. *Journal of Quaternary Science*, **17**, 723-730.

WASTEGÅRD, S. 2008. Tephra and Cryptotephra from European and North Atlantic marine sequences. Linkages with the Greenland record. *Final RESOLuTION Workshop. Rapid Climatic and Environmental Shifts during OIS 2 and 3 – Linking High-resolution Terrestrial, Ice Core and Marine Archives*, Ecole Pratique des Hautes Etudes, Haut Carré, Bordeaux 1 University.

WASTEGÅRD, S. & RASMUSSEN, T. L. 2001. New tephra horizons from Oxygen Isotope Stage 5 in the North Atlantic: correlation potential for terrestrial, marine and ice-core archives. *Quaternary Science Reviews*, **20**, 1587-1593.

WASTEGÅRD, S. & DAVIES, S. M. 2009. An overview of distal tephrochronology in northern Europe during the last 1000 years. *Journal of Quaternary Science*, **24**, 500-512.

WASTEGÅRD, S., BJÖRCK, S., GRAUERT, M. & HANNON, G. E. 2001. The Mjáuvøtn Tephra and other Holocene tephra horizons from the Faroe Islands: a link between the Icelandic source region, the Nordic Seas, and the European continent. *The Holocene*, **11**, 101-109.

WASTEGÅRD, S., RASMUSSEN, T. L., KUIJPERS, A., NIELSEN, T. & VAN WEERING, T. C. E. 2006. Composition and origin of ash zones from Marine Isotope Stages 3 and 2 in the North Atlantic. *Quaternary Science Reviews*, **25**, 2409-2419.

North Atlantic marine radiocarbon reservoir ages through Heinrich event H4: a new method for marine age model construction

JESPER OLSEN[1,4]*, TINE L. RASMUSSEN[2] & PAULA J. REIMER[3]

[1] *14CHRONO Centre for Climate, the Environment and Chronology, School of Geography, Archaeology and Palaeoecology, Queen's University Belfast, Belfast BT7 1NN, UK*

[2] *CAGE–Centre for Arctic Gas Hydrate, Environment and Climate, UiT, the Arctic University of Tromsø, N9037 Tromsø, Norway*

[3] *14CHRONO Centre for Climate, the Environment and Chronology, School of Geography, Archaeology and Palaeoecology, Queen's University Belfast, Belfast BT7 1NN, UK*

[4] *Present address: Department of Physics and Astronomy, Aarhus University, DK-8000 Aarhus, Denmark*

**Corresponding author (e-mail: jesper.olsen@phys.au.dk)*

Abstract: Cooling and sinking of dense saline water in the Norwegian–Greenland Sea is essential for the formation of North Atlantic Deep Water. The convection in the Norwegian–Greenland Sea allows for a northward flow of warm surface water and southward transport of cold saline water. This circulation system is highly sensitive to climate change and has been shown to operate in different modes. In ice cores the last glacial period is characterized by millennial-scale Dansgaard–Oeschger (D–O) events of warm interstadials and cold stadials. Similar millennial-scale variability (linked to D–O events) is evident from oceanic cores, suggesting a strong coupling of the atmospheric and oceanic circulations system. Particularly long-lasting cold stadials correlate with North Atlantic Heinrich events, where icebergs released from the continents caused a spread of meltwater over the northern North Atlantic and Nordic seas. The meltwater layer is believed to have caused a stop or near-stop in the deep convection, leading to cold climate. The spreading of meltwater and changes in oceanic circulation have a large influence on the carbon exchange between atmosphere and the deep ocean and lead to profound changes in the ^{14}C activity of the surface ocean. Here we demonstrate marine ^{14}C reservoir ages (R) of up to *c.* 2000 years for Heinrich event H4. Our R estimates are based on a new method for age model construction using identified tephra layers and tie-points based on abrupt interstadial warmings.

Over the last two decades numerous palaeoclimate reconstructions and numerical climate models have demonstrated close coupling between major atmospheric and oceanographic changes occurring over the course of Dansgaard–Oeschger (D–O) events in the North Atlantic during the last glacial period (e.g. Berger *et al.* 2010; Dansgaard *et al.* 1993; Waelbroeck *et al.* 2011). D–O events as seen in ice cores consist of a warm interstadial followed by a cold stadial (Dansgaard *et al.* 1993). Palaeoceanographic research has shown that the major reorganizations of the thermohaline circulation in the North Atlantic Ocean are related to large changes in sea ice coverage, sea surface temperature and the strength in the northward flow of warm Atlantic surface water the North Atlantic Drift (e.g. Bond *et al.* 1993; Rasmussen & Thomsen 2004; Rasmussen *et al.* 1996; Vautravers & Shackleton 2006). In particular, episodic ice rafting events known as North Atlantic Heinrich events correlate with particularly cold and long-lasting Greenland stadials (Bond *et al.* 1993). Six major events termed H6–H1 occurred during the mid–late Weichselian glacial 60 000–15 000 years BP. During Heinrich events icebergs released from the Laurentide and Fennoscandian Ice Sheets caused spreading of meltwater and sea ice over a large region of the North Atlantic Ocean (Bond *et al.* 1993; Broecker *et al.* 1992; Heinrich 1988). Associated with these events it is expected that the marine radiocarbon reservoir ages of the North Atlantic Ocean will show large deviations from the assumed 400 year reservoir age offset (e.g. Austin & Kroon 2001; Austin *et al.* 2011; Singarayer *et al.* 2008; Thornalley *et al.* 2011). A 400 year reservoir age correction between the 'global' ocean and contemporaneous atmospheric ^{14}C values is adopted, in want of a better estimate, by the international

From: Austin, W. E. N., Abbott, P. M., Davies, S. M., Pearce, N. J. G. & Wastegård, S. (eds) 2014. *Marine Tephrochronology*. Geological Society, London, Special Publications, **398**, 95–112.
First published online February 6, 2014, http://dx.doi.org/10.1144/SP398.2

radiocarbon calibration IntCal working group (Reimer *et al.* 2009).

To overcome the problems with variations in reservoir ages, methods other than radiocarbon are needed for independent age control of marine sediments. Tephrostratigraphy offers an important alternative and can provide direct links between marine records and the Greenland ice cores and other archives such as lake sediments (Davies *et al.* 2008, 2012; Eiríksson *et al.* 2011; Gronvold *et al.* 1995; Ruddiman & Glover 1972), which can be used to test and construct independent marine age models. However, even though tephrostratigraphy offers one of the only ways of independent age control of marine sediments, it cannot, at present, stand alone. Often only a few tephra layers are found in marine sediment cores and some tephra layers are still to be identified and dated in the Greenland ice cores, for example, only a few tephra horizons from the deglaciation period *c.* 15–28 cal kyr BP have been identified (e.g. Blockley *et al.* 2012; Davies *et al.* 2012 and references therein). Refining and extending the tephrostratigraphic framework is an ongoing task and, until an extended tephrochronological framework is available, dating of marine sediment cores must be aided by other methods (e.g. Austin *et al.* 2012).

Here we present the distribution pattern of the polar planktic foraminifera species *Neogloboquadrina pachyderma* sinistral (s), planktic oxygen isotopes measured on *N. pachyderma* s and magnetic susceptibility from the LINK15 core taken in the Faroe–Shetland Channel in the North Atlantic Ocean (Fig. 1). The magnetic susceptibility record has been published previously in Nielsen *et al.* (2007). We use the evidence from the climate proxies to construct a regional event stratigraphy, which is linked to the NGRIP ice core stratotype with the aid of three tephra layers. To construct age models we have developed a method that takes into account the age and depth errors involved as well as incorporating differential dates between Greenland interstadials (GI), which are known with high precision in the NGRIP ice core (typically around ±40 to ±90 years, Table 1). An age model for core ENAM93-21 using our new approach and based on tephra layers and tie-points identified by Rasmussen *et al.* (1996) are also presented.

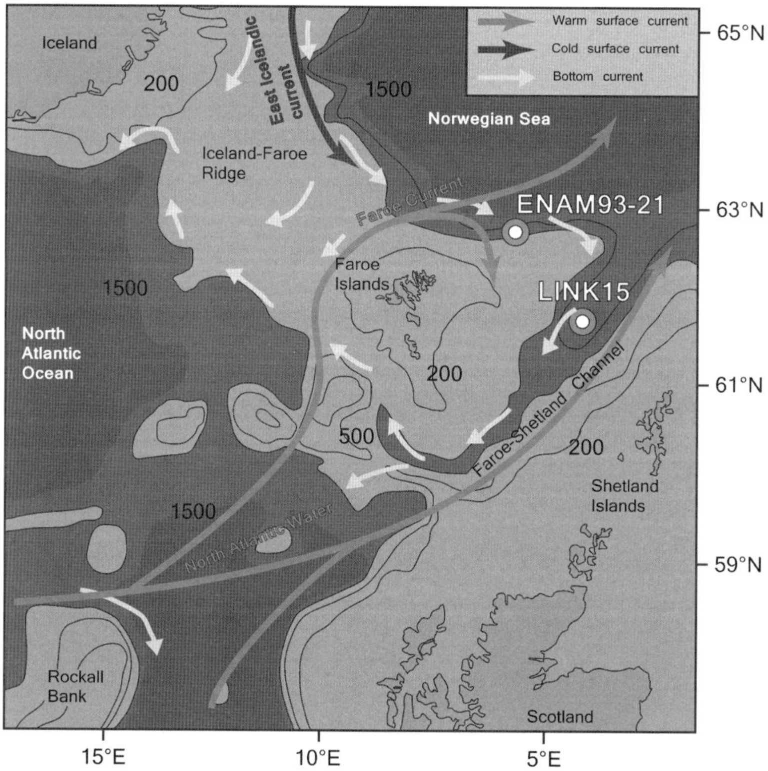

Fig. 1. Location map showing the ENAM93-21 and LINK15 sites with major surface and bottom currents indicated (for details see main text).

Table 1. *LINK15 and ENAM93-21 identified tephra and GI onset tie-points*

Event	NGRIP age (b2k)	Age difference (years)	ENAM93-21 depth (cm)	LINK15 depth (cm)
Saksunarvatn	10 347 ± 45		30.3 ± 0.9	
Vedde	12 171 ± 114			10.0 ± 1.7
GI1	14 680 ± 93		162.5 ± 1.3	21.8 ± 0.6
GI2	23 340 ± 298	8660 ± 205	277.5 ± 1.3	75.9 ± 2.5
Fugloyarbanki	26 740 ± 390		313.8 ± 0.4	94.0 ± 1.3
GI3	27 780 ± 416	4440 ± 118	332.5 ± 1.3	106.4 ± 0.6
GI4	28 900 ± 449	1120 ± 33	352.5 ± 1.3	126.0 ± 1.0
GI5	32 500 ± 566	3600 ± 117	422.5 ± 1.3	160.6 ± 0.7
GI6	33 740 ± 606	1240 ± 40	455.0 ± 2.5	177.8 ± 0.6
GI7	35 480 ± 661	1740 ± 55	522.5 ± 1.3	200.5 ± 0.3
FMAZ III	38 122 ± 723		627.5 ± 2.5	231.0 ± 0.7
GI8	38 220 ± 724	2740 ± 63	632.5 ± 4.5	232.3 ± 1.2
GI9	40 160 ± 790	1940 ± 66	676.5 ± 2.1	260.5 ± 0.8
GI10	41 460 ± 817	1300 ± 27	730.0 ± 1.3	281.0 ± 0.3
GI11	43 340 ± 868	1880 ± 51	782.7 ± 1.4	310.5 ± 0.3
GI12	46 860 ± 956	3520 ± 88	903.5 ± 1.3	
GI13	49 280 ± 1015	2420 ± 59	958.0 ± 1.3	
NAAZ II	55 380 ± 1184		1060.0 ± 5.0	
GI14	54 220 ± 1150	4940 ± 135	1018.5 ± 1.3	
GI15	55 800 ± 1196	1580 ± 46	1088.0 ± 1.3	

Absolute ages are from Andersen *et al.* (2006), Rasmussen *et al.* (2006) and Svensson *et al.* (2008).

Further, we evaluate the marine radiocarbon reservoir age *R* using different calendar age estimates around the transition from Heinrich event H4 to the following interstadial GI8. This transition is dated to 38220 ± 724 b2k in the Greenland NGRIP ice core by layer counting (the GICC05 timescale Rasmussen *et al.* 2006; Svensson *et al.* 2008).

Location

The LINK15 (61°45′ N, 2°40′ W) core was retrieved from the Faroe–Shetland Channel at a water depth of 1602 m (Fig. 1). The northward flowing Atlantic Water gives rise to the Faroe Current, that flows west and north of the Faroe Islands. It is classified as Modified North Atlantic Water. North Atlantic Water flows into the Nordic seas in the eastern part of the Faroe–Shetland Channel. Colder and less saline Arctic Intermediate Water and Modified East Icelandic Water circulate below the Modified North Atlantic Water at depths of 400–600 m. Polar surface water of the East Icelandic Current flows towards the Faroe Islands from the NW (Fig. 1). It is separated from the warmer water of the Faroe Current by the Iceland–Faroe Front. The area has been central for our understanding of the coupling between oceanography and climate during the last glacial period (Moros *et al.* 2004; Rasmussen *et al.* 2002, 1996; Rasmussen & Thomsen 2004; Rasmussen & Thomsen 2008).

Methods

Core LINK15 was sampled in 0.7 cm-thick slices each 4 cm except for the interval 180–300 cm, which was sampled every 2 cm. The samples were dried, weighed and wet-sieved over mesh-sizes of 63 and 100 μm. The dry residues were weighed. Approximately 300 specimens of planktic foraminifera were picked from the >100 μm size fraction, counted and identified, and the relative abundance of species calculated. Rhyolitic and basaltic tephra grains were counted separately in the >100 μm fraction. The number of tephra grains per gram dry weight sediment was calculated. Stable isotope measurements of the planktic foraminifera species *N. pachyderma* s were performed at Woods Hole Oceanographic Institution using standard procedures (see Ostermann & Curry 2000). The samples were accelerator mass spectrometry (AMS) ^{14}C dated using monospecific samples of *N. pachyderma* s. Around 1000 generally well-preserved specimens of similar size were picked. The samples were analysed at the ^{14}CHRONO Centre facility at Queen's University Belfast. Pre-treatment of the foraminifera specimens followed the approach by Nadeau *et al.* (2001). About 12 mg of sample was transferred into a septum-sealed vial. Then 200 μl of a 15% hydrogen peroxide (H_2O_2) solution was added and the tube placed in an ultrasonic bath for *c.* 3 min to remove organic surface contaminants and to provide a mild leach of the foraminifera material. Subsequently, hydrogen peroxide was decanted using a fine syringe and rinsed repeatedly in Milli-Q water. While still damp the samples were evacuated and 1–2 ml of 85% phosphoric acid was added to hydrolyse the samples to CO_2 at 80 °C until complete dissolution. Samples were then converted into graphite for AMS analysis on a Fe catalyst using the hydrogen reduction method (Vogel *et al.* 1987). ^{14}C background corrections were based on core-specific *N. pachyderma* s samples (>50 kcal yr BP) measured together with ordinary LINK15 samples with a small sample background correction for samples <0.5 mg C.

Tie-points to NGRIP

Sea surface temperature proxies of numerous marine sediment cores exhibit a pattern typical for the D–O events upon which a regional event stratigraphy can be constructed. Using tephrochronology this regional event stratigraphy can then be compared with the North Atlantic stratotype, the NGRIP ice cores (e.g. Lowe *et al.* 2008). Importantly the regional event stratigraphy can be linked to the NGRIP stratotype, providing age estimates for age model construction. However, linking the regional event stratigraphy to the NGRIP stratotype comes at a price, namely assuming synchronicity between the atmospheric variations observed by the Greenland ice cores and major oceanographic reorganizations as observed by sea surface temperature proxies. Hence using the tie-point method leads or lags cannot be quantified.

The transitions from stadials (Greenland stadials, GS) into interstadials (GI) in the Greenland ice cores are abrupt and well defined (Andersen *et al.* 2006; Svensson *et al.* 2008) and thus useful as tie-points. These abrupt transitions into warmer periods are chosen as the best markers to correlate LINK15 to NGRIP. The $\delta^{18}O_{NPS}$ values along with the relative abundance of the polar planktic foraminifera *N. pachyderma* s are interpreted as a reflection of the sea surface temperature and therefore offer the most appropriate tie-points for the GS to GI transitions in LINK15 (Fig. 2). However, owing to the northern location of LINK15 and proximity to the Iceland–Faroe Front, the climate conditions were generally cold during the late Pleistocene. The relative abundance of *N. pachyderma* s (range between 78 and 99%) does not show large fluctuations with D–O cycles as observed further south (Fig. 2, Rasmussen & Thomsen 2008). The magnetic susceptibility (MS) signal is interpreted to represent a signal of bottom current strength and is linked to the strength of the North Atlantic Current (Kissel *et al.* 1999; Moros *et al.* 2004; Rasmussen *et al.* 1996, 1998). Therefore the MS record offers a good proxy for tie-points to NGRIP. The $\delta^{18}O_{NPS}$ and %NPS tie-points are therefore complemented by the use of MS (Fig. 2). Because the sample resolution in LINK15 (and marine cores in general) does not match the sample resolution in NGRIP, each tie-point is associated with a non-negligible error in the depth domain. Errors in depth are assumed to be constrained by neighbouring samples points with 2σ and are incorporated into the age model construction (see the Appendix; Fig. 2).

Tephrostratigraphy plays a critical role in validating correlations between marine cores and the Greenland ice cores and serves as a sound basis for testing and building chronological models (Austin *et al.* 2012; Davies *et al.* 2012). Three identified tephra layers in LINK15 and other cores from the Faroe Islands margin (Nielsen *et al.* 2007; Rasmussen *et al.* 2003) provide additional tie-points as well as an independent check on the LINK15 to NGRIP correlation (Fig. 2). In particular, it is reassuring that FMAZ-III position at the GIS8 transitions is in accordance with the FMAZ-III stratigraphical level in LINK15 (Fig. 2). Tephra layers and tie-points for ENAM93-21 follow Rasmussen *et al.* (1996) and Davies *et al.* (2010). Identified tephras and tie-points are provided in Table 1.

Chronology and ^{14}C reservoir ages

The chronology of the LINK15 and ENAM93-21 record is constructed using three approaches. The first approach is based on identified tephra layers to anchor to the NGRIP ice core (referred to as tephra model), whereas the second approach is based on using abrupt warming events as tie-points to NGRIP (referred to as the tie-point model). The third model applies the identified tephra layers to anchor to NGRIP and then further uses the relative age difference between GI events to match the identified tie-points into the model. No assumptions on the absolute age of each tie-point are made in the model (referred to as the combined model). The differential dating method makes use of the precise age difference between onsets of GIs (see the Appendix; Table 1). All age model construction methods takes into account the depth d with an uncertainty σ_d and associated with an age a with error σ_a (Table 1). To construct a robust chronology, errors are included in the age model construction using a statistical approach, where errors are given as probability distributions (see the Appendix). The resulting age models are shown in Figures 3–5 and Table 2. Around the H4 event/GI8 interstadial the accumulation rate is *c.* 0.13 mm/yr, which with a sample resolution of 7 mm corresponds to an age resolution of *c.* 50 years per sample.

Using five ^{14}C dates on planktic foraminifera, changes in the ^{14}C reservoir age are inferred

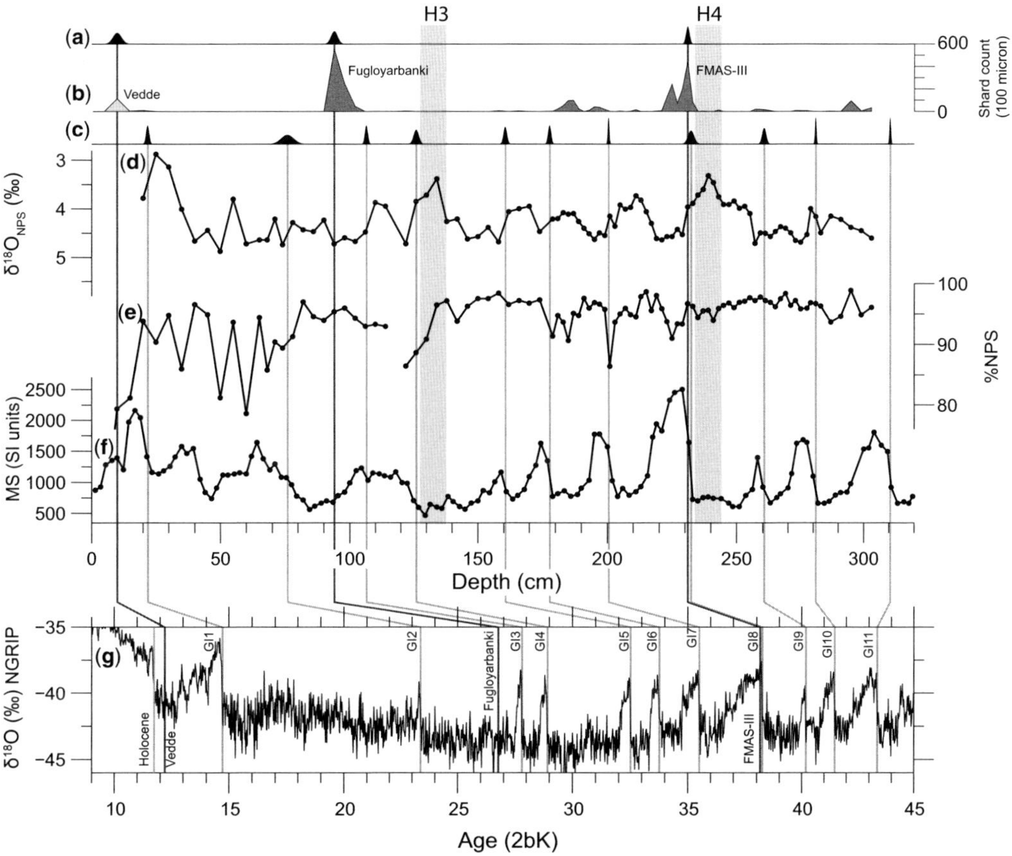

Fig. 2. Tie-points for LINK15 to the NGRIP ice core are shown together with identified tephra layers. Abrupt warming defining GI1–11 in NGRIP is used to link LINK15 surface temperature proxies percentage *N. pachyderma* s (abbreviated NPS) and $\delta^{18}O_{NPS}$ values with the magnetic susceptibility (MS) record. (**a**) Depth probability distributions for identified tephra layers. (**b**) Tephra shard counts in number/gram dry weight sediment. (**c**) Depth probability distributions for tie-points. (**d**) $\delta^{18}O_{NPS}$ (‰ v. PDB). (**f**) Relative abundance of *N. pachyderma* s. (**e**) MS. (**g**) NGRIP $\delta^{18}O$ values (‰ v. SMOW).

using the combined age model (Table 3). The contemporaneous atmospheric ^{14}C content is estimated using the radiocarbon calibration curve (IntCal09) and the Bahamas speleothem ^{14}C record (Figs 6 & 7; Hoffmann *et al.* 2010; Reimer *et al.* 2009; see the Appendix for details on reservoir age calculations).

Discussion

A robust chronological framework is essential for reconstructing long-term climate variability. Therefore it is of paramount importance to adopt sound methods for constructing age models. For the Holocene one of the most frequently used methods for obtaining absolute dates is radiocarbon dating and several methods exist for constructing age models (e.g. Blaauw & Christen 2005; Ramsey 2008). Moreover the INTIMATE community has developed protocols on the adoption of regional event stratigraphies, which can be linked to other archives with the purpose of investigating leads and lags within the climate system (e.g. Austin *et al.* 2012; Lowe *et al.* 2008). The INTIMATE protocol stresses the need to use independent age information such as tephrochronology, optical stimulated luminesence (OSL) or radiocarbon dates to secure an independent chronological framework for each site. Furthermore, the INTIMATE protocol makes the point that correlation using tie-points between local (core specific) event stratigraphies and the Greenland stratotype should be avoided (Lowe *et al.* 2008). However, the future objective of INTIMATE is to extend correlations of different archives further

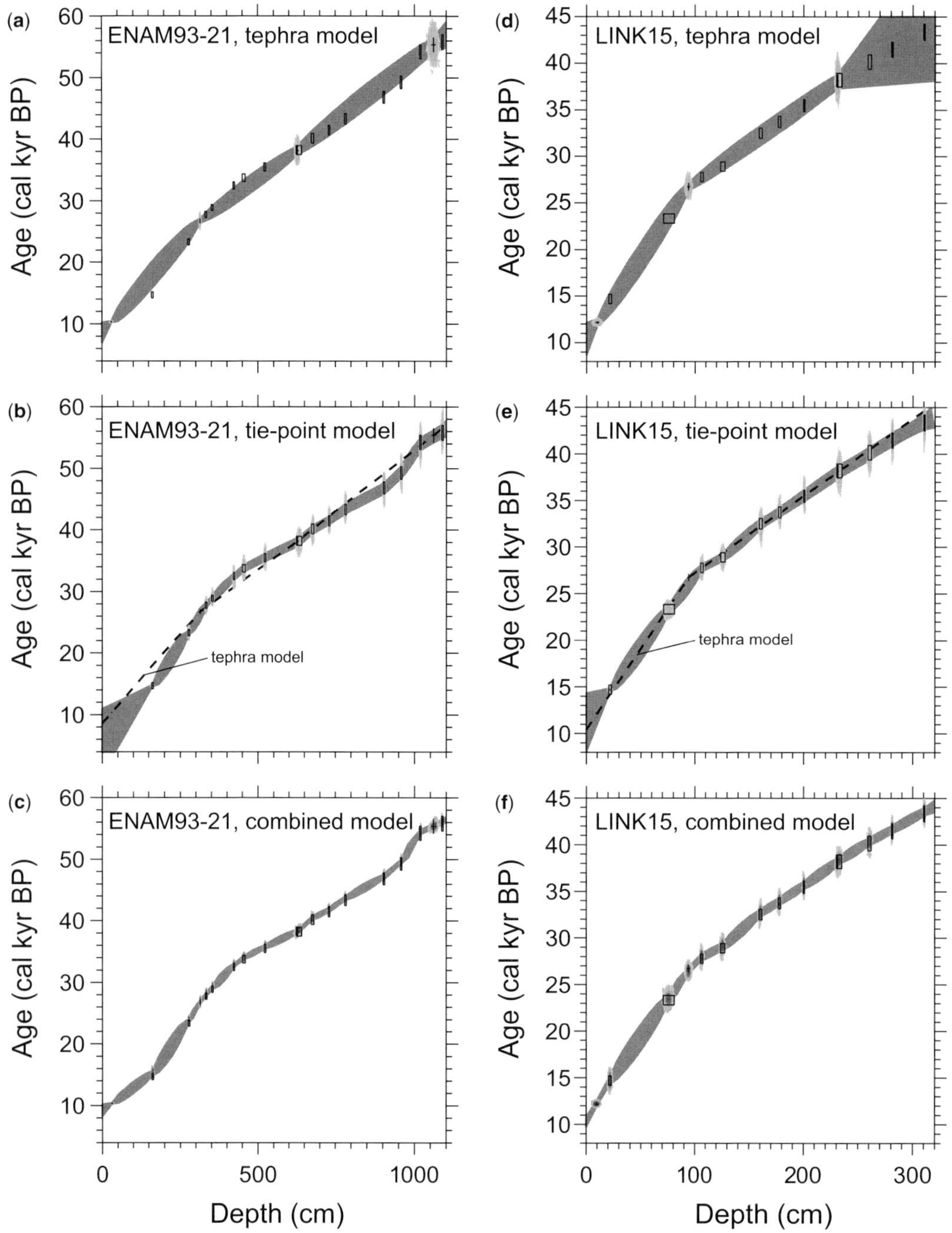

Fig. 3. Age model construction based on either tephra or tie-points, or tephra and tie-points. (**a**) ENAM93-21 tephra-based age model. (**b**) ENAM93-21 tie-point-based age model. The tephra-based age model is shown by a stippled line. (**c**) ENAM93-21 combined age model. (**d**) LINK15 tephra-based model (**e**) LINK15 tie-point-based age model. The tephra-based age model is shown by a stippled line. (**f**) LINK15 combined age model.

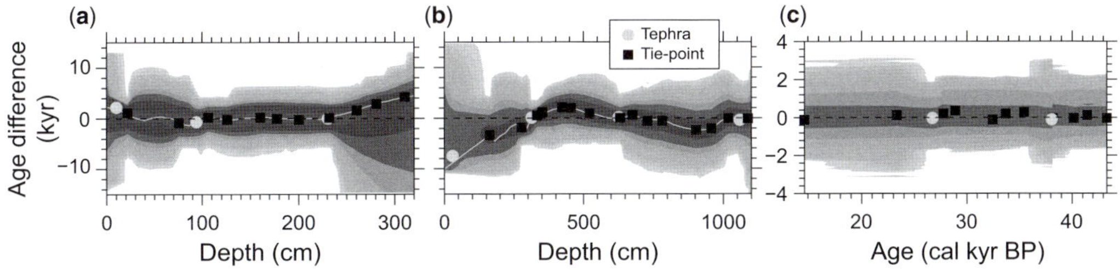

Fig. 4. Age deviations between different age model approaches. (**a**) LINK15 age difference between tephra and tie-point age models. (**b**) ENAM93-21 age difference between tephra and tie-point age models. (**c**) Age difference between LINK15 and ENAM93-21 combined age models.

back in time (to 60 cal kyr BP), which imposes challenges to the INTIMATE protocol in general and in particular to the marine chronologies.

Age models: leads or lags?

Identification of leads or lags is challenging in the marine realm because of the generally unknown marine ^{14}C reservoir effect. One of the best possibilities of testing leads or lags between North Atlantic marine records and Greenland ice cores around the H4–GI8 transition is the FMAZ-III tephra horizon, because this tephra layer is positioned some 100 years after the onset of GI8 (Rasmussen *et al.* 2003; Andersen *et al.* 2006; Svensson *et al.* 2008; Davies *et al.* 2010). In both ENAM93-21 and LINK15, H4 is found below the FMAZ-III tephra and hence confirms the NGRIP stratigraphy. However the interpretations of both the GI8 onset as well as the position of the FMAZ-III tephra layer are not unproblematic, as can be seen by the large and overlapping depth uncertainties (Fig. 2, Table 1). Hence, considering the general age and stratigraphic errors involved in constructing age models within the marine realm, leads or lags in the order of hundreds of years are difficult to disentangle. Additionally, marine tephra deposits may be dislocated in time owing to the different transport mechanisms involved, that is, ice transported tephra or direct atmospheric fall-out. In addition, tephra can be affected by bioturbation and bottom current activities during and after deposition. Whereas direct atmospheric fall-out of tephra provides a secure chronological event, a significant time lag may be introduced of up to 400 years or more in marine deposits (Brendryen *et al.* 2011).

In an attempt to investigate possible leads or lags an age model based only on tephra horizon anchors to NGRIP has been constructed and is compared with an age model constructed using

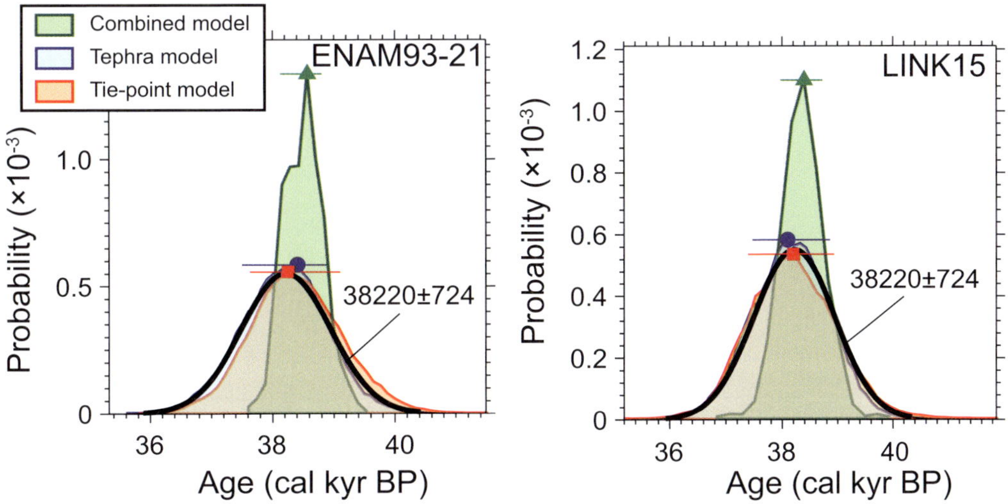

Fig. 5. Modelled GI8 age probability distributions of ENAM93-21 and LINK15 compared with NGRIP age information.

Table 2. *GI onset age model output*

Event	M	NGRIP deviation	WM	NGRIP deviation	68.2% Confidence interval	65.4% Confidence interval	O
LINK15, tie model							
GI1	14684	4 (0.0)	14693	13 (0.1)	14 555–14 815 68.9%	14 364–14 998 95.4%	91.0%
GI2	23320	−20 (−0.1)	23388	48 (0.2)	22 933–23 719 68.6%	22 459–24 167 95.5%	95.0%
GI3	27749	−31 (−0.1)	27841	61 (0.1)	27 350–28 173 68.4%	26 917–28 546 95.4%	100.0%
GI4	28913	13 (0.0)	29057	157 (0.3)	28 495–29 360 68.4%	28 059–29 756 95.5%	100.1%
GI5	32470	−30 (−0.1)	32449	−51 (−0.1)	31 960–33 039 68.5%	31 412–33 511 95.5%	99.9%
GI6	33779	39 (0.1)	33735	−5 (0.0)	33 226–34 330 68.5%	32 715–34 833 95.4%	100.2%
GI7	35544	64 (0.1)	35524	44 (0.1)	34 892–36 123 68.4%	34 328–36 837 95.4%	100.1%
GI8	38140	−80 (−0.1)	38092	−128 (−0.2)	37 454–38 845 68.4%	36 768–39 452 95.4%	100.1%
GI9	40045	−115 (−0.1)	40150	−10 (0.0)	39 346–40 723 68.3%	38 637–41 421 95.5%	100.1%
GI10	41571	111 (0.1)	41529	69 (0.1)	40 890–42 323 68.2%	40 152–43 009 95.4%	100.2%
GI11	43466	126 (0.1)	43462	122 (0.1)	42 539–44 233 68.6%	41 907–45 117 95.5%	100.0%
		7 ± 71 (χ^2: 0.1 ≤ 18.3)		**29 ± 75** (χ^2: 0.3 ≤ 18.3)			
ENAM93-21, tie model							
GI1	14686	6 (0.1)	14685	5 (0.1)	14 549–14 798 68.5%	14 415–14 967 95.4%	94.6%
GI2	23332	−8 (0.0)	23395	55 (0.2)	23 027–23 648 68.4%	22 708–23 961 95.5%	99.6%
GI3	27790	10 (0.0)	27844	64 (0.2)	27 388–28 206 68.4%	27 001–28 581 95.5%	100.1%
GI4	28950	50 (0.1)	28847	−53 (−0.1)	28 556–29 323 68.5%	28 183–29 755 95.4%	100.1%
GI5	32377	−123 (−0.2)	32284	−216 (−0.4)	31 871–32 938 68.3%	31 293–33 452 95.4%	99.8%
GI6	33745	5 (0.0)	33742	2 (0.0)	33 176–34 320 68.4%	32 669–34 853 95.5%	100.2%
GI7	35530	50 (0.1)	35457	−23 (0.0)	34 883–36 161 68.4%	34 331–36 801 95.5%	100.2%
GI8	38161	−59 (−0.1)	38391	171 (0.2)	37 495–38 868 68.3%	36 709–39 484 95.4%	100.1%
GI9	40027	−133 (−0.2)	40116	−44 (−0.1)	39 379–40 739 68.3%	38 621–41 332 95.4%	100.2%
GI10	41534	74 (0.1)	41483	23 (0.0)	40 823–42 178 68.4%	40 217–42 871 95.4%	100.2%
GI11	43461	121 (0.1)	43443	103 (0.1)	42 543–44 240 68.3%	41 889–45 134 95.5%	100.1%
GI12	46768	−92 (−0.1)	46941	81 (0.1)	45 901–47 674 68.3%	45 043–48 496 95.4%	100.2%
GI13	49404	124 (0.1)	49311	31 (0.0)	48 437–50 337 68.4%	47 566–51 350 95.4%	100.2%
GI14	53941	−279 −0.2)	54185	−35 (0.0)	52 920–54 999 68.5%	51 932–55 972 95.4%	100.1%
GI15	56099	299 (0.3)	55663	−137 (−0.1)	54 939–57 097 68.3%	54 056–58 311 95.4%	100.2%
		3 ± 130 (χ^2: 0.3 ≤ 23.7)		**2 ± 92** (χ^2: 0.2 ≤ 23.7)			

LINK15, tephra model							
GI1	14213	−467 (−5.0)	13693	−987 (−10.6)	12 780–14 879 68.5%	12 340–16 529 95.4%	14.1%
GI2	23611	271 (0.9)	24111	771 (2.6)	22 511–25 251 68.4%	20 751–26 139 95.5%	38.2%
GI3	27780	0 (0.0)	27652	−128 (−0.3)	27 021–28 322 68.4%	26 514–29 217 95.5%	94.0%
GI4	29402	502 (1.1)	29190	290 (0.6)	28 404–30 219 68.4%	27 648–31 266 95.5%	81.2%
GI5	32258	−242 (−0.4)	32271	−229 (−0.4)	31 129–33 291 68.4%	30 125–34 423 95.4%	85.6%
GI6	33680	−60 (−0.1)	33688	−52 (−0.1)	32 580–34 773 68.4%	31 534–35 844 95.5%	89.4%
GI7	35556	76 (0.1)	35532	52 (0.1)	34 567–36 590 68.5%	33 517–37 584 95.4%	94.5%
GI8	38183	−37 (−0.1)	38194	−26 (0.0)	37 380–38 923 68.5%	36 686–39 732 95.5%	99.4%
GI9	40514	354 (0.4)	38959	−1201 (−1.5)	37 638–41 466 68.5%	36 791–46 052 95.4%	73.7%
GI10	42209	749 (0.9)	39344	−2116 (−2.6)	37 655–43 436 68.3%	36 760–51 229 95.4%	48.4%
GI11	44649	1309 (1.5)	40164	−3176 (−3.7)	37 630–46 404 68.6%	36 529–58 680 95.5%	27.8%
		223 ± 473 (χ^2: 30.8 ≤ 18.3)		**− 618 ± 1019** (χ^2: 142.4 ≤ 18.3)			
ENAM93-21, tephra model							
GI1	17993	3313 (35.6)	17775	3095 (33.3)	15 581–20 101 68.4%	13 795–22 298 95.4%	2.0%
GI2	24660	1320 (4.4)	25559	2219 (7.4)	23 850–26 375 68.6%	21 720–26 971 95.4%	20.4%
GI3	27446	−334 (−0.8)	27151	−629 (−1.5)	26 620–27 915 68.6%	26 147–29 028 95.5%	87.6%
GI4	28172	−728 (−1.6)	27781	−1119 (−2.5)	27 026–28 809 68.3%	26 481–30 238 95.5%	65.8%
GI5	30729	−1771 (−3.1)	30532	−1968 (−3.5)	29 174–31 963 68.4%	28 110–33 486 95.5%	40.5%
GI6	31910	−1830 (−3.0)	31792	−1948 (−3.2)	30 346–33 290 68.4%	29 108–34 750 95.4%	45.9%
GI7	34365	−1115 (−1.7)	34468	−1012 (−1.5)	33 019–35 890 68.3%	31 492–37 084 95.4%	69.3%
GI8	38374	154 (0.2)	38245	25 (0.0)	37 620–39 090 68.5%	36 923–39 902 95.4%	99.6%
GI9	40126	−34 (0.0)	39630	−530 (−0.7)	38 648–41 104 68.4%	37 748–42 843 95.5%	92.8%
GI10	42259	799 (1.0)	42089	629 (0.8)	40 343–43 656 68.4%	39 104–45 678 95.5%	81.0%
GI11	44348	1008 (1.2)	43945	605 (0.7)	42 338–46 054 68.5%	40 794–48 057 95.4%	78.7%
GI12	49159	2299 (2.4)	49143	2283 (2.4)	47 259–51 126 68.4%	45 350–52 936 95.5%	61.3%
GI13	51332	2052 (2.0)	51493	2213 (2.2)	49 640–53 281 68.5%	47 632–54 836 95.5%	64.6%
GI14	53742	−478 (−0.4)	54105	−115 (−0.1)	52 373–55 372 68.3%	50 602–56 720 95.5%	95.7%
GI15	56519	719 (0.6)	56123	323 (0.3)	54 693–57 837 68.6%	53 238–60 125 95.4%	93.9%
		358 ± 1440 (χ^2: 34.5 ≤ 23.7)		**271 ± 1528** (χ^2: 36.8 ≤ 23.7)			

(*Continued*)

Table 2. *GI onset age model output* (*Continued*)

Event	M	NGRIP deviation	WM	NGRIP deviation	68.2% Confidence interval	65.4% Confidence interval	O
LINK15, combined model							
GI1	14743	63 (0.7)	14672	−8 (−0.1)	14 312–15 169 68.4%	13 868–15 610 95.4%	46.8%
GI2	23519	179 (0.6)	23575	235 (0.8)	23 057–23 979 68.9%	22 552–24 484 95.5%	91.3%
GI3	27930	150 (0.4)	27945	165 (0.4)	27 549–28 278 68.6%	27 204–28 671 95.5%	99.8%
GI4	29085	185 (0.4)	29110	210 (0.5)	28 702–29 439 68.6%	28 338–29 849 95.4%	99.8%
GI5	32666	166 (0.3)	32656	156 (0.3)	32 266–33 024 68.5%	31 911–33 411 95.5%	100.2%
GI6	33917	177 (0.3)	33995	255 (0.4)	33 509–34 257 68.5%	33 197–34 661 95.5%	100.2%
GI7	35664	184 (0.3)	35717	237 (0.4)	35 259–36 008 68.6%	34 946–36 423 95.5%	100.2%
GI8	38351	131 (0.2)	38380	160 (0.2)	37 952–38 689 68.5%	37 637–39 127 95.5%	100.2%
GI9	40351	191 (0.2)	40362	202 (0.3)	39 959–40 716 68.6%	39 616–41 126 95.5%	100.2%
GI10	41658	198 (0.2)	41693	233 (0.3)	41 268–42 024 68.5%	40 916–42 435 95.5%	100.2%
GI11	43545	205 (0.2)	43460	120 (0.1)	43 149–43 917 68.6%	42 805–44 335 95.5%	100.2%
		166 ± 38 (χ^2: 1.6 ≤ 18.3)		**179 ± 71** (χ^2: 1.6 ≤ 18.3)			
ENAM93-21, combined model							
GI1	15251	571 (6.1)	15112	432 (4.6)	14 833–15 742 68.4%	14 536–15 945 95.6%	18.50%
GI2	23799	459 (1.5)	23891	551 (1.8)	23 421–24 100 68.6%	23 203–24 444 95.5%	88.1%
GI3	28043	263 (0.6)	27891	111 (0.3)	27 712–28 361 68.8%	27 480–28 641 95.5%	100.2%
GI4	29166	266 (0.6)	28862	−38 (−0.1)	28 814–29 476 68.8%	28 616–29 785 95.4%	100.2%
GI5	32745	245 (0.4)	32831	331 (0.6)	32 408–33 091 69.0%	32 180–33 364 95.4%	100.2%
GI6	33990	250 (0.4)	33892	152 (0.3)	33 624–34 313 68.6%	33 447–34 748 95.5%	100.2%
GI7	35755	275 (0.4)	35516	36 (0.1)	35 368–36 005 68.3%	35 174–36 347 95.5%	100.2%
GI8	38492	272 (0.4)	38556	336 (0.5)	38 156–38 817 68.7%	37 914–39 170 95.5%	100.2%
GI9	40431	271 (0.3)	40641	481 (0.6)	40 051–40 748 68.5%	39 835–41 052 95.5%	100.2%
GI10	41774	314 (0.4)	41765	305 (0.4)	41 399–42 074 68.6%	41 174–42 367 95.4%	100.2%
GI11	43658	318 (0.4)	43675	335 (0.4)	43 304–43 956 68.7%	43 043–44 276 95.5%	100.2%
GI12	47211	351 (0.4)	47381	521 (0.5)	46 842–47 479 68.6%	46 593–47 841 95.5%	100.2%
GI13	49651	371 (0.4)	49639	359 (0.4)	49 309–49 965 68.6%	49 051–50 291 95.4%	100.2%
GI14	54422	202 (0.2)	54254	34 (0.0)	54 049–54 776 68.5%	53 699–55 073 95.5%	100.2%
GI15	56111	311 (0.3)	56058	258 (0.2)	55 691–56 409 68.2%	55 400–56 775 95.4%	100.2%
		316 ± 91 (χ^2: 1.4 ≤ 23.7)		**280 ± 178** (χ^2: 1.8 ≤ 23.7)			

Deviation column provides the age difference in years and in parentheses the deviation in terms of the standard deviation. O provides the probability distribution overlap between the NGRIP age and the modelled age. All ages in BP (before present where present is 1950 AD). M, Probability distribution mode; WM, weighted mean of probability distribution.

Table 3. *^{14}C reservoir estimation*

Laboratory code	Depth (cm)	^{14}C age (BP)	Age, M (cal yr BP)	Age, WM (cal yr BP)	68.2% Confidence interval	95.4.2% Confidence interval	*R*, M (^{14}C years)	*R*, WM (^{14}C years)	68.2% Confidence interval	95.4.2% Confidence interval
Bahamas ^{14}C calibration dataset										
UB-16323	229	34 524 ± 350	37 993	37 986	37 609–38 356 (68.5%)	37 255–38 783 (95.5%)	866	1047	257–1464 (68.2%)	7–2218 (95.4%)
UB-17235	231	34 700 ± 341	38 235	38 160	37 776–38 493 (68.5%)	37 490–38 921 (95.4%)	926	1018	351–1460 (68.2%)	8–2058 (95.4%)
UB-17236	233	35631 ± 379	38 315	38 394	38 017–38 742 (68.3%)	37 704–39 154 (95.4%)	1651	1653	1093–2278 (68.2%)	513–2801 (95.4%)
UB-16315	235	35512 ± 398	38 477	38 530	38 162–38 910 (68.6%)	37 800–39 304 (95.5%)	1490	1448	842–2081 (68.3%)	237–2598 (95.4%)
UB-17237	237	34403 ± 332	38 718	38 549	38 259–38 956 (68.3%)	37 832–39 199 (95.5%)	302	589	13–746 (68.3%)	3–1361 (95.4%)
INTCAL09 ^{14}C calibration dataset										
UB-16323	229	34524 ± 350	37 993	38 010	37 611–38 387 (68.5%)	37 262–38 808 (95.5%)	1309	1291	874–1742 (68.2%)	380–2166 (95.4%)
UB-17235	231	34700 ± 341	38 235	38 182	37 790–38 524 (68.5%)	37 488–38 953 (95.5%)	1409	1374	975–1859 (68.3%)	401–2263 (95.4%)
UB-17236	233	35631 ± 379	38 315	38 402	38 015–38 748 (68.4%)	37 696–39 171 (95.4%)	2295	2092	1737–2738 (68.3%)	1092–2991 (95.4%)
UB-16315	235	35512 ± 398	38 477	38 546	38 156–38 920 (68.5%)	37 797–39 342 (95.4%)	2074	1854	1379–2567 (68.2%)	734–2932 (95.4%)
UB-17237	237	34403 ± 332	38 557	38 593	38 257–38 967 (68.4%)	37 858–39 349 (95.5%)	846	810	199–1220 (68.2%)	6–1614 (95.4%)

M, Probability distribution mode; WM, weighted mean of probability distribution.

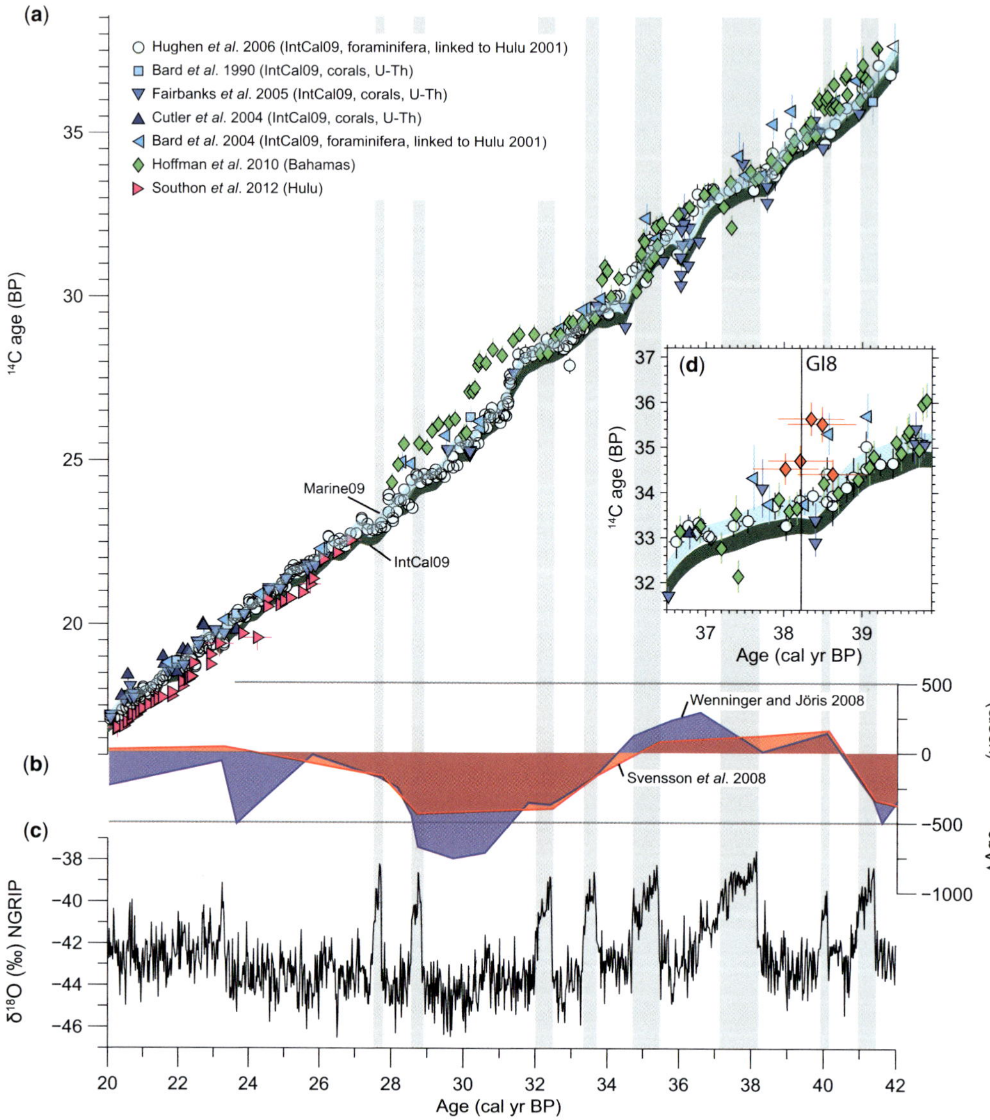

Fig. 6. Radiocarbon calibration and age scale (Hulu, GICC05) differences. (**a**) The IntCal09 and Marine09 radiocarbon calibration curves (Reimer *et al.* 2009) together with the accompanying datasets (Bard *et al.* 1990, 2004; Cutler *et al.* 2004; Fairbanks *et al.* 2005; Hughen *et al.* 2006). Radiocarbon from the Bahamas speleothem and Hulu cave stalagmite are also shown (Hoffmann *et al.* 2010; Southon *et al.* 2012). (**b**) Age differences between the Hulu and GICC05 timescales (Svensson *et al.* 2008; Weninger & Jöris 2008). (**c**) NGRIP $\delta^{18}O$ values. (**d**) Insert shows a zoom on the GI8 onset. Also plotted are the LINK15 ^{14}C data. Grey-shaded areas indicate interstadials.

identified LINK15 abrupt warming correlated to onsets of interstadials (Fig. 3, Table 2). The average difference between the tephra and tie-point models amounts to −577 and 627 years for the ENAM93-21 and LINK15 cores, respectively (Fig. 4). The tephra age models of both are, however, without any detailed structure owing to the small number of identified tephra layers explaining the large discrepancy between the tephra and tie-point age models. Nonetheless, the FMAZ-III tephra layer and the GI8 onset provide a secure comparison ranging between −37 and 154 years for tie-point and tephra models of both ENAM93-21 and LINK15 (Table 2), suggesting an insignificant lead or lag between the North Atlantic Ocean and the NGRIP ice core. Likewise, the combined age

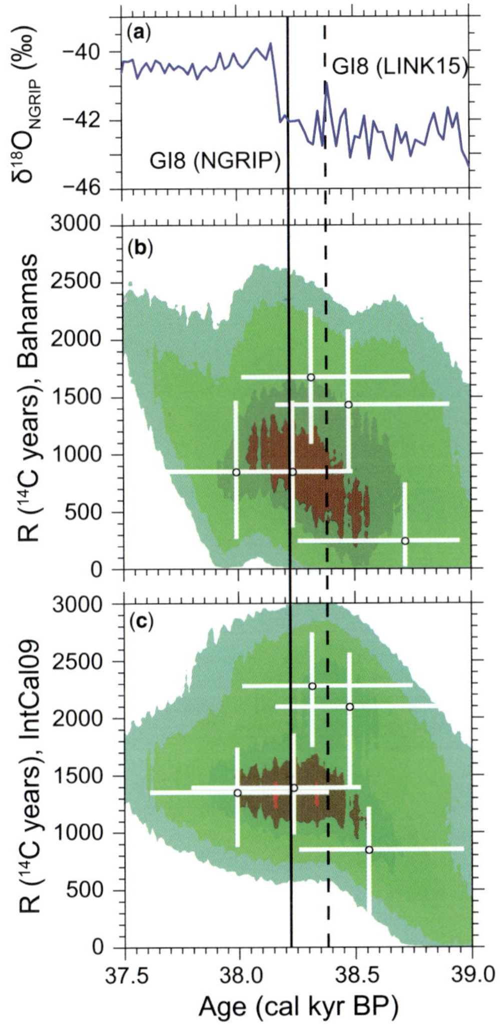

Fig. 7. ^{14}C reservoir ages (R) measured using *N. pachyderma*. The start of interstadial GI8 is shown for NGRIP ice core (black) and LINK15 (stippled black). (**a**) δ^{18}O from the NGRIP ice core. (**b**) Reservoir ages R calculated using the combined age model and the Bahamas ^{14}C calibration dataset. (**c**) Reservoir ages R calculated using the combined age model and the IntCal09 ^{14}C calibration dataset.

models estimate the difference between the onsets of interstadials in the North Atlantic Ocean and in NGRIP to an average of 166 ± 38 and 316 ± 91 years. Hence the age differences in either age model relative to the GICC05 age scale are well within age uncertainties. The overlap between age model probability distributions and the GI onsets is overall about 100%, that is, the complete model age probability distribution is within the GICC05 error boundaries (Fig. 5, Table 2). Thus within the present age uncertainties it is difficult to quantify leads or lags between North Atlantic Ocean cores and the NGRIP stratigraphy.

^{14}C Reservoir variability

Applying radiocarbon in the marine environment beyond *c.* 30 kcal yr BP requires careful considerations for several reasons. Numerous North Atlantic records document that major oceanographic circulation changes accompany the D–O transition between stadials and interstadials (e.g. Bond *et al.* 1993; Rasmussen *et al.* 1996; Rasmussen & Thomsen 2004; Waelbroeck *et al.* 2011). These changes are associated with weakening or strengthening of the North Atlantic Drift as well as large fluctuations in the sea ice-covered areas of the Nordic Seas and North Atlantic Ocean. Such changes are potentially accompanied by large variability of the marine reservoir age and hence may obscure radiocarbon calibration.

The ^{14}C radiocarbon calibration curve (Marine09; Reimer *et al.* 2009) is constructed using numerous coral records and ocean cores (Cariaco basin and Iberian margin) (Fig. 6). The coral records are independently dated using U–Th, while the non-varved foraminifera records are dated using tie-points to the Hulu cave speleothems (Wang *et al.* 2001) by assuming synchroneity of the D–O events observed in the marine environment and Hulu cave (Bard *et al.* 2004; Hughen *et al.* 2006). This may be problematic because the Hulu cave timescale deviates from the GICC05 age scale with age differences up to *c.* 500 years (Fig. 5; Svensson *et al.* 2008; Weninger & Jöris 2008). Hence, using the radiocarbon calibration curve to calibrate ^{14}C samples may artificially introduce age offset between the ^{14}C dated record and onsets of interstadials (Fig. 5). Additionally, the Hulu cave GICC05 age deviations complicate calculation of marine reservoir effects beyond 30 kcal yr BP as this requires knowledge of the absolute age. This is illustrated by the LINK15 ^{14}C reservoir age values, which depend on the choice of atmospheric calibration dataset with the IntCal09 calibration curve, generally producing higher R values than the Bahamas speleothem dataset (Fig. 7; Reimer *et al.* 2009; Hoffmann *et al.* 2010).

However, the Hulu cave and the GICC05 timescale show minimal age deviations around 38 kcal yr BP (Fig. 6). Fortunately this is around the age of the FMAZ-III tephra fall-out, offering a secure anchor from the North Atlantic marine records to the NGRIP ice core. The LINK15 ^{14}C-dated samples show that the transition into the warming is marked by a reduction of the reservoir age with onset of warming (strengthening of the

North Atlantic Drift and deep convection) in the LINK15 core independent of the chosen ^{14}C calibration dataset (Fig. 6). This confirms that variability in ocean circulation may be associated with large changes in ^{14}C reservoir ages. However, the low ^{14}C reservoir age at the beginning of H4 indicates that ^{14}C reservoir age may vary considerably and on a short timescale (Fig. 7). Furthermore, the very high ^{14}C reservoir ages (*c.* 1500–2000 ^{14}C years) prior to the onset of GI8 during the latter part of H4 confirm that the ^{14}C reservoir age can reach exceptionally high values during Heinrich events (e.g. Voelker 2000; Reimer *et al.* 2009).

Summary and conclusion

A new approach to construct tie-point age models taking into account both depth and age errors is suggested. Tie-points are identified using abrupt changes in $\delta^{18}O$ values, decreases in relative abundance of *N. pachyderma* s and increase in magnetic susceptibility values to mark the abrupt warming, that is, the start of interstadials. The age model approach uses the ages of identified tephra layers to anchor the chronology to the NGRIP GICC05 age scale, whereas no *a priori* ages for the tie-points are assumed. Tie-points are fitted into the chronology by assuming that their relative time difference equals the relative time difference observed between NGRIP interstadials onsets.

Using the FMAZ-III tephra layer, positioned around 100 years after the onset of GI8, leads and lags have been estimated. Based on different age model approaches using the ENAM93-21 and LINK15 marine cores, the lead/lag between the North Atlantic Ocean and the NGRIP ice core has been found to range between −37 and 154 years. The estimated age differences between ENAM93-21 and LINK15 at the start of interstadial 8 are much smaller than the general age uncertainties (around ±725 years) and we therefore conclude that with present knowledge it is difficult to disentangle leads or lags between the NGRIP ice core and the North Atlantic Ocean. However we note that the stratigraphical position of FMAZ-III and GI8 of both ENAM93-21 and LINK15 confirms the correlation to the NGRIP stratigraphy.

The radiocarbon reservoir age *R* has been estimated using the Bahamas ^{14}C dataset and the IntCal09 calibration curve. The estimated *R* appears to be very variable and range between 300 and 1650 or 850 and 2300 ^{14}C years depending of the chosen ^{14}C dataset. The IntCal09 calibration consistently yields the highest *R* values. However, independent of the ^{14}C dataset, the *R* values appear to have a similar trend with highest values during the H4 event and lower *R* values before and after. The start of interstadial 8 appears to be associated with a significant lowering of the *R* values.

The study of LINK15 forms part of the project 'PaleoCIRCUS' financed by the Mohn Foundation and the University of Tromsø. The project is also part of CAGE supported by the Research Council of Norway, Grant No. 223259/F5. D. Ostermann, Woods Hole Oceanographic Institution, supervised the stable isotope measurements.

Appendix

Model configuration

Markov Chain Monte Carlo algorithm

The model is driven by the Metropolis–Hastings sampler (see Martinez & Martinez 2002) by obtaining a candidate matrix $\boldsymbol{Y}$ from the proposal distributions $q_i(y_i)$. The candidate point $\boldsymbol{Y}$ is then accepted as the next state of the chain with a probability given by α:

$$R(t) = \frac{P_{model}(Y_t)}{P_{model}(Y_{t-1})} \text{ with } \alpha = \min\{1, R(t)\}. \quad \text{(A1)}$$

If the candidate matrix $\boldsymbol{Y}$ is not accepted, then the chain does not move and the state t is set to $\boldsymbol{X}_t = \boldsymbol{X}_{t-1}$. The candidate matrix $\mathbf{Y}$ is accepted if

$$U(0, 1) \leq \alpha \quad \text{(A2)}$$

where $U(0,1)$ is a uniform random number between 0 and 1. With $\alpha = 0$ all candidates are accepted opposed to $\alpha = 1$, where only candidates with unity probability are accepted. The model probability may have five contributions – time, depths, differential dates, differential depth and accumulation rate limits – and is expressed as

$$P_{model} = P_T \cdot P_D \cdot P_{\Delta_T} \cdot P_{\Delta_D} \cdot P_R. \quad \text{(A3)}$$

The calculation of each of the probabilities P_T, P_D, P_{Δ_T}, P_{Δ_D} and P_R will be discussed below. The model performance and convergence is monitored by accepted P_{model} values (see Fig. 3c, e). In particular the model algorithm takes a number of iterations, typical 10 000, before finding its 'equilibrium' state. This so-called burn-in time is disregarded in the resulting model.

For each iteration (typical 10^6 iterations in total) age values for input event points as well as at interpolated depths are stored ($\boldsymbol{M}_{raw} = n_d \times n_{iterations}$). The resulting age model is taken as the mode of the resulting age probability distributions and errors are estimated by 68.2% (*c.* 1σ) and 94.5% (*c.* 2σ) confidence intervals.

Defining events and associated probabilities

Events are labelled in time t_i from the start a to the bottom b with associated depths d_i such that:

$$t_i < t_{i+1} \text{ for all } i \text{ such that } a < t_i < b \quad \text{(A4)}$$

and

$$d_i > d_{i+1} \text{ for all } i \text{ such that } d_a > d_i > d_b. \quad (A5)$$

Using the above, each modelled age depth relationship can be given probabilities as

$$P_T(t) = \begin{cases} \prod_{i=a}^{b} P_T(t_i) & t_a < \cdots < t_i \cdots < t_b. \\ 0 & \text{otherwise} \end{cases} \quad (A6)$$

Similarly, the probability for depths is given as

$$P_D(t) = \begin{cases} \prod_{i=a}^{b} P_D(d_i) & d_a > \cdots > d_i \cdots > d_b. \\ 0 & \text{otherwise} \end{cases} \quad (A7)$$

Each event is associated with an age $t_i{}^*$ and with an error $\sigma_i{}^{t*}$. Similarly, the event depth is expressed as $d_i{}^*$ with an error $\sigma_i{}^{d*}$. The errors $\sigma_i{}^{t*}$ and $\sigma_i{}^{d*}$ are assumed to be normally distributed. Hence the probabilities $P_T(t_i)$ (A6) and $P_D(d_i)$ (A7) are proportional to

$$P_T(t_i) \propto \exp\left(-\frac{(t_i - t_i^*)^2}{2\sigma_i^{t*2}}\right) \quad (A8)$$

and

$$P_D(d_i) \propto \exp\left(-\frac{(d_i - d_i^*)^2}{2\sigma_i^{d*2}}\right). \quad (A9)$$

Differential dating

The errors introduced, for example, when counting seasonal layers in the NGRIP ice core are cumulative and therefore increase with depth. However, the age difference between two events can be known with a high precision. For example, the age difference between the onset of GI7 and the onset of GI6 is 1740 ± 55 years (Table 1). Such differential ages may further constrain the age model. Suppose the age difference between event t_x and t_y, expressed as Δt_{xy}, is known to be $\tau \pm \sigma_\tau$, then the probability of Δt_{xy} being equal to $\tau \pm \sigma_\tau$ is

$$P_\Delta(\Delta t_{xy}) \propto \exp\left(-\frac{(\Delta t_{xy} - \tau_{xy})^2}{2\sigma_{xy}^{\tau\,2}}\right). \quad (A10)$$

Thus the total probability P_Δ by taking all differential age pairs into account is given by

$$P_\Delta = \prod_{xy=1}^{N} P_\Delta(\Delta t_{xy}). \quad (A11)$$

Similarly, the stratigraphy can be checked by requiring that the depth difference Δ_D defined by succeeding depth points (d_i and d_{i+1}) fulfils

$$P_{\Delta_D}(d_i - d_{i+1}) \propto \exp\left(-\frac{(d_i - d_{i+1} - \Delta d_{xy})^2}{2\sigma_{xy}^{d\,2}}\right) \quad (A12)$$

with $\Delta d_{xy} = d_i^* - d_{i+1}^*$ and $(\sigma_{xy}{}^{d*})^2 = (\sigma_i{}^{d*})^2 + (\sigma_{i+1}{}^{d*})^2$. Hence the total probability is given by:

$$P_{\Delta_D} = \prod P_{\Delta_D}(d_i - d_{i+1}). \quad (A13)$$

Interpolation

Probability distributions for endpoints are estimated by calculating the mean accumulation rate for the first three (top) and last three (bottom) tie-points. Uncertainties for top and bottom ages are provided using the standard deviation for all accumulation rates. Interpolated depth (d_p) and age (t_p) between events (defined by d_i, t_i) are estimated such that:

$$\begin{aligned} d_i > d_p > d_{i+1} \\ t_i < t_p < t_{i+1}. \end{aligned} \quad (A14)$$

Constraining accumulation rates

The age and depth constraints (A4) and (A5) may result in unrealistic rapid changes in the sedimentation rate for overlapping age and depth probability distributions. To avoid this, an optional additional requirement constraining allowed sedimentation rates is added using the gamma probability distribution:

$$P_R = \prod \frac{1}{\theta^k} \frac{1}{\Gamma(k)} r_i^{k-1} e^{-(r_i/\theta)} \quad (A15)$$

where r is the accumulation rate, k the shape parameter and θ the scale parameter. k and θ are determined using the tie-point data, for example, identified tephra layers:

$$\theta = \frac{VAR}{\mu} \text{ and } k = \frac{\mu}{\theta} - 1 \quad (A16)$$

where VAR is the accumulation rate variance and μ the average accumulation rate.

Model setup

Table A1 shows the model setup for the constructed tie-point, tephra and combined age models for ENAM93-21 and LINK15.

Table A1. *Age model options used*

Model	Constraints used	Data used
Tie-point model	(A4), (A5), (A8), (A9), (A12), (A15)	GI onset tie-points
Tephra layer model	(A4), (A5), (A8), (A9), (A12), (A15)	Identified tephra layers
Combined model	(A4), (A5), (A8), (A9), (A12), (A15)	Identified tephra layers
	(A5), (A8), (A9), (A11), (A12), (A15)	GI onset tie-points

Calculating reservoir ages

Radiocarbon reservoir ages are expressed as the quantities $R(t)$ and $\Delta R(t)$ (Stuiver & Braziunas 1993). $R(t)$ is determined as the difference between a measured marine ^{14}C age, $^{14}C_M(t)$, and the contemporaneous atmospheric ^{14}C age, $^{14}C_{ATM}(t)$, derived from the atmospheric calibration curve (IntCal09: Reimer *et al.* 2009) at the calibrated age t of the deposit. The calibrated age t may be given by the age–depth model, or from the ^{14}C content of a terrestrial sample at a similar depth. Therefore,

$$R(t_i) = {}^{14}C_M(t_i) - {}^{14}C_{ATM}(t_i) \quad (A17)$$

ΔR, in contrast, expresses the difference between the measured marine ^{14}C age, $^{14}C_M(t)$, and the global marine calibration curve, $^{14}C_{MAR}(t)$ (Marine09: Reimer *et al.* 2009), yielding

$$R(t_i) = {}^{14}C_M(t_i) - {}^{14}C_{MAR}(t_i). \quad (A18)$$

The parameters t_i, $^{14}C_M$, $^{14}C_{ATM}$ and $^{14}C_{MAR}$ are all associated with errors. The errors on $^{14}C_M$, $^{14}C_{ATM}$ and $^{14}C_{MAR}$ can be assumed to be normally distributed. Hence $^{14}C_M$ is sampled using normally distributed data, $N(\mu,\sigma)$, with μ and σ given by the radiocarbon measurement. Likewise $^{14}C_{ATM}(t_i)$ and $^{14}C_{MAR}(t_i)$ are sampled as normally distributed data with μ and σ given by the radiocarbon mean value and error given by each calibration curve at the time t_i.

However, the probability distributions for the calibrated ages t_i cannot be assumed normally distributed and are therefore sampled from a general probability distribution by providing a probability density function. Calibrated ages t_i may also be provided as saved mcmc output (i.e. the $model_{raw}$ matrix from above or *ex* from depositional models in OxCal using the Markov chain Monte Carlo command). Probability density functions can then be calculated from the N calculated reservoir ages (R(t), ΔR(t)) and, from these PDFs, 68.4% (*c.* 1σ) and 95.4% (*c.* 2σ) confidence intervals can be deduced.

References

Andersen, K. K., Svensson, A. *et al.* 2006. The Greenland ice core chronology, 2005, 15–42 ka. Part 1: constructing the time scale. *Quaternary Science Reviews*, **25**, 3246–3257.

Austin, W. E. N. & Kroon, D. 2001. Deep sea ventilation of the northeastern Atlantic during the last 15,000 years. *Global Planetary Change*, **30**, 13–31.

Austin, W. E. N., Telford, R. J., Ninnemann, U. S., Brown, L., Wilson, L. J., Small, D. P. & Bryant, C. L. 2011. North Atlantic reservoir ages linked to high Younger Dryas atmospheric radiocarbon concentrations. *Global Planetary Change*, **79**, 226–233.

Austin, W. E. N., Hibbert, F. D., Rasmussen, S. O., Peters, C., Abbott, P. M. & Bryant, C. L. 2012. The synchronization of palaeoclimatic events in the North Atlantic region during Greenland Stadial 3 (ca 27.5 to 23.3 kyr b2k). *Quaternary Science Reviews*, **36**, 154–163.

Bard, E., Hamelin, B., Fairbanks, R. G. & Zindler, A. 1990. Calibration of the ^{14}C timescale over the past 30,000 years using mass spectrometric U–Th ages from Barbados corals. *Nature*, **345**, 405–410.

Bard, E., Rostek, F. & Ménot-Combes, G. 2004. Radiocarbon calibration beyond 20,000 ^{14}C yr B.P. by means of planktonic foraminifera of the Iberian Margin. *Quaternary Research*, **61**, 204–214.

Berger, W. H., Schulz, M. & Wefer, G. 2010. Quaternary oceans and climate change: lessons for the future? *International Journal of Earth Sciences*, **99**, S171–S189.

Blaauw, M. & Christen, J. A. 2005. Radiocarbon peat chronologies and environmental change. *Applied Statistics*, **54**, 805–816.

Blockley, S. P. E., Lane, C. S. *et al.* 2012. Synchronisation of palaeoenvironmental records over the last 60,000 years, and an extended INTIMATE event stratigraphy to 48,000 b2k. *Quaternary Science Reviews*, **36**, 2–10.

Bond, G., Broecker, W., Johnsen, S., McManus, J., Labeyrie, L., Jouzel, J. & Bonani, G. 1993. Correlations between climate records from North-Atlantic sediments and Greenland ice. *Nature*, **365**, 143–147.

Brendryen, J., Haflidason, H. & Sejrup, H. P. 2011. Non-synchronous deposition of North Atlantic Ash Zone II in Greenland ice cores, and North Atlantic and Norwegian Sea sediments: an example of complex glacial-stage tephra transport. *Journal of Quaternary Science*, **26**, 739–745.

Broecker, W., Bond, G., Klas, M., Clark, E. & McManus, J. 1992. Origin of the northern Atlantic's Heinrich events. *Climate Dynamics*, **6**, 265–273.

Cutler, K. B., Gray, S. C. *et al.* 2004. Radiocarbon calibration to 50 kyr BP with paired ^{14}C and ^{230}Th dating of corals from Vanuatu and Papua New Guinea. *Radiocarbon*, **46**, 1127–1160.

Dansgaard, W., Johnsen, S. J. *et al.* 1993. Evidence for general instability of past climate from a 250-kyr ice-core record. *Nature*, **364**, 218–220.

Davies, S. M., Wastegård, S., Rasmussen, T. L., Svensson, A., Johnsen, S. J., Steffensen, J. P. & Andersen, K. K. 2008. Identification of the Fugloyarbanki tephra in the NGRIP ice core: a key tie-point for marine and ice-core sequences during the last glacial period. *Journal of Quaternary Science*, **23**, 409–414.

Davies, S. M., Wastegård, S. *et al.* 2010. Tracing volcanic events in the NGRIP ice-core and synchronising North Atlantic marine records during the last glacial period. *Earth and Planetary Science Letters*, **294**, 69–79.

Davies, S. M., Abbott, P. M., Pearce, N. J. G., Wastegård, S. & Blockley, S. P. E. 2012. Integrating the INTIMATE records using tephrochronology: rising to the challenge. *Quaternary Science Reviews*, **36**, 11–27.

Eiríksson, J., Knudsen, K. L. *et al.* 2011. Coupling of palaeoceanographic shifts and changes in marine reservoir ages off North Iceland through the last millennium. *Palaeogeography, Palaeoclimatology, Palaeoecology*, **302**, 95–108.

FAIRBANKS, R. G., MORTLOCK, R. A. *ET AL.* 2005. Radiocarbon calibration curve spanning 0 to 50,000 years BP based on paired ^{230}Th/^{234}U/^{238}U and ^{14}C dates on pristine corals. *Quaternary Science Reviews*, **24**, 1781–1796.

GRONVOLD, K., OSKARSSON, N., JOHNSEN, S. J., CLAUSEN, H. B., HAMMER, C. U., BOND, G. & BARD, E. 1995. Ash layers from Iceland in the Greenland GRIP ice core correlated with oceanic and land sediments. *Earth and Planetary Science Letters*, **135**, 149–155.

HEINRICH, H. 1988. Origin and consequences of cyclic ice rafting in the Northeast Atlantic-Ocean during the past 130,000 years. *Quaternary Research*, **29**, 142–152.

HOFFMANN, D. L., BECK, J. W., RICHARDS, D. A., SMART, P. L., SINGARAYER, J. S., KETCHMARK, T. & HAWKESWORTH, C. J. 2010. Towards radiocarbon calibration beyond 28 ka using speleothems from the Bahamas. *Earth and Planetary Science Letters*, **289**, 1–10.

HUGHEN, K., SOUTHON, J., LEHMAN, S., BERTRAND, C. & TURNBULL, J. 2006. Marine-derived C-14 calibration and activity record for the past 50,000 years updated from the Cariaco Basin. *Quaternary Science Reviews*, **25**, 3216–3227.

KISSEL, C., LAJ, C., LABEYRIE, L., DOKKEN, T., VOELKER, A. & BLAMART, D. 1999. Rapid climatic variations during marine isotopic stage 3: magnetic analysis of sediments from Nordic Seas and North Atlantic. *Earth and Planetary Science Letters*, **171**, 489–502.

LOWE, J. J., RASMUSSEN, S. O., BJORCK, S., HOEK, W. Z., STEFFENSEN, J. P., WALKER, M. J. C. & YU, Z. C. 2008. Synchronisation of palaeoenvironmental events in the North Atlantic region during the Last Termination: a revised protocol recommended by the INTIMATE group. *Quaternary Science Reviews*, **27**, 6–17.

MARTINEZ, W. L. & MARTINEZ, A. R. 2002. *Computational Statistics Handbook with Matlab*. Chapman & Hall/CRC, Boca Raton, FL.

MOROS, M., MCMANUS, J. F. *ET AL.* 2004. Quartz content and the quartz-to-plagioclase ratio determined by X-ray diffraction: a proxy for ice rafting in the northern North Atlantic? *Earth and Planetary Science Letters*, **218**, 389–401.

NADEAU, M. J., GROOTES, P. M., VOELKER, A., BRUHN, F., DUHR, A. & ORIWALL, A. 2001. Carbonate 14C background: does it have multiple personalities? *Radiocarbon*, **43**, 169–176.

NIELSEN, T., RASMUSSEN, T. L., CERAMICOLA, S. & KUIJPERS, A. 2007. Quaternary sedimentation, margin architecture and ocean circulation variability around the Faroe Islands, North Atlantic. *Quaternary Science Reviews*, **26**, 1016–1036.

OSTERMANN, D. R. & CURRY, W. B. 2000. Calibration of stable isotopic data: an enriched delta O-18 standard used for source gas mixing detection and correction. *Paleoceanography*, **15**, 353–360.

RAMSEY, C. B. 2008. Deposition models for chronological records. *Quaternary Science Reviews*, **27**, 42–60.

RASMUSSEN, T. L. & THOMSEN, E. 2004. The role of the North Atlantic Drift in the millennial timescale glacial climate fluctuations. *Palaeogeography, Palaeoclimatology, Palaeoecology*, **210**, 101–116.

RASMUSSEN, T. L. & THOMSEN, E. 2008. Warm Atlantic surface water inflow to the Nordic seas 34–10 calibrated ka B.P. *Paleoceanography*, **23**, PA1201.

RASMUSSEN, T. L., THOMSEN, E., LABEYRIE, L. & VAN WEERING, T. C. E. 1996. Circulation changes in the Faeroe–Shetland Channel correlating with cold events during the last glacial period (58–10 ka). *Geology*, **24**, 937–940.

RASMUSSEN, T. L., THOMSEN, E. & VAN WEERING, T. C. E., 1998. Cyclic changes in sedimentation on the Faeroe Drift 53–9 kyr BP related to climate variations. *In*: STOKER, M., EVANS, D. & CRAMP, R. (eds) *Geological Processes on Continental Margins: Sedimentation, Mass-Wasting and Stability*. Geological Society, London, Special Publications, **129**, 255–267.

RASMUSSEN, T. L., BÄCKSTRÖM, D. *ET AL.* 2002. The Faroe–Shetland Gateway: Late Quaternary water mass exchange between the Nordic seas and the northeastern Atlantic. *Marine Geology*, **188**, 165–192.

RASMUSSEN, T. L., WASTEGÅRD, S., KUIJPERS, A., VAN WEERING, T. C. E., HEINEMIEIER, J. & THOMSEN, E. 2003. Stratigraphy and distribution of tephra layers in marine sediment cores from the Faeroe Islands, North Alantic. *Marine Geology*, **199**, 263–277.

RASMUSSEN, S. O., ANDERSEN, K. K. *ET AL.* 2006. A new Greenland ice core chronology for the last glacial termination. *Journal of Geophysical Research*, **111**, D06102.

REIMER, P. J., BAILLIE, M. G. L. *ET AL.* 2009. Intcal09 and Marine09 radiocarbon age calibration curves, 0–50,000 years cal BP. *Radiocarbon*, **51**, 1111–1150.

RUDDIMAN, W. F. & GLOVER, L. K. 1972. Vertical mixing of ice-rafted volcanic ash in North-Atlantic sediments. *Geological Society of the American Bulletin*, **83**, 2817–2836.

SINGARAYER, J. S., RICHARDS, D. A., RIDGWELL, A., VALDES, P. J., AUSTIN, W. E. N. & BECK, J. W. 2008. An oceanic origin for the increase of atmospheric radiocarbon during the Younger Dryas. *Geophysical Research Letters*, **35**, L14707.

SOUTHON, J., NORONHA, A. L., CHENG, H., EDWARDS, R. L. & WANG, Y. 2012. A high-resolution record of atmospheric ^{14}C based on Hulu Cave speleothem H82. *Quaternary Science Reviews*, **33**, 32–41.

STUIVER, M. & BRAZIUNAS, T. F. 1993. Modeling atmospheric ^{14}C influences and ^{14}C ages of marine samples to 10,000 BC. *Radiocarbon*, **35**, 137–189.

SVENSSON, A., ANDERSEN, K. K. *ET AL.* 2008. A 60 000 year Greenland stratigraphic ice core chronology. *Climate of the Past*, **4**, 47–57.

THORNALLEY, D. J. R., BARKER, S., BROECKER, W. S., ELDERFIELD, H. & MCCAVE, I. N. 2011. The deglacial evolution of North Atlantic deep convection. *Science*, **331**, 202–205.

VAUTRAVERS, M. J. & SHACKLETON, N. J. 2006. Centennial-scale surface hydrology off Portugal during marine isotope stage 3: insights from planktonic foraminiferal fauna variability. *Paleoceanography*, **21**, PA3004.

VOELKER, A. H. L. 2000. Radiocarbon levels in the Iceland Sea from 25–53 kyr and their link to the Earth's magnetic field intensity. *Radiocarbon*, **42**, 437–452.

VOGEL, J. S., SOUTHON, J. R. & NELSON, D. E. 1987. Catalyst and binder effects in the use of filamentous graphite for AMS. *Nuclear Instruments and Methods in Physics Research, Section B*, **29**, 50–56.

WAELBROECK, C., SKINNER, L. C. ET AL. 2011. The timing of deglacial circulation changes in the Atlantic. *Paleoceanography*, **26**, PA3213.

WANG, Y. J., CHENG, H., EDWARDS, R. L., AN, Z. S., WU, J. Y., SHEN, C.-C. & DORALE, J. A. 2001. A high-resolution absolute-dated Late Pleistocene monsoon record from Hulu Cave, China. *Science*, **294**, 2345–2348.

WENINGER, B. & JÖRIS, O. 2008. A ^{14}C age calibration curve for the last 60 ka: the Greenland–Hulu U/Th timescale and its impact on understanding the Middle to Upper Paleolithic transition in Western Eurasia. *Journal of Human Evolution*, **55**, 772–781.

Last millennium dispersal of air-fall tephra and ocean-rafted pumice towards the north Icelandic shelf and the Nordic seas

GUDRÚN LARSEN*, JÓN EIRÍKSSON & ESTHER R. GUDMUNDSDÓTTIR

Institute of Earth Sciences, University of Iceland, Askja, Sturlugata 7, IS-101 Reykjavík, Iceland

**Corresponding author (e-mail: glare@hi.is)*

Abstract: The paper focuses on dispersal of airborne tephra that may have reached the marine environment north of Iceland during the last millennium, particularly the shelf off north Iceland. Many of these tephra horizons may extend into the Nordic Seas and the Arctic. The tephrochronology of Iceland after the settlement in the late ninth century AD relates to volcanic events that have been dated with documentary records as well as ice cores. The relevant eruptions for long-distance transport of tephra have been explosive or partly explosive, and mostly deposited from volcanic plumes. However, other methods of transport into the area north of Iceland are also considered. These include rafting of pumice from offshore eruptions elsewhere around Iceland, leading to instantaneous flotation of tephra, and river-rafted pumice from inland areas of heavy tephra fall, as well as potential contributions from volcanogenic glacial bursts.

Airborne tephra in 25–30 explosive and partly explosive eruptions had the potential to reach the north Icelandic shelf and beyond during the time slice considered here. Four instances of ocean-rafted pumice off the north coast are known. Tephra from 15 of these eruptions has been identified in marine cores from the north Icelandic shelf and eastern Norwegian sea.

The last millennium represents a particularly convenient time slice for the application of tephrochronology as a high-precision dating tool in the Iceland region, both within terrestrial Iceland and in offshore marine sediments. Exact eruption dates, with a precision to a year, month, day and even the hour, are available from contemporary or near-contemporary sources (e.g. documentary annals) and these are often complemented by descriptions of the explosive eruptions or explosive phases, as well as the areas affected by tephra fall (e.g. Thorarinsson, 1958, 1967, 1974). The proximity to the tephra-producing volcanoes provides the potential of small and average sized tephra layers to increase the resolution of the tephrochronology within terrestrial Iceland and also within the shelf area off north Iceland (e.g. Thorarinsson 1958, 1967, 1974, Larsen *et al.* 1998, 2002; Haflidason *et al.* 2000; Eiríksson *et al.* 2004; Thordarson & Larsen 2007, Gudmundsdóttir *et al.* 2012*a*), as compared with more distal areas where relatively few and, usually, large tephra layers form the tephrochronological framework (e.g. Persson 1971; Dugmore *et al.* 1995; Pilcher *et al.* 1996; van den Bogaard & Schmincke 2002; Wastegård & Davies 2009; Swindles *et al.* 2011).

The compilation of northward-dispersed tephra during the last 1000 years was originally carried out to provide a background for the dating of marine sediments in gravity cores, multicores and box cores collected in 1999 and 2006 on the north Icelandic shelf and investigated as part of the Millennium project (European climate of the last Millennium, European Commission Sixth Framework Programme 2006–2009). The tephrochronological age models for the marine records were used in the Millennium Project for high-resolution reconstructions of the marine palaeoclimate of the last millennium (Eiríksson *et al.* 2011; Ran *et al.* 2011; Sicre *et al.* 2011; Knudsen *et al.* 2012; Wanamaker *et al.* 2012; Cunningham *et al.* 2013). The tephra compilation is based on available published and unpublished data. Eruptions postdating the coring of 1999 and 2006 are not included.

Volcanism in Iceland and on the Icelandic shelf is confined to numerous volcanic systems (Fig. 1a). About 30 such volcanic systems have been defined on land while those on the shelf are less well known (Sæmundsson 1978, 1979; Jakobsson 1979; Jakobsson *et al.* 2008; Jóhannesson & Sæmundsson 1998). During the last millennium at least 12 volcanic systems have contributed basaltic and silicic tephra, most of them land-based (e.g. Thordarson & Larsen 2007; Larsen & Eiríksson 2008). The number of submarine volcanic systems active during this time period remains, however, uncertain (e.g. Jakobsson *et al.* 2008 and references therein).

The major producers of basaltic tephra during the last millennium are the Grímsvötn, Veidivötn–Bárdarbunga and Katla volcanic systems, all partly

From: Austin, W. E. N., Abbott, P. M., Davies, S. M., Pearce, N. J. G. & Wastegård, S. (eds) 2014. *Marine Tephrochronology*. Geological Society, London, Special Publications, **398**, 113–140.
First published online March 3, 2014, http://dx.doi.org/10.1144/SP398.4

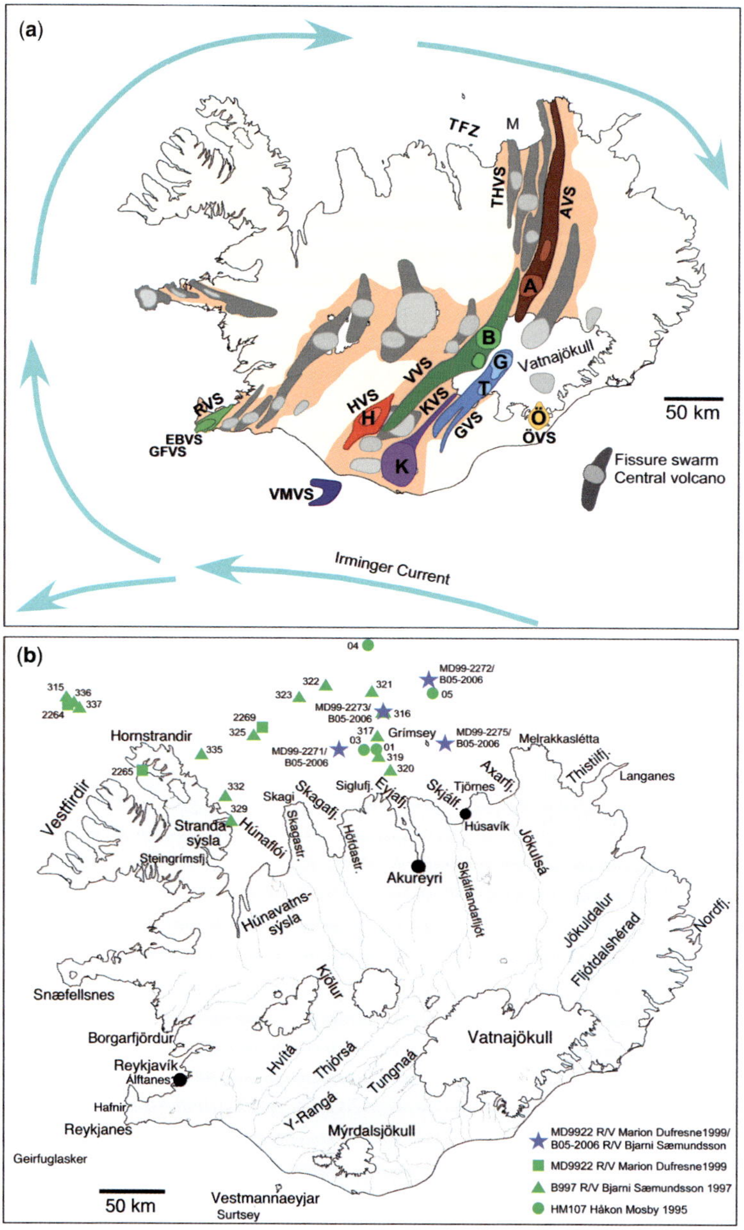

Fig. 1. (**a**) The volcanic systems of Iceland. The major producers of tephra during last millennium are shown in colour, which also reflects the different chemical characteristics of these systems. Adapted from Jóhannesson & Sæmundsson (1998). A, Askja; B, Bárdarbunga; G, Grímsvötn; H, Hekla; K, Katla; Ö, Öræfajökull; EBVS, Eldeyjarbodi Volcanic System; GVS, Grímsvötn Volcanic System; GFVS, Geirfuglasker Volcanic System; RVS, Reykjanes Volcanic System; THVS, Theistareykir Volcanic System; VMVS, Vestmannaeyjar Volcanic System; VVS, Veidivötn–Bárdarbunga Volcanic System; TFZ, Tjörnes Fracture Zone; M, Mánáreyjar. (**b**) Location of marine sediment cores on the north Icelandic shelf. Millennium core sites are marked by stars. Also shown are place names referred to in the text.

ice-covered, as well as the partly submarine Reykjanes and Vestmannaeyjar systems. Several submarine volcanic systems on the Reykjanes ridge have erupted during this time slice and possibly two volcanic systems on the fracture zone off north Iceland. The major producers of silicic tephra are the Hekla, Öræfajökull and Askja volcanic systems (Thordarson & Larsen 2007; Larsen & Eiríksson 2008, and references therein).

The type of volcanic activity that has the potential to disperse basaltic tephra tens or hundreds of kilometres from the source is hydromagmatic explosive activity, also referred to as phreatomagmatic and surtseyan activity. This includes eruptions in subglacial, submarine and subaquous settings and in areas of high ground-water. Some of these eruptions may be classified as mixed eruptions if the eruption style changes to effusive during some stage or persists on parts of the initially active volcanic fissure (e.g. Thordarson & Höskuldsson 2008 and references therein).

Tephra of silicic and intermediate composition (rhyolitic, dacitic, andesitic) is generated by explosive eruptions or explosive phases of the Plinian or subplinian type in a dry environment. At volcanoes that are capped by ice or perennial snow, the opening phase may be phreatoplinian, generating much finer-grained tephra than the Plinian and subplinian eruptions/phases. Many of these eruptions produce substantial lava flows as well and belong to the category of mixed eruptions (e.g. Thordarson & Höskuldsson 2008 and references therein).

Sources of information on volcanic eruptions of the last millennium

Sources of information about explosive and partly explosive eruptions in Iceland fall mainly into two categories. The first is written descriptions of volcanic eruptions or phenomena that could be caused by volcanic eruptions. These are mentioned or described in various documents (e.g. annals, reports from officials, letters) from the twelfth century onwards and may include date of the eruption (year, month, day, hour), location of the volcano, description of the type of activity and course of events, duration of the activity, type of products, direction or area of tephra fall, and areas of damage (e.g. Thorarinsson 1958, 1967, 1974, 1975, and references therein). This is the case for many eruptions of the Hekla and Katla volcanoes, which lie close to farming areas in south Iceland, and some eruptions within the Vatnajökull ice cap.

For eruptions in more remote areas such as the central highlands or Vatnajökull ice cap, the information may consist only of the eruption year and a few words about eruption cloud or fire, glow in the sky, tephra fall, darkness, flood or other phenomena that people thought were signs of volcanic activity. The location of the eruption site may be vague or attributed to the wrong volcano. Also, names that differ from the present ones may have been used for mountains, ice caps and districts at the time of writing.

In general, however, the date of an eruption is likely to be more accurate than the location of the volcano because the calendar was common knowledge although geographical knowledge was often poor. Volcanic eruptions were as much news then as they are today and were definitely worth mentioning in the annals.

Some sightings of phenomena that were interpreted as signs of volcanic eruptions may or may not have been caused by volcanic activity. Some accounts of tephra fall may also be false – as late as in the 2004 Grímsvötn eruption, reports of tephra fall were spurred by dust storms (Larsen, unpublished data).

The second category is volcanic products, which in this case are tephra layers, that have been identified, mapped and/or chemically fingerprinted. Tephra layers have many macroscopic and microscopic characteristics that can be used to identify, correlate, map and trace tephra layers to source – and to build a local or regional tephra stratigraphy (e.g. Thorarinsson 1958, 1967; Larsen & Thorarinsson 1977; Larsen *et al.* 2001; Alloway *et al.* 2006; Larsen 2010; Lowe 2011; Óladóttir *et al.* 2011, 2012). Many volcanic systems in Iceland have chemical characteristics that allow correlation of their product to the source, and can support or verify correlations of tephra layers in the field (e.g. Jakobsson 1979; Larsen *et al.* 1998; Jóhannsdóttir 2007; Óladóttir *et al.* 2011; Gudmundsdóttir *et al.* 2012*a*).

Many descriptions of eruptions and tephra fall are detailed enough to allow correlation of certain tephra layers to definite dated eruptions, thereby providing securely dated tephra horizons within the local or regional tephra stratigraphy. Such marker horizons facilitate further correlation of tephra layers to dated eruptions. Dated pairs or series of tephra layers can also serve as marker horizons. The first such series in the Icelandic tephrochronology consisted of the H-1766, K-1721 and H-1693 tephra layers (Thorarinsson 1944).

Aeolian dust layers are, normally, easily recognizable for the trained eye. In Iceland this dust is mostly volcanic glass, but differs from fresh tephra in being inhomogenous both in grain morpology and composition, and the layers are not laterally extensive. No such layers are considered in the following sections.

Transport of volcanic material: air-fall, ocean-rafting, jökulhlaups

The most common transport of primary tephra is by an eruption cloud carried downwind from the volcano, covering distances of tens to thousands of kilometres. Deposition by *air-fall* has delivered by far the most tephra to the terrestrial and marine environment outside Iceland. In large Plinian eruptions, such as the prehistoric Hekla-4 eruption, detectable tephra deposits covered an area of several hundred thousand square kilometres (e.g. Davies *et al.* 2010; Swindles *et al.* 2011). In smaller Plinian, subplinian or phreatomagmatic eruptions the fallout area may be confined to narrow sectors of limited area, although exceptions such as the 2010 Eyjafjallajökull eruption are known (Gudmundsson *et al.* 2012). Numerous eruptions of Icelandic volcanoes have delivered tephra to sediments off the north shore of Iceland in Holocene time (e.g. Haflidason *et al.* 2000; Eiríksson *et al.* 2000; Andrews *et al.* 2002; Kristjánsdóttir *et al.* 2007; Gudmundsdóttir *et al.* 2012*a*, b). The number of eruptions that have dispersed tephra off the north coast during the last millennium is not precisely known but will certainly exceed the 30+ eruptions listed below. It should be noted that fires, such as the Surtsey Fires 1963–1967, are counted as a single eruption.

Air-fall tephra deposits originating from Icelandic volcanoes during the last millennium have been reported from numerous sediment cores on the north Iceland shelf (e.g. Kristjánsdóttir 1999; Eiríksson *et al.* 2000, 2004; Jónsdóttir 2001; Larsen *et al.* 2002; Kristjánsdóttir *et al.* 2007; Gudmundsdóttir *et al.* 2011, 2012*a*, *b*) and from two cores in the Nordic seas (Sejrup *et al.* 2010, 2011). Occurrences in terrestrial sediments in the North Atlantic region (e.g. Wastegård & Davies 2009 and references therein; D'Andrea *et al.* 2012) and in glacial ice (Palais *et al.* 1991; Isaksson *et al.* 2005; Kekonen *et al.* 2005) are, however, beyond the scope of this paper.

Pumice rafts have been known to float for hundreds of kilometres on the ocean before becoming waterlogged and sinking (e.g. Newton 1999). In the Hekla eruptions of 1970, 1947 and 1766, pumice rafts were formed off the south coast of Iceland as a result of offshore tephra fall (1947) or seaward transport by rivers (1970, 1766) and transported clockwise by near-shore currents and a branch of the Irminger current along the west and north coast of Iceland during the next three months (see references in the following sections). The Askja 1875 pumice was carried 130 km by a glacial river, Jökulsá á Fjöllum, to the north shore and then at least 70 km at sea to Grímsey Island. Although no reports of ocean-transported pumice off north Iceland exist for other eruptions, the presence of such pumice cannot be excluded. In the Öræfajökull eruption in 1362, pumice rafts were reported off west Iceland and rounded pumice on salt-marshes in west Iceland has been correlated to eruption off Reykjanes in 1226/1227 (see references in the following sections).

Jökulhlaups (glacial bursts) are the third transport mechanism that may have delivered volcanic glass to the sediments off north Iceland. Spring floods in the river Jökulsá á Fjöllum carry a load of volcanogenic material into Axarfjördur, and silt-coloured water is known to extend as far out as Grímsey. Large floods in this river (e.g. 1729, 1726, 1684/1685 and 1681; see references in the following sections) and in Skjálfandafljót River, caused by subglacial eruptions within Vatnajökull ice cap, are likely to have extended at least as far and are therefore potential contributors of volcanogenic material to the shelf sediments. Although unlikely to consist of homogenous glass grains, a resemblance to a primary ash cannot be ruled out.

Explosive and partly explosive eruptions 2005–870: an overview

Some 185 tephra-producing eruptions are known to have occurred since the late ninth century, of which about 170 happened during the last 1000 years (Thordarson & Larsen 2007; Larsen & Eiríksson 2008). Figure 2a, b summarizes the current knowledge on frequency of explosive and partly explosive eruptions in Iceland for this period. The frequency per 100 years is shown in Figure 2a. Eruption frequency was exceptionally low at the beginning of the AD 2005–1000 time interval, the period under consideration in the Millennium project, with only four explosive or partly explosive eruptions known in the eleventh century (Óladóttir *et al.* 2011). The eighteenth and ninteenth centuries experienced the highest frequency of explosive and partly explosive eruptions on record, but most of these eruptions were small (e.g. Thorarinsson 1974).

The majority, or about 120 of the explosive and partly explosive eruptions, can be assigned a date based on written sources. Not all tephra layers can be correlated to a documentary date; there are more tephra layers than records. Those tephra layers that cannot be correlated to an eruption year are assigned a date calculated using soil or sediment accumulation rate (SAR), or based on ice thickness between securely dated tephra horizons. These are often accurate to ± 5 or ± 10 years, which explains the 20 year binning used in Figure 2b. Furthermore, not all eruptions mentioned in documents can be fitted to a tephra layer, and about 30 eruptions are labelled 'description only' in Figure 2b. Even in the last century there are several eruptions that are classified as reported only because no corresponding volcanic products have been found.

An apparent general increase in eruption frequency with time in Figure 2 is an artefact, reflecting

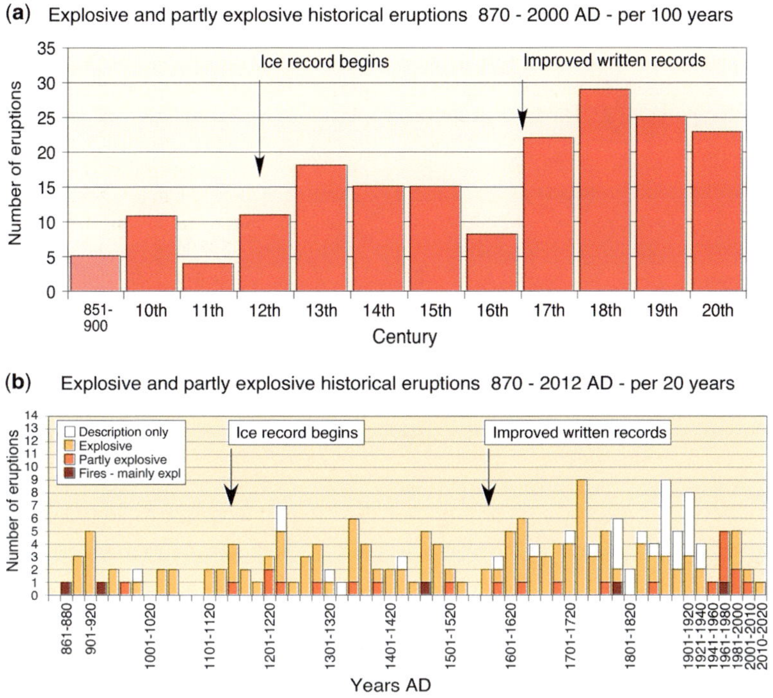

Fig. 2. Frequency of explosive and partly explosive eruptions between AD 870 and AD 2005, binned (**a**) per 100 years and (**b**) per 20 years. Adapted from Thordarson & Larsen (2007) and Larsen & Eiríksson (2008). Roughly 170 tephra-producing eruptions are known during last millennium, of which about 30 are 'reported only' eruptions within Vatnajökull ice cap. The frequency per 20 years varies from zero to nine eruptions, with peaks spaced 120–160 years apart. An apparent general increase in eruption frequency with time is probably an artefact, reflecting improved knowledge. Tephra layers in Vatnajökull ice go back to the twelfth century, greatly adding to the knowledge of volcanic eruptions within the Vatnajökull ice cap (first arrow). Written descriptions, including those of volcanic eruptions, improved in the seventeenth century (second arrow). Information based on Jónsson (1945), Thorarinsson (1950, 1958, 1963, 1965*a*, *b*, 1966, 1967, 1970, 1974), Jóhannesson (1977, 1983, 1984, 1987), Jóhannesson & Einarsson (1988), Sæmundsson (1991), Steinthorsson (1978), Larsen (1982, 1984, 2000), Larsen *et al.* (1992, 1998), Sigurgeirsson (1995), Thordarson & Self (1993), Thordarson *et al.* (2003), Grönvold *et al.* (1995), Zielinski *et al.* (1995), Eiríksson *et al.* (2004), Lacasse *et al.* (2004), Sigmundsson & Gudmundsson (2004), Óladóttir *et al.* (2011) and Larsen and Gudmundsson (unpublished data).

improved documentary records. Tephra layers in Vatnajökull ice go back to the twelfth century (first arrow). Written descriptions improve in the seventeenth century (second arrow). The highs (15–30 eruptions/100 years) and lows (2–8 eruptions/100 years) revealed by the 100 year bins represent actual changes in eruption frequency. The sixteenth-century low and the seventeenth-century rise in eruption frequency is, for instance, seen in all three archives used to compile the graphs, written documents, soil sections and glacier ice. Frequency per 20 years varies from 0 to 9 eruptions, with peaks spaced 120–160 years apart (Larsen *et al.* 1998).

In the following sections semi-standardized information on explosive and partly explosive eruptions is compiled, including the location, time and length of eruptions if known, type of explosive activity, dispersal and volume of the tephra together with selected references. The dispersal maps are based on contemporary descriptions and on ground data (thickness measurements or mass per unit area). Areas of documented tephra fall are shown and potential dispersal off north Iceland is indicated. River and/or ocean transport of pumice and volcanogenic sediments is indicated by arrows. Potential rafting of pumice from submarine eruptions is also indicated by an arrow and a question mark.

Note that the dispersal maps provided in this report do not necessarily include all known areas of deposition for the tephra layers on land because the emphasis is on the north Icelandic shelf. In the

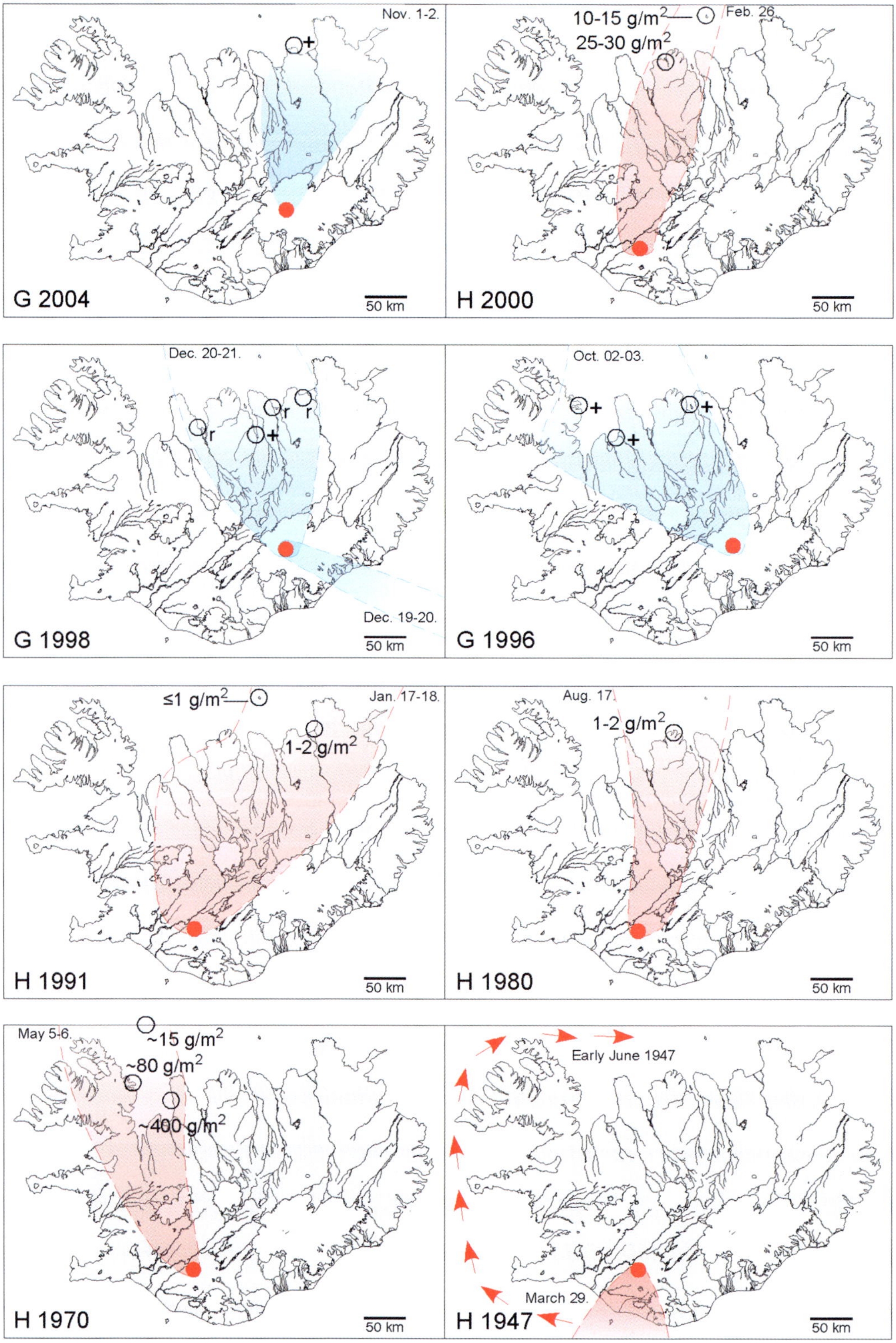
Nov. 1-2.
G 2004
50 km
10-15 g/m2
25-30 g/m2
Feb. 26.
H 2000
50 km
Dec. 20-21.
r
r
r
Dec. 19-20.
G 1998
50 km
Oct. 02-03.
G 1996
50 km
≤1 g/m2
Jan. 17-18.
1-2 g/m2
H 1991
50 km
Aug. 17.
1-2 g/m2
H 1980
50 km
May 5-6.
~15 g/m2
~80 g/m2
~400 g/m2
H 1970
50 km
Early June 1947
March 29.
H 1947
50 km

following sections, a small tephra layer refers to 0.01– 0.05 km^3, moderate 0.05–0.1 km^3, average 0.1–0.5 km^3 and fairly large 0.5–1 km^3 as freshly fallen.

Tephra producing eruptions 2005–1901

Verified and reported explosive or partly explosive eruptions in the period 2005–1901 are at least 24, counting the Surtsey Fires as a single eruption. Of them 17 eruptions have been verified. Airborne tephra from at least 11 eruptions was carried northwards across Iceland and onto the shelf off north Iceland (Figs 3 & 4). Below are short descriptions of these eruptions together with relevant references. Also included are those that delivered large quantities of pumice into the sea through tephra fall or by eruption at sea (Fig. 5). Volcanogenic jökulhlaups into Skjálfandaflói and Axarfjördur are included (Fig. 6) as flood water may be expected to extend tens of kilometres into the sea as freshwater plumes.

Grímsvötn 2004

Eruption site: Grímsvötn caldera, Grímsvötn volcanic system. Eruption began around 21:50 on 1 November 2004 and lasted for five days.

The subglacial opening stage was followed by subaerial explosive activity and the only product was basaltic tephra. Maximum height of eruption plume was close to 12 km, based on corrected records of the weather radar of the Icelandic Meteorological Office. Tephra dispersal was towards NNE, north, NNW and minor tephra fall reached the coast of NE Iceland. The tephra layer was a small one; the volume of tephra deposited outside the crater was close to 24 million m^3 or 0.024 km^3. In north Iceland 6–100 g m^{-2} of tephra were deposited 40–60 km inland (based on Sigmundsson & Gudmundsson 2004; Oddsson 2007; Oddsson *et al.* 2012; Jude-Eton *et al.* 2012).

Hekla 2000

Eruption site: Hekla central volcano, Hekla volcanic system. Eruption began at 18:19 on 26 February 2000 and lasted 12 days.

The subplinian opening stage emitted basaltic–andesitic tephra, followed by mixed explosive and effusive activity. Maximum height of the eruption plume was 11–12 km according to the weather radar of the Icelandic Meteorological Office. The main tephra-producing phase lasted a few hours and tephra was carried NNE onto the shelf off north Iceland along a narrow sector crossing Grímsey Island, where 10–15 g m^{-2} was deposited. The resulting tephra layer was a small one, with calculated volume of 10 million m^3 or 0.01 km^3. Near-vent deposits that accumulated during the following activity are not included in this value (based on Haraldsson *et al.* 2002; Lacasse *et al.* 2004; Höskuldsson *et al.* 2007).

Grímsvötn 1998

Eruption site: Grímsvötn caldera, Grímsvötn volcanic system. Eruption began around 9.30 a.m. on 18 December 1998, and lasted until 28 December.

Subaerial explosive explosive activity emitting basaltic tephra, apparently without a preceding subglacial stage, was continuous for the first two days and fluctuating after that. Maximum height of eruption cloud observed from aircraft was 9–10 km a.s.l. Main tephra dispersal was towards ESE but minor tephra fall reached the coast of north Iceland on 22 December. The volume of the north trending tephra sector is not known (based on Gudmundsson *et al.* 2000; Sigmarsson *et al.* 2000).

Gjálp 1996

Eruption site: Gjálp in Vatnajökull, Grímsvötn volcanic system. Eruption began around 10 p.m. on 30 September 1996 and lasted until the 13 or 14 October.

The subglacial opening stage was followed by subaerial explosive activity, the only product being basaltic–andesitic tephra. Maximum height of eruption cloud observed from aircraft was about 9 km a.s.l. Main tephra dispersal was towards NW and minor tephra fall reached the coast of north and NW Iceland. The volume of the tephra layer was estimated at about 0.01 km^3 (based on Gudmundsson *et al.* 1997, 2004; Sigmarsson *et al.* 2000; Steinthorsson *et al.* 2000).

Hekla 1991

Eruption site: Hekla central volcano, Hekla volcanic system. Eruption began at 5 p.m. on 17 January 1991 and lasted until 11 March.

The Plinian opening stage emitted basaltic–andesitic tephra and was followed by mixed

Fig. 3. Relvant eruptions in the period 2012–1940. Red, Hekla; blue, Grímsvötn. Arrows indicate routes of ocean rafted pumice. Small circles with numbers indicate locations where mass of deposited tephra is known near or off the north coast, ' + ' and 'r' indicate confirmed and reported traces of tephra, respectively. Estimated mass of *c.* 300 g m^{-2} is assigned to localities where footprints were traceable after light tephra fall, in accordance with Thorarinsson (1955).

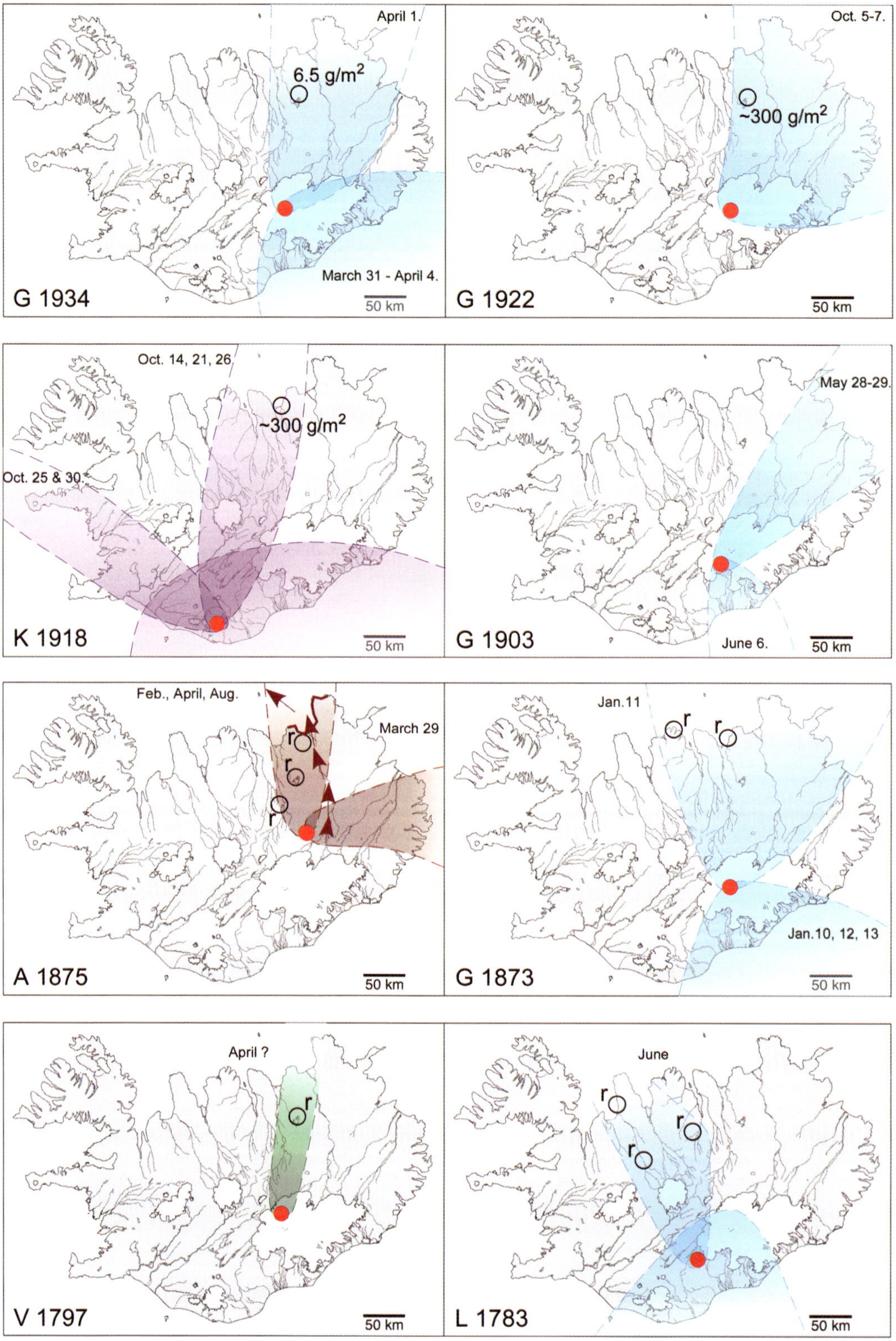
April 1.
6.5 g/m2
March 31 - April 4.
G 1934
50 km
Oct. 5-7.
~300 g/m2
G 1922
50 km
Oct. 14, 21, 26
~300 g/m2
Oct. 25 & 30.
K 1918
50 km
May 28-29.
G 1903
June 6.
50 km
Feb., April, Aug.
March 29
r
r
r
A 1875
50 km
Jan.11
r
r
Jan.10, 12, 13
G 1873
50 km
April ?
r
V 1797
50 km
June
r
r
r
L 1783
50 km

explosive and effusive activity. Maximum observed height of the eruption cloud was about 12 km a.s.l. The main tephra-producing phase lasted a few hours; tephra was carried NNE onto the shelf off north Iceland along a sector with axis east of Grímsey Island, where about 1 g m^{-2} was deposited. The tephra layer was small, the calculated volume being 0.02 km^3 (based on Gudmundsson *et al.* 1992; Larsen *et al.* 1992).

Hekla 1980

Eruption site: Hekla central volcano, Hekla volcanic system. Eruption began at *c.* 1.27 p.m. on 17 August 1980 and ended on 20 August.

The Plinian opening phase emitted basaltic–andesitic tephra and was followed by mixed explosive and effusive activity. Maximum observed height of the eruption cloud was 15 km a.s.l. The main tephra-producing phase lasted 4–5 h. The tephra was carried due north to the coast, where 1–2 g m^{-2} were deposited and onto the shelf off north Iceland along a sector mainly west of Grímsey Island. The tephra layer was of moderate size, the calculated volume being about 0.06 km^3. Activity was resumed in April 1981 and lasted 7 days. Weak explosive activity in several short-lived phases dispersed tephra up to 40 km from source. Maximum cloud height was only 6.6 km a.s.l. as observed by radar (based on Grönvold *et al.* 1983).

Hekla 1970

Eruption site: flanks of Hekla central volcano, Hekla volcanic system. Eruption began between 9.20 and 9.25 p.m. on 5 May and lasted 2 months.

The Plinian/subplinian opening phase emitted basaltic–andesitic tephra from two separate fissures and was followed by mixed explosive and effusive activity. Maximum observed height of the eruption cloud was about 16 km a.s.l. The main tephra-producing phase lasted 4–5 h. The tephra was carried NNW to the coast where up to 160 g m^{-2} were deposited and onto the shelf off north and NW Iceland where up to 15 g m^{-2} were deposited at about 66°50′N. The tephra layer was a moderate one, the calculated volume being about 0.07 km^3.

Rafts of pumice were detected along the NW coast of Iceland (Raudisandur) in May 1970 and off and along the north coast in late June and July 1970 (Strandir to Skjálfandaflói). This was assumed to be Hekla tephra that was transported down to the sea by rivers in south Iceland during and after the Hekla eruption (based on Thorarinsson 1970; Thorarinsson & Sigvaldason 1972).

Surtsey Fires 1963–1967

Eruption site: Vestmannaeyjar archipelago, Vestmanneyjar volcanic system. The eruption was first observed on 14 November 1963 and lasted intermittently until 4 June 1967.

The eruption episode beginning on 14 November 1963 consisted of several events. The eruption sites where permanent or temporary islands were built above sea-level were Surtsey in 1963–1967, Syrtlingur in 1965 and Jólnir in 1966. In all instances the activity was submarine for some time before explosive activity broke the surface of the sea. At Surtsey the explosive activity continued for 4.5 months and built a double tephra cone before the activity turned effusive. At Syrtlingur and Jólnir the explosive activity died down after 5 and 7 months and both islands disappeared below sea-level shortly after.

The tephra produced at these eruption sites is estimated to have exceeded 0.6 km^3. Airborne tephra was not widely spread but reached Heimaey, 20 km to the NE. Rafts of pumice were frequently observed around the eruption sites and such pumice repeatedly reached the shores of Reykjanesskagi, 80–120 km to the WNW. Sea-transported pumice was reported on beaches in SW Iceland (Álftanes) and NE Iceland (Melrakkaslétta) in 1965 and in north Iceland (Höfdaströnd) in late summer 1966. It was assumed that these rafts were the product of the Surtsey Fires (based on Morgunbladid 1965, 6 July, p. 2; Morgunbladid 1966, 10 September, p. 32; Thorarinsson 1965*a*, 1966, 1968; Gudmundsson & Ingólfsson 1968; Jakobsson & Moore 1982; Thordarson & Sigmarsson 2009).

Hekla 1947

Eruption site: Hekla central volcano, Hekla volcanic system. Eruption began around 6.40 a.m. on 29 March and lasted 13 months.

The Plinian opening phase emitted andesitic to basaltic–andesitic tephra and was followed by mixed explosive and effusive activity. Maximum observed height of the eruption cloud during the

Fig. 4. Relevant eruptions in the period 1940–1780. Blue, Grímsvötn; green, Veidivötn–Bárdarbunga; violet, Katla; brown, Askja. Arrows indicate routes of river and ocean rafted pumice. Small circles with numbers indicate locations where mass of deposited tephra is known near or off the north coast, and 'r' indicates reported traces of tephra. Estimated mass of *c.* 300 g m^{-2} is assigned to localities where footprints were traceable after light tephra fall, in accordance with Thorarinsson (1955).

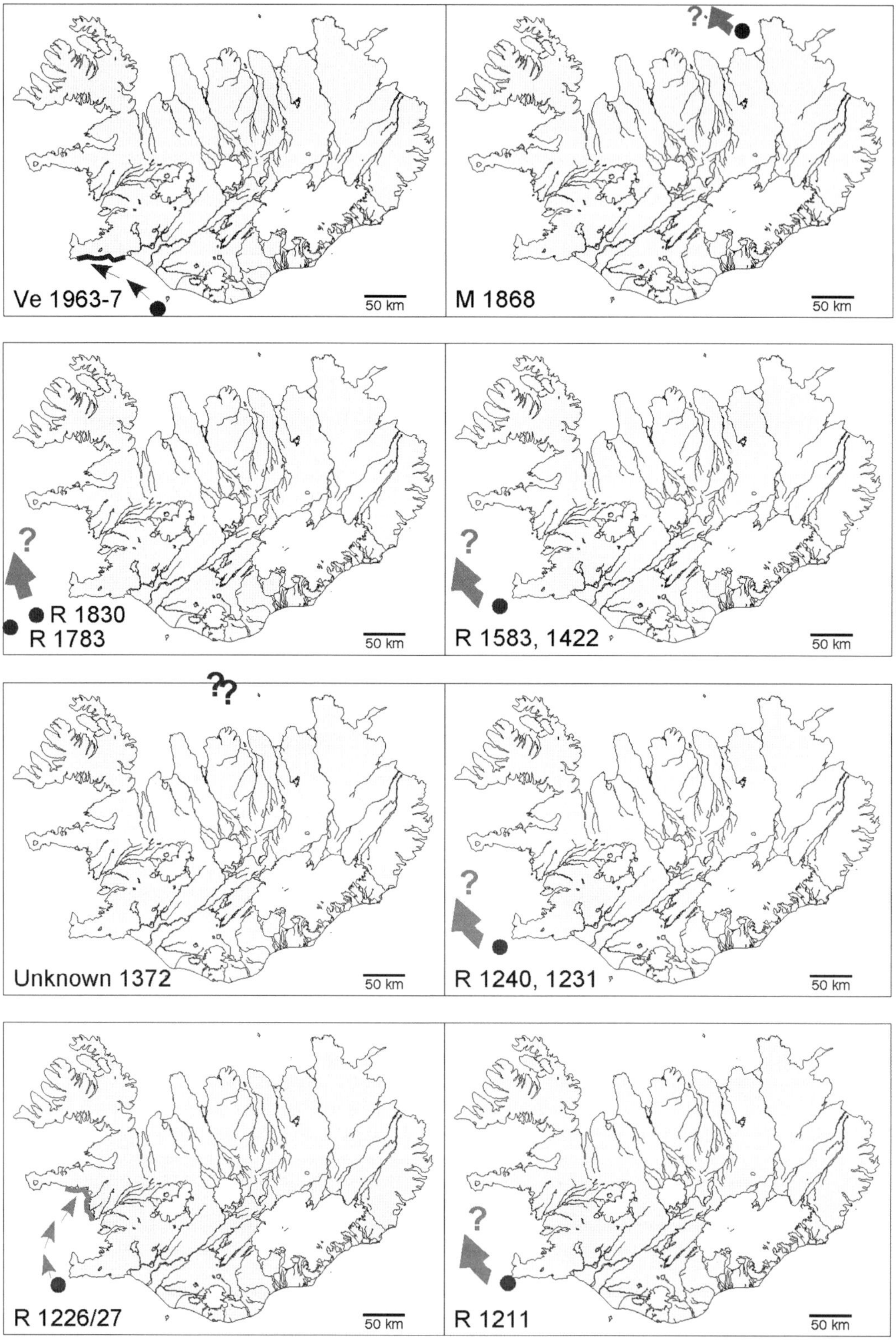
Ve 1963-7
50 km
M 1868
50 km
R 1830
R 1783
50 km
R 1583, 1422
50 km
Unknown 1372
50 km
R 1240, 1231
50 km
R 1226/27
50 km
R 1211
50 km

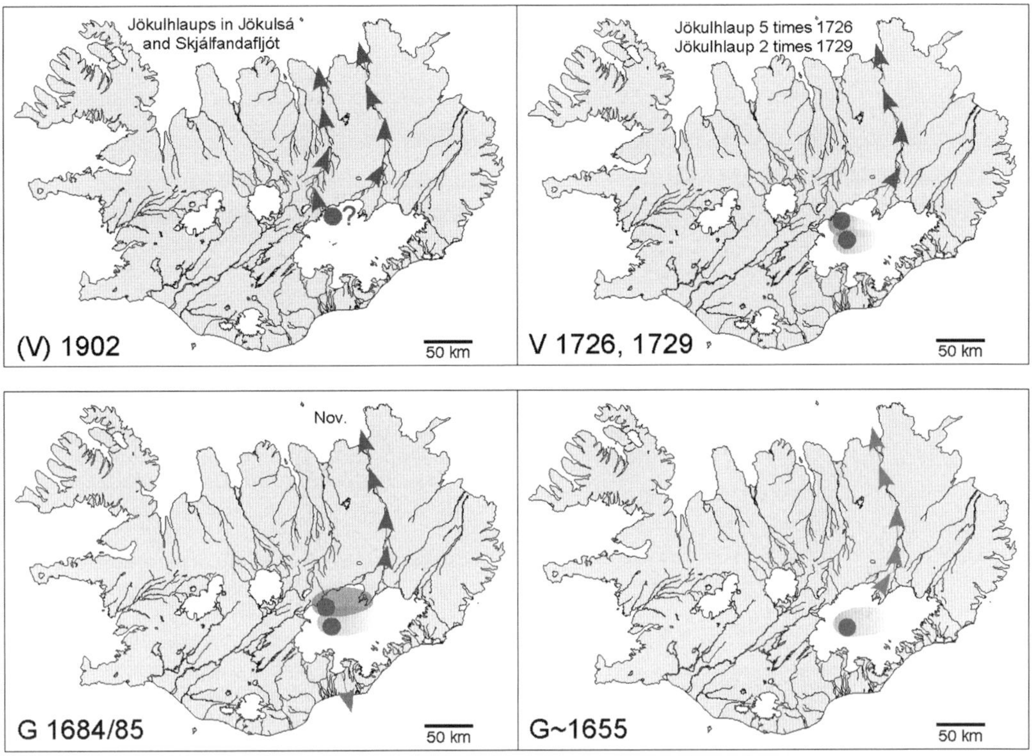

Fig. 6. Routes of relevant northbound jökulhlaups caused by subglacial eruptions during the last millennium.

opening phase was about 30 km a.s.l. The main tephra-producing phase lasted about an hour, but strong explosive activity continued into the afternoon and was significant during the next two days. The tephra produced during these first days was carried due south and to the European mainland. Currents carried some of the tephra that was deposited into the sea westwards along the coast. The tephra layer was an average one, the calculated volume being about 0.18 km^3.

Weaker explosive activity continued intermittently during the next 2–3 months and tephra was carried towards west and north on several occasions but so far no information exists on tephra fall in west or north Iceland or off the north coast. However, rafts of pumice and ash that resembled shoals of herring when seen from a distance were observed off north Iceland in June 1947. It was assumed that these rafts came from the tephra that fell offshore on March 29 (based on Thorarinsson 1954, 1967; Morgunbladid 1947, 14 June, p. 12).

Grímsvötn 1934

Eruption site: Grímsvötn caldera, Grímsvötn volcanic system. The eruption was first observed on 30 March and lasted until the middle of April.

Presumably the eruption had a subglacial opening stage that was followed by subaerial explosive activity. The only known product was basaltic tephra. The maximum observed height of eruption cloud was about 12.8 km a.s.l. Tephra dispersal was towards SE, east and north. In north Iceland 6 g m^{-2} to perhaps 300 g m^{-2} of tephra were deposited 30–60 km inland. Tephra fall in the coastal areas and offshore must have been considerably less. The

---

Fig. 5. Relevant submarine eruptions during the last millennium. Arrows indicate known or possible transport of ocean rafted pumice. A branch of the Irminger current and the coastal currents flow clockwise along the south, west and north coasts. The strength of the coastal currents changes seasonally (J. Ólafsson pers. comm.) which may partly explain why pumice from some eruptions, but not other similar ones, reaches the seas off NW and north Iceland).

estimated volume of tephra is 0.01–0.02 km^3 (based on Áskelsson 1936; Thorarinsson 1974).

Grímsvötn 1922

Eruption site: Grímsvötn caldera, Grímsvötn volcanic system. The eruption began on 2 October and lasted at least until 23 October.

Presumably the eruption had a subglacial opening stage that was followed by subaerial explosive activity. The only known product was basaltic tephra. Tephra dispersal was mainly towards the north, NE and east from 5 to 7 October, across north and NE Iceland and off the coast. Minor tephra fall (dust) was reported in NW Iceland. The tephra volume erupted in October was at least 0.015 km^3.

A predominantly effusive eruption that began in November on the Askja volcanic system may have contributed slightly to the tephra deposits in NE Iceland later on (based on Thorkelsson 1923; Thorarinsson 1974).

Katla 1918

Eruption site: Katla caldera in Mýrdalsjökull, Katla volcanic system. The eruption began on 12 October and lasted for three weeks.

Presumably the eruption had a short subglacial opening stage that was soon followed by strong hydromagmatic explosive activity producing basaltic tephra. Explosive activity was most intense during the first days and continued intermittently throughout the eruption. The tephra was dispersed widely and tephra fall reached north Iceland on 14 October and also later during the eruption. On 26 October tephra deposition in Akureyri and Húsavík village on the north coast of Iceland may have exceeded 300 g m^{-2} (footprints traceable). The tephra layer was fairly large, the estimated volume being about 0.7 km^3 (based on Morgunbladid 1918, 15 October, p. 1; Morgunbladid 1918, 27 October, p. 1; Sveinsson 1919; Eggertsson 1919; Thorarinsson 1955, 1975; Larsen 2000).

Grímsvötn 1903

Eruption site: Thordarhyrna, Grímsvötn volcanic system. The eruption was first observed on 28 May and continued intermittently for over 7 months.

Presumably the eruption had a subglacial opening stage that was followed by subaerial explosive activity. The only known product was basaltic tephra. The maximum height of eruption cloud is not known. Tephra production was most intense during the first two days and the tephra was carried towards the NE, over NE and east Iceland and off the coast. Tephra fall was reported by a ship 30 miles east of the coast, off Langanes. Tephra was also dispersed towards SSE. The tephra layer was an average one, the estimated volume on land being about 0.1 km^3 (based on Thoroddsen 1924; Mouritzen & Noe-Nygaard 1950; Thorarinsson 1974).

Summary 2005–1901 and occurrences outside Iceland

Tephra from 11 eruptions was carried northwards across Iceland and onto the shelf off north Iceland in the period 1900–2005 (Figs 3 & 4). These are the G 2004, H 2000, G 1998, Gjálp 1996, H 1991, 1980, 1970, G 1934, 1922, K 1918 and G 1903, which had a northeasterly trend. Pumice rafts attributed to the H 1970, H 1947 and the Surtsey 1963–1967 eruptions were reported off north Iceland.

All the northbound Hekla tephra layers are small, between 0.01 and 0.07 km^3. Where data exists on tephra deposition offshore in the Hekla eruptions the mass is 1–15 g m^{-2}, and north of 66°N the range is 1–80 g m^{-2}. Tephra from Grímsvötn eruptions forms thinner 'tephra veils' in north Iceland than the tephra from Hekla eruptions, partly because of the different type of explosive activity and in some cases because the main tephra dispersal was in other directions. In the 1918 Katla eruption, tephra fall in north Iceland may have exceeded 300 g m^{-2} in Húsavík (footprints traceable) at 66°N. This eruption was the largest of the twentieth century.

Tephra from two eruptions, H 1947 and K 1918, has been identified in marine sediments (Table 1; Sejrup *et al.* 2011; Gudmundsdóttir *et al.* 2012*a*). H 1947 and G 1903 have also been found in terrestrial archives outside Iceland (Wastegård & Davies 2009). The tephra from H 1947 eruption identified in marine sediments on the north Icelandic shelf is most likely due to pumice rafting (Fig. 4).

The best candidates for twentieth-century isochrones in the marine realm north of Iceland are probably the H 2000, 1970 and K 1918 tephra.

Tephra-producing eruptions 1900–1501

The documentation of volcanic events in the period AD 1900–1501 improves after AD 1600 and is reasonably good for the last two centuries (Fig. 2). Eruption frequency was exceptionally low between AD 1600 and 1500, with only eight or nine explosive or partly explosive eruptions on record. This is, however, not caused by poor documentation because only a few tephra layers are found in soil sections and in Vatnajökull ice during this period.

Table 1. *Tephra layers from the last millennium in marine sediments on the north Icelandic shelf*

Age (cal BP)	Volcanic system	Tephra markers	MD99-2269	HM107-04	MD99-2271	HM107-03	MD99-2273 & B05-2006-MC03	HM107-01	B997-319GGC	B997-319	BB97-316	B997-320	B997-321	MD99-2275 & B05-2006-MC04
3	**Hekla**	**H 1947**					×							×
70	Veidiv–Bárdarb or Reykjanes*													×
75	Veidiv–Bárdarb?													×
111	Veidiv–Bárdarb*													×
119	Veidiv–Bárdarb or Reykjanes													×
153	**Veidiv–Bárdarb**	**V 1797**												×
175	Veidiv–Bárdarb													×
184	**Hekla**	**H 1766**					×							
184	**Veidiv–Bárdarb**	**V 1766**					×							
215	Veidiv–Bárdarb*													×
233	**Veidiv–Bárdarb**	**V 1717**					×							×
257	**Hekla**	**H 1693**					×							
310	Grímsvötn	G ~ 1640					×							
336	Veidiv–Bárdarb													×
406	Hekla*													×
439	Veidiv–Bárdarb?	V ~ 1511												×
450	Grímsvötn	G ~ 1500					×							
470	Grímsvötn	G ~ 1480					×							
473	**Veidiv–Bárdarb**	**V ~ 1477**												×
539	Veidiv–Bárdarb	V ~ 1410												×
580	**Unspecified system on Kolbeinsey ridge**	**KOL 1372**	×			×		×			×			
650	**Hekla**	**H 1300**			×		×							×
850	**Hekla**	**H 1104**	×	×	×	×	×	×	×	×		×	×	×

Absolutely dated tephra layers or tephra markers are in bold. Not all tephra layers shown in the table are discussed in the paper. Asterisks indicate tephra horizons that are not robust. Adapted from Gudmundsdóttir *et al.* (2012*b*).

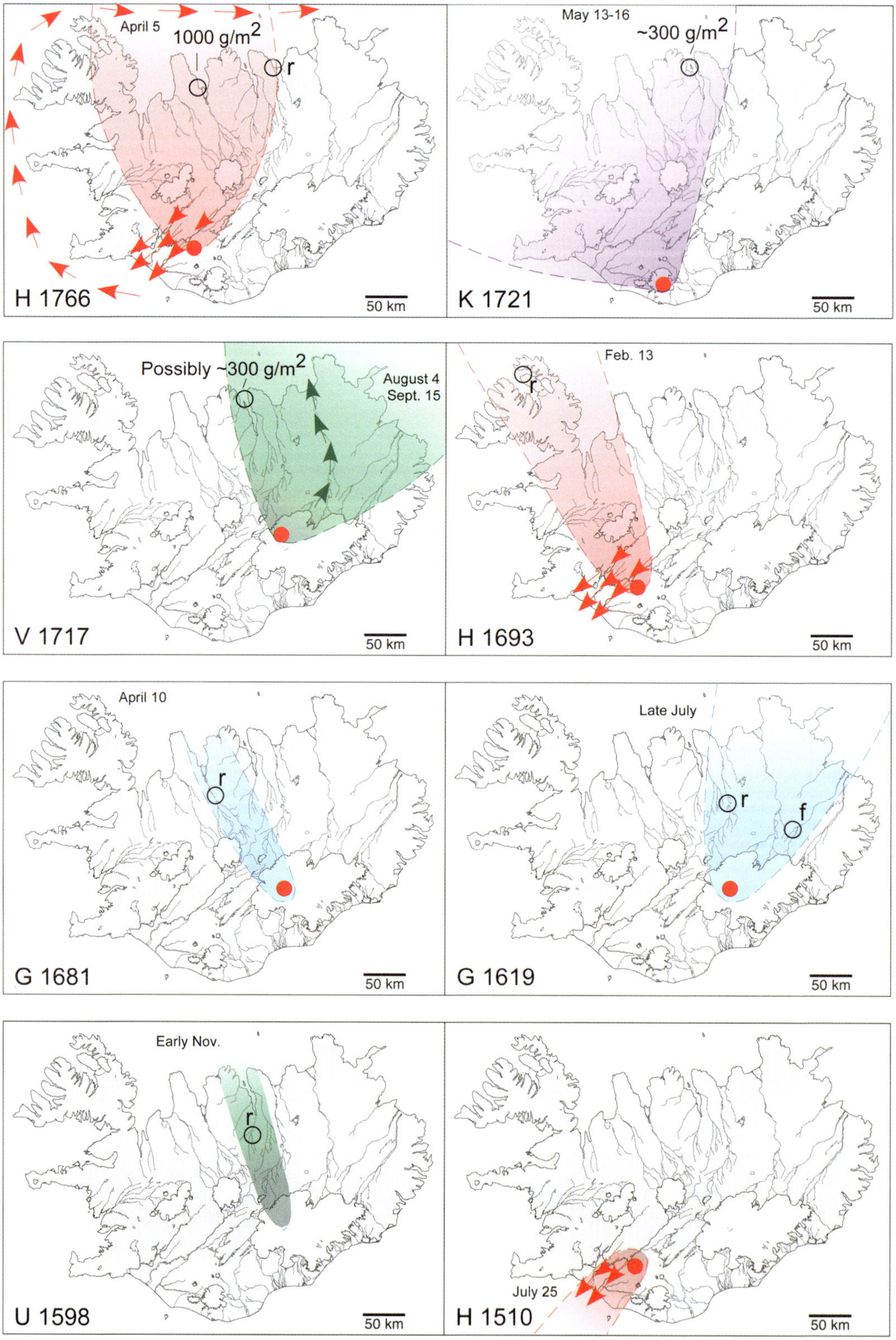
April 5
1000 g/m2
r
H 1766
50 km
May 13-16
~300 g/m2
K 1721
50 km
Possibly ~300 g/m2
August 4
Sept. 15
V 1717
50 km
Feb. 13
r
H 1693
50 km
April 10
r
G 1681
50 km
Late July
r
f
G 1619
50 km
Early Nov.
r
U 1598
50 km
July 25
H 1510
50 km

Verified and reported explosive and partly explosive eruptions in the period 1900–1501 are at least 84, counting the 1783–1785 Skaftá Fires as a single eruption. Over 20 eruptions are 'reported only', which means that corresponding tephra layers have not been found. The tephra layers of the remaining 63 eruptions range from small to large (0.01–1 km^3). Of them 16 have a fairly wide dispersal: Askja 1875; Hekla 1845, 1766, 1693, 1636, 1597, 1510; Katla 1755, 1721, 1660, 1625; Grímsvötn 1873, 1783 (Skaftá Fires), 1619; and Veidivötn–Bárdarbunga 1717.

Airborne tephra from at least 11 eruptions was carried northwards across Iceland and potentially onto the shelf off north Iceland during this period (Figs 4 & 7). Below are short descriptions together with relevant references. Also included are eruptions that delivered large quantities of pumice into the sea through tephra fall or by river transport and eruptions at sea (Fig. 5). Volcanogenic jökulhlaups in Jökulsá á Fjöllum are included (Fig. 6) because flood water is known to extend tens of kilometres into the sea at Axarfjördur.

Reykjanes ridge 1879

Eruption site: Geirfuglasker volcanic system? Eruption began on 30 May and lasted intermittently until mid-June.

Submarine eruption occurred SW of Reykjanes, possibly 30 km offshore. The eruption plume was seen from Hafnir village and in mid-June a minor tephra fall was reported there. The eruption was apparently a small one, the eruption site was never visited and pumice is not mentioned. Rafting of pumice by ocean currents (branch of Irminger current or coastal currents flowing clockwise along the west and north coast) towards the north can, however, not be excluded, given the location of the eruption site (based on Thorarinsson 1965*b*; Jakobsson 1974; Jakobsson *et al.* 2008).

Askja 1875

Eruption site: Askja volcanic system, central volcano and fissure swarm. Activity at Askja central volcano in Dyngjufjöll began in January 1875.

A minor (basaltic) tephra fall from Dyngjufjöll was reported in Kelduhverfi, Axarfjördur, some time before mid-February. The phreatoplinian and Plinian eruption on 29 March emitted close to 1.8 km^3 of silicic tephra which was carried towards the east, the tephra fall reaching Scandinavia within 24 h. The eruption left thick deposits of silicic pumice in the uppermost part of the drainage basins of Jökulsá á Fjöllum and its tributary river Kreppa, which transported substantial quantities of pumice into Axarfjördur. The pumice washed up onto the shore from Tjörnes in the west to Thistilfjördur in the east. In the spring of 1875 pumice-rafts were reported north of Grímsey Island. Jökulsá still transports small amounts of pumice from the 1875 eruption into Axarfjördur. Minor tephra fall was reported in Mývatnssveit on 5 April and in Bárdardalur on 14 August, either from Askja or from the predominantly effusive eruption episodes on the Sveinagjá fissure swarm (based on Nordanfari 14 1875; Ísafold II, 9 1875; Jónsson 1945; Thorarinsson 1963; Sparks *et al.* 1981; Carey *et al.* 2010; Hartley &Thordarson 2012; and references in these papers).

Grímsvötn 1873

Eruption site: Grímsvötn central volcano, most probably the Grímsvötn caldera. Eruption began on 8 or 9 January 1873.

Presumably the eruption had a subglacial opening stage that was followed by subaerial explosive activity. The only known product was basaltic tephra. Tephra production was most intense between 9 and 13 January, in particular on 10 January, but the eruption lasted intermittently until August. Tephra dispersal was towards the north and NE on 9, 10 and 11 January, with tephra fall in north and NE Iceland (e.g. Siglufjördur, Húsavík) and, presumably, off the coast. Tephra was also carried to the SE and east. The tephra volume erupted in January is estimated at several hundred million cubic metres (average to fairly large), and the 1873 eruption is one of the largest Grímsvötn eruption on record (based on Thorarinsson 1974; and references therein).

Mánáreyjar area 1868

Likely eruption site: Theistareykir volcanic system north of Mánáreyjar Islands (possibly a separate system, Mánáreyjar). Activity was reported in January 1868.

Fig. 7. Relevant eruptions in the period 1780–1501. Red, Hekla; blue, Grímsvötn; green, Veidivötn–Bárdarbunga; violet, Katla. Arrows indicate routes of ocean rafted pumice, river rafted pumice and jökulhlaup routes. Small green dots indicate core locations. Small circles with numbers indicate locations where mass of deposited tephra is known near or off the north coast; 'f' and 'r' indicate confirmed and reported traces of tephra, respectively. Estimated mass of *c.* 300 g m^{-2} is assigned to localities where footprints were traceable after light tephra fall, in accordance with Thorarinsson (1955).

Descriptions of the activity north of Mánáreyjar are scarce, only mentioning 'fire' rising from the sea. Neither tephra nor pumice rafts are mentioned. It is, however, included here because the eruption site may have been 40 km or less from the coring sites around Grímsey Island (based on Thorarinsson 1965*b*; Nordanfari 7 1868; Jakobsson *et al.* 2008; Magnúsdóttir & Brandsdóttir 2011).

Reykjanes ridge 1830

Eruption site: Eldeyjarbodi volcanic system? Eruption began on 13 March and lasted intermittently for a year.

Submarine eruption occurred SW of Reykjanes, possibly 55–60 km offshore. Large eruption plume was evident from Reykjavík in late March. Great quantities of scoria and pumice were formed at the eruption site. Rafting of pumice by ocean currents (branch of Irminger current or coastal currents flowing clockwise along the west and north coast) towards the north is possible but no reports are known (based on Thorarinsson 1965*b*; Jakobsson *et al.* 2008).

Veidivötn–Bárdarbunga 1797

Eruption site(s): Veidivötn–Bárdarbunga volcanic system, subglacial part below NW Vatnajökull. Some activity was reported in winter (possibly April) 1797 and observed in July–August 1797.

Presumably the eruption had a subglacial opening phase but without an ensuing jökulhlaup. The observed activity in July–August was in the NW corner of Vatnajökull, apparently a rather small hydromagmatic eruption featuring three separate plumes. Minor tephra fall in NE Iceland apparently occurred earlier in 1797 (based on Thorarinsson 1974; and references therein).

Reykjanes Ridge 1783

Eruption site: Gullhóll volcanic system, Reykjanes Ridge. Activity was first reported on 1 May 1783.

Submarine eruption occurred SW of Reykjanes, about 150 km offshore. Surtseyjan activity gradually built a small island, given the name Nýey, which disappeared within a year. Huge amounts of floating pumice were reported west of Reykjanes in May 1783, extending over an area 20–25 miles wide. Most authors assume that these are Danish nautical miles, 7.4 km, in which case the pumice covered a 150–200 km-wide area. In case these were 1.85 km-long nautical miles, the area is 37–47 km wide. The rafts were so thick that ships were slowed down. Rafting of pumice by ocean currents further towards the north is possible but no reports are known (based on Thoroddsen 1925; Thorarinsson 1965*b*; Jakobsson 1974, Jakobsson *et al.* 2008; and references in these papers).

Lakagígar 1783–85

Eruption site: Grímsvötn volcanic system, at the Lakagígar fissure and Grímsvötn central volcano. Eruption began on 8 June 1783 and continued until May 1785.

The atmospheric pollution and haze resulting from the large Skaftáreldar fissure eruption in 1783–1784 was heavily felt in north and NE Iceland, but tephra fall is only reported in locations in central and north Iceland. Tephra fell on Kjölur, central Iceland, on 24 July 1783. In Skagaströnd minor tephra fall (greyish sand) is reported in 1783 but no date is given. Official reports from Húnavatnssýsla and Strandasýsla mention that fine sand had fallen 'here and there' without further specifications, presumably in the summer 1783. Official reports from Eyjafjördur area mention that fine black sand accompanied the haze the southern part of the district in the summer 1783 (based on Gunnlaugsson *et al.* 1984; Thordarson 1991; Thordarson & Self 1993, 2003; and references in these papers).

Hekla 1766

Eruption site: Hekla volcanic system, Hekla central volcano. Eruption began on 5 April 1766 and lasted intermittently until April 1768.

The Plinian opening stage, lasting 5–6 h, emitted andesitic to basaltic–andesitic tephra and was followed by mixed explosive and effusive activity. Tephra was carried towards the north and tephra fall was reported in NW and north Iceland from Strandasýsla to the western part of north Thingeyjasýsla. Tephra was probably deposited on the shelf off north Iceland along a sector from Hornstrandir in the west to Skjálfandaflói in the east. The axis of thickness crossed the Skagafjördur district and its continuation on the shelf probably lay some 50 km west of Grímsey Island. It is likely that some air-fall tephra was deposited to the east of Grímsey Island as well. Assuming that contemporary reports on tephra thickness in Skagafjördur are reliable, the minimum mass deposited there per square metre was 1000 g. The darkness caused by the tephra fall was so intense that two fishing boats collided at sea. The tephra layer has a calculated volume of 0.4 km^3.

The tephra sector crosses the rivers Y-Rangá and Thjórsá just north of Hekla. So much pumice was deposited into Y-Rangá that it was temporarily dammed and both rivers carried huge quantities of pumice to the sea off the south coast. The pumice rafts were too thick for rowing boats to pass through. Coastal currents transported the pumice rafts westwards and then clockwise along the coast to the shelf off north Iceland. Using the 1947 Hekla eruption as an analog, this is likely to have occurred within 3–4 months from the beginning of the eruption. It is possible that sea-transported grains of pumice and ash have been deposited in marine sediments outside (and within) the areas affected by the tephra fall (based on Thorarinsson 1967; and references therein).

Unknown 1766 (V 1766)

Eruption site: unknown – presumably within western part of Vatnajökull ice cap. Eruption began in July, duration unknown.

An eruption was observed on 24 July in the direction of Vatnajökull ice cap. A bluish haze was reported extending from Vestmannaeyjar in the south to Borgarfjördur in the west, accompanied with minor ash fall. No further information exists on the tephra fall but several jökulhlaups in Thjórsá River were reported during the summer 1766, probably via its tributaries Tungnaá or Kaldakvísl rivers. The tephra has been identified in Vatnajökull ice and has the chemical characteristics of the Veidivötn–Bárdarbunga system (based on Thorarinsson 1974; Steinthorsson 1978; Larsen *et al.* 1998).

Veidivötn–Bárdarbunga 1729 and 1726

Likely eruption site(s): Veidivötn–Bárdarbunga volcanic system, subglacial part below NW Vatnajökull, in 1729 and 1726. Simultaneous activity may have occurred somewhere on the Grímsvötn volcanic system in 1726.

Jökulhlaups, almost certainly of volcanogenic origin, were reported in Jökulsá á Fjöllum in winter and late summer 1729 and five times in late winter–early spring 1726. Reports on the volcanic activity within Vatnajökull are poor but and tephra fall appears to have been confined to the ice cap. However, the reports on the jökulhlaups are considered reliable. Jökulsá discharges into Axarfjördur and at least one of those jökulhlaups carried pumice down to the sea.

During spring floods in Jökulsá á Fjöllum the silty water extends tens of kilometres out into the Axarfjördur Bay, reportedly even as far as Grímsey Island. Melt-water in volcanogenic jökulhlaups is likely to behave similarly and these floods are therefore included here. However, the pumice and sand-size material are not necessarily primary material from the eruptions that cause the jökulhlaups and may consist partly or mostly of remobilized material from older deposits along their course (based on Thorarinsson 1950; Steinthorsson 1978; Ísaksson 1985; Larsen *et al.* 1998; and references in these papers).

Katla 1721

Eruption site: Katla caldera in Mýrdalsjökull, Katla volcanic system. Eruption began on 11 May and lasted until autumn.

A short subglacial phase was followed by hydromagmatic explosive eruption producing basaltic tephra and a large jökulhlaup. Explosive activity was most intense during the first few days and the tephra was dispersed towards the west, NW and north. Tephra fall reached north Iceland on 13–16 May and may have exceeded 300 g m^{-2} in the northern part of Eyjafjördur (footprints traceable). The eastern limit of the tephra fall was most likely east of Grímsey Island. A large jökulhlaup discharged into the sea at the south coast but pumice rafting similar to that observed in Hekla eruptions is not reported. The estimated volume of the airborne tephra is about 0.3 km^3 (based on Thorarinsson 1955, 1975; Larsen 2000, 2010; and references in these papers).

Veidivötn–Bárdarbunga 1717

Eruption site: Veidivötn–Bárdarbunga volcanic system, subglacial part below NW Vatnajökull. The eruption occurred in summer 1717.

Presumably the eruption had a subglacial opening phase, followed by hydromagmatic explosive activity producing basaltic tephra and generating a substantial jökulhlaup. Tephra was dispersed to the north, NE and ENE and tephra fall was reported from Eyjafjörður in the west to the Fljótsdalshérad in the east. The tephra has been identified in soil and analysed. In the northern part of Eyjafjördur it may have exceeded 300 g m^{-2} (footprints traceable). The offshore extension is not known but the tephra fall in Eyjafjördur indicates that the western limit of the tephra fall area may lie some 50 km west of Grímsey Island.

The ensuing jökulhlaup followed the course of Jökulsá á Fjöllum and discharged into Axarfjördur (based on Jónsson 1945; Thorarinsson 1950; Larsen 1982).

Hekla 1693

Eruption site: Hekla volcanic system, Hekla central volcano. Eruption began on 13 February 1693 and lasted at least 7 months.

The Plinian opening stage, emitting andesitic to basaltic–andesitic tephra, was most intense during the first half an hour and tephra fall was mostly over in about 12 h. Tephra was carried towards the NW and tephra fall was reported in NW and north

Iceland, having an eastern limit at Skagaströnd, Húnavatnssýsla. The axis of thickness crossed the Hrútafjördur area. Tephra was probably deposited on the shelf off NW and north Iceland along a sector from Dýrafjördur in NW Iceland to Húnaflói in the north. The eastern limit of tephra fall on the shelf probably lay more than 100 km west of Grímsey Island. The tephra layer has a calculated volume of 0.3 km^3.

The tephra sector crossed the rivers Y-Rangá and Thjórsá just NW of Hekla and both rivers carried huge quantities of pumice to the sea off the south coast. The sea around the Vestmannaeyjar archipelago was, reportedly, completely covered by ash. It is likely, although no reports are known, that coastal currents transported the ash and pumice rafts westwards and then clockwise along the coast towards the shelf off north Iceland. It is therefore possible that sea-transported grains of ash and pumice were deposited in marine sediments outside (and within) the areas affected by the tephra fall (based on Thorarinsson 1967; and references therein).

Grímsvötn 1684/1685, 1681

Likely eruption site(s): Grímsvötn volcanic system, subglacial part. Simultaneous activity may have occurred somewhere on the Veidivötn–Bárdarbunga volcanic system.

Reports on the activity within Vatnajökull in this period are scant. Minor tephra fall was reported in Skagafjördur on 10 April 1681. It may have reached the coast, but deposition further north is unlikely. Grímsvötn are mentioned as the source but the source could be elsewhere within Vatnajökull. Jökulhlaups in Jökulsá á Fjöllum in 1684/1685 were also attributed to an eruption in Grímsvötn (based on Jónsson 1945; Thorarinsson 1974; and references therein).

Hekla 1636

Eruption site: Hekla volcanic system, Hekla central volcano. Eruption began on 8 or 15 May 1636 and lasted at least a year.

The opening phase may have been subplinian rather than Plinian and emitted basaltic–andesitic tephra towards the NE. Minor air-fall may have reached the shelf off east Iceland. The tephra layer has a calculated volume of 0.18 km^3.

The axis of thickness crossed Tungnaá River about 30 km NE of Hekla, one of the main tributaries of Thjórsá River. Substantial quantities of pumice must therefore have been rafted into Thjórsá River and into the sea off the south shore. One annal claims that drowning of men on a ferry on Thjórsá River was caused by the pumice. Although no reports describe pumice rafts off the coasts, the deposition of sea-transported grains of ash and pumice in marine sediments west and north of Iceland cannot be excluded (based on Thorarinsson 1967; and references therein; Larsen *et al.* 1999).

Grímsvötn 1619

Eruption site: Grímsvötn volcanic system, most probably the Grímsvötn caldera. Eruption began in July 1619.

The eruption presumably had a subglacial opening stage that was followed by subaerial explosive activity. The only known product was basaltic tephra. Tephra production was most intense in late July. Tephra dispersal was towards the north and NE with tephra fall in north and NE Iceland, as reported in Bárdardalur, and in Jökuldalur the tephra has been identified in soil and analysed. The tephra fall may have extended offshore east of Grímsey Island. Tephra fall also reached the Faroe Islands and Norway and the 1619 eruption is thought to be among the largest Grímsvötn eruptions, but no volume estimates exist (based on Thorarinsson 1974, 1981; Larsen 1982, Larsen *et al.* 1998; and references in these papers).

Unknown 1598

Likely eruption site: within Vatnajökul, but the location is unknown. The beginning and duration of this eruption are unknown.

Minor tephra fall was reported in Eyjafjördur in November 1598. It may have reached the coast and possibly a limited area offshore, to the west of Grímsey Island (based on Jónsson 1945; Thorarinsson 1974).

Reykjanes Ridge 1583

Eruption site: Reykjanes Ridge. The location is not known but may have been in the vicinity of Eldey.

Reports of this eruption are uncertain, but fire was seen at sea from a merchant ship. It is not known whether an island was formed. Rafting of pumice from a submarine eruption is possible but in this case not very likely (based on Thoroddsen 1925; Thorarinsson 1965*b*; and references therein).

Hekla 1510 and unknown ~1510

Eruption site: Hekla central volcano, Hekla volcanic system. Eruption began around 25 July 1510 but its duration is not known. An unknown eruption site in north Iceland was active around 1510, the beginning and duration not known.

Reports on this Hekla eruption are very scant, but tephrochronological studies indicate that its Plinian opening phase was similar to but more intense than that of the Hekla 1947 eruption. The andesitic to basaltic–andesitic tephra was carried towards SSW and great quantities must have been deposited at sea off the south coast. The rivers affected by the tephra fall may have added substantial quantities. The tephra layer has a calculated volume of about 0.32 km^3. It is possible that currents carried some of the tephra that was deposited offshore in clockwise fashion along the Icelandic coast, as happened in 1947. Given the magnitude of the eruption, some rafts of pumice and ash may have reached the shelf off north Iceland and deposition of grains in the marine sediments there cannot be excluded.

An eruption in the vicinity of the Herdubreid and Trölladyngja mountains, occurring at the same time as the Hekla eruption, is mentioned in a single annal. No further information exists but the eruption cloud must have been visible in populated areas. The eruption site could lie within the Askja volcanic system or adjacent volcanic systems such as Veidivötn–Bárdarbunga (based on Jónsson 1945; Thorarinsson 1954, 1967; and references therein).

Summary 1900–1501 and occurrences outside Iceland

Tephra from at least 11 eruptions was carried northwards across Iceland and onto the shelf off north Iceland in the period 1900–1501 but in five cases the tephra dispersal was limited (Figs 4 & 7). Air-fall tephra from the G 1873, H 1766, K 1721, V 1717 and G 1619 eruptions is most likely to have been deposited offshore, but three minor tephra falls may also have reached some parts of this area in 1875, 1797 and 1598.

The north-bound tephra layers have volumes up to 0.4 km^3. The two most distinct tephra layers in north Iceland from this period are the H 1766 and V 1717 tephra layers. At the north coast the mass per square metre is thought to be close to 300 g (footprints traceable).

Tephra from hydromagmatic Katla and Grímsvötn eruptions forms thinner 'tephra veils' in north Iceland than the tephra from Plinian Hekla eruptions, partly because of the different type of explosive activity and partly because the main tephra dispersal was towards other directions. In the 1721 Katla eruption tephra fall in Eyjafjördur, north Iceland, may have exceeded 300 g m^{-2} (footprints traceable).

Tephra from the eruptions V 1797, H 1766, an unknown eruption V 1766, V 1717 and an unknown eruption ~1511 has been reported in marine sediments on the north Icelandic shelf (Gudmundsdóttir *et al.* 2012*b* and references therein). Silicic tephra from Askja 1875 eruption has been identified in sediments off the western coast of Norway (Sejrup *et al.* 2011) and in terrestrial environments in Norway, Sweden and Germany (Wastegård & Davies 2009). It has not been identified in the marine sediments on the north Icelandic shelf but basaltic tephra is found at the 1875 level there, having the chemical characteristics of the Veidivötn–Bárdarbunga system.

Tephra fall from Lakagígar eruption 1783–1785 was observed in north Iceland and has been reported in glacier ice in Greenland and Svalbard (Fiacco *et al.* 1994; Kekonen *et al.* 2005; Abbott & Davies 2012).

Ocean-rafted pumice from the eruptions in Askja 1875 and Hekla 1766 is known to have reached the north Icelandic shelf. To what extent this resulted in deposition of volcanic material is not known. Although other occurrences have not been reported, such a possibility should not be entirely disregarded.

Tephra-producing eruptions 1500–1001

Contemporary reports mentioning volcanic eruptions in this period are more scarce and less detailed than in the following centuries. In many cases the only knowledge about an eruption is a tephra layer revealed by tephrochronological research. Eruption years indicated by '~' may be based on age calculations using SAR between tephra layers having a documented eruption year.

Amongst the 61 explosive or partly explosive eruptions known between 1001 and 1500 are seven Hekla eruptions with tephra volumes ranging from 0.04 to 2 km^3, the 1362 Öræfajökull eruption and the Veidivötn eruption with about 10 km^3 of tephra each, and nine Katla eruptions with volumes ranging up to 0.5 km^3. Some 30 eruptions occurred on the ice-covered parts of the Grímsvötn and Veidivötn volcanic systems. No volume estimates exist for these, but most of them are thought to be of small to moderate (0.01–0.1 km^3) volume.

Airborne tephra from at least seven eruptions was carried northwards across Iceland and potentially onto the shelf off north Iceland during this period (Fig. 8). Below are short descriptions together with relevant references. Also included are eruptions that delivered large quantities of pumice into the sea through tephra fall or by river transport and eruptions at sea (Fig. 5).

Grímsvötn ~1500

Eruption site: Grímsvötn volcanic system. No reports mention this eruption. The tephra has been

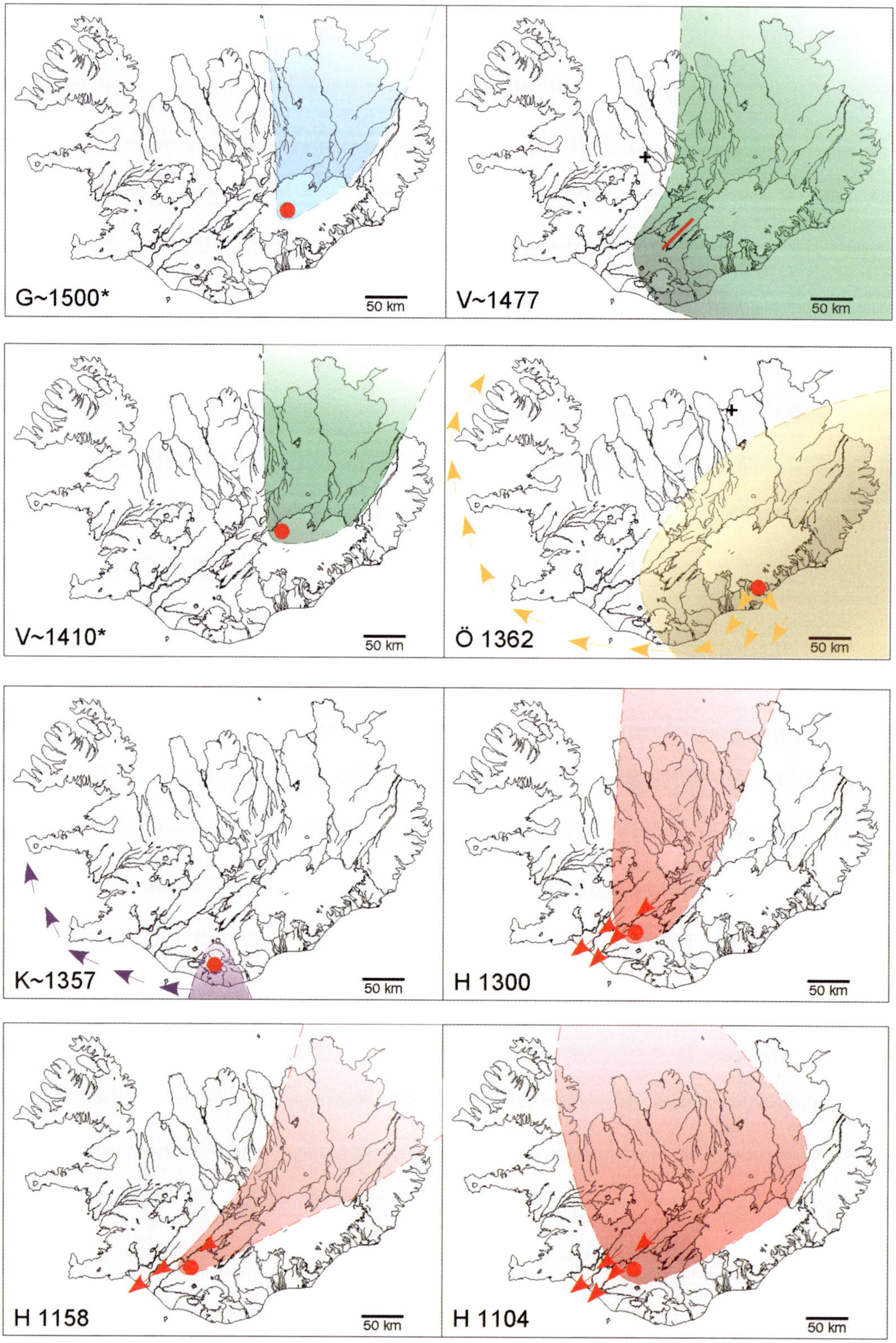
G~1500*
50 km
V~1477
50 km
V~1410*
50 km
Ö 1362
50 km
K~1357
50 km
H 1300
50 km
H 1158
50 km
H 1104
50 km

identified in soil and geochemically fingerprinted. Its age was calculated using SAR between tephra layers of known age.

The activity during this eruption is presumed to be typical for Grímsvötn eruption with a subglacial opening stage followed by subaerial explosive activity. The only known product was basaltic tephra. Tephra dispersal was towards the north and NE with tephra fall in north and NE Iceland, between Bárdardalur in the west and Jökuldalur in the east. The tephra fall may have extended offshore east of Grímsey Island. The volume of tephra has not been estimated (based on Larsen 1982; Larsen *et al.* 1998).

Veidivötn ~1477

Eruption site: Veidivötn–Bárdarbunga volcanic system, Veidivötn volcanic fissure on southwestern part of the system. Tephra fall documented in north Iceland in February 1477 is correlated to the early stages of the eruption. Simultaneous activity may have occurred on the subglacial part below NW Vatnajökull and also on the Grímsvötn volcanic system.

The Veidivötn eruption was a predominantly phreatomagmatic eruption producing over 10 km^3 of basaltic tephra. The tephra, which has been geochemically fingerprinted and mapped in detail, was carried to the north, NE and east. The 1 cm isopach extends offshore, indicating extensive deposition on the shelf off north, NE and east Iceland. The tephra sector has a sharp western limit on land indicating that its boundary on the shelf lies some tens of kilometres west of Grímsey Island.

Some subglacial activity below NW Vatnajökull, generating jökulhlaup in River Jökulsá á Fjöllum, may have occurred simultaneously with the eruption on the southwestern part. The presence of subordinate tephra component having Grímsvötn composition in part of the Veidivötn tephra layer indicates simultaneous or near-simultaneous eruption on that volcanic system (based on Benjamínsson 1982; Larsen 1982; Larsen *et al.* 2002).

Katla ~1440

Eruption site: Katla volcanic system. No reports mention this eruption. The tephra has been identified in soil and geochemically fingerprinted. Its age was calculated using SAR between tephra layers of known age.

The activity during this eruption is presumed to be typical for Katla eruption with a subglacial opening stage followed by explosive hydromagmatic activity. The only known product is basaltic tephra. The known tephra dispersal was towards the north. The tephra fall may have extended offshore towards Grímsey Island. The volume of this tephra layer is estimated as small to moderate (based on Larsen 2000).

Reykjanes Ridge 1422

Eruption site: Reykjanes Ridge. The location was off Reykjanes, but is not known.

This eruption is briefly mentioned in an old annal as fire at sea forming an island. Pumice rafting is not mentioned but is possible (Thoroddsen 1925; Thorarinsson 1965*b*; and references therein).

Veidivötn–Bárdarbunga ~1410

Eruption site: Veidivötn–Bárdarbunga volcanic system. No reports mention this eruption. The tephra has been identified in soil and geochemically fingerprinted. Its age was calculated using SAR between tephra layers of known age.

The activity is presumed to be typical for an eruption on the ice-covered part of this system, beginning with a subglacial opening stage followed by explosive hydromagmatic activity. The only known product was basaltic tephra. The known tephra dispersal was towards the north and NE with tephra fall in north and NE Iceland, between Bárdardalur in the west and Jökuldalur in the east. The tephra fall may have extended offshore east of Grímsey Island (based on Larsen 1982; Larsen *et al.* 1998).

Unknown 1372

Eruption site: Kolbeinsey Ridge? The location is not known.

Reports of this eruption are uncertain. A new island was reportedly seen NW of Grímsey Island from several localities in north Iceland but tephra fall is not mentioned. Rafting of pumice off north Iceland, in the case of an eruption there, is likely (based on Thoroddsen 1925; Thorarinsson 1965*b*; and references therein).

Fig. 8. Relvant eruptions in the period 1500–1001. Red, Hekla; blue, Grímsvötn; green, Veidivötn–Bárdarbunga; violet, Katla; yellow, Öræfajökull. Arrows indicate routes of ocean rafted pumice, river rafted pumice and jökulhlaup routes. Small green dots indicate core locations. Small circles with numbers indicate locations where mass of deposited tephra is known near or off the north coast; ' + ' and 'r' indicate confirmed and reported traces of tephra, respectively.

Öræfajökull 1362

Eruption site: Öræfajökull volcanic system, caldera of the central volcano. The eruption is described in the old annals and began in the spring, but the eruption year differs slightly between annals. The tephra has been identified in soil, mapped and geochemically fingerprinted.

The Öræfajökull eruption was Plinian with a phreatoplinian opening phase, producing one of the largest tephra layers of the last millennium. The thickness axis of the tephra trends towards the SW but tephra fall reached NE Iceland and possibly the north coast. Jökulhlaups carrying pumice, and possibly pyroclastic surges or flows, discharged into the sea at the south shore. Pumice rafts thick enough to slow down ships were reported off the western fjords. Such rafts are likely to have continued into the seas off north Iceland but no reports exist. Estimates of tephra volume differ but are up to 10 km^3 (based on Thorarinsson 1958; Jónsson *et al.* 2007; Karma *et al.* 2008).

Katla ~1357

Eruption site: Katla volcanic system, Katla caldera. The eruption is briefly mentioned in annals. The eruption site is misplaced but the tephra fall area is correctly placed. The tephra has been identified in soil, mapped and geochemically fingerprinted.

The activity during this eruption is presumed to be typical for Katla eruption with a subglacial opening stage followed by explosive hydromagmatic activity. The only known product was basaltic tephra. The known tephra dispersal was towards the south across Mýrdalur and off the south coast, which lies about 30 km south of the Katla caldera. Therefore, the tephra deposited offshore was relatively coarse and formed pumice rafts that were transported westwards and were reported as far as west as the Snæfellsnes peninsula. Pumice rafts may have travelled further north but no reports exist. The tephra volume is estimated at about 0.3 km^3 (based on Thorarinsson 1975; Einarsson *et al.* 1980; Larsen 2000).

Hekla 1341

Eruption site: Hekla volcanic system, Hekla central volcano. Eruption began on 19 May 1341 and lasted at least 2 months.

The Plinian opening stage was intense and emitted andesitic tephra that was mostly carried towards SE. This part of the tephra layer has been identified and mapped. Some tephra fall was reported in west and NW Iceland between Borgarfjördur and Skagi, but this tephra has not yet been identified in soil sections. No pumice rafting is mentioned in connection with this eruptions and is considered unlikely. Tephra volume is estimeated as 0.18 km^3 (based on Thorarinsson 1967; Larsen *et al.* 1999).

Hekla 1300

Eruption site: Hekla volcanic system, Hekla central volcano. Eruption began on 12 July 1300, and lasted nearly 12 months.

The Plinian opening stage was violent and emitted andesitic tephra that was carried towards the north. Tephra fall was reported from Skagafjördur to Axarfjördur and it was particularly stated that people in the area did not dare to go out on boats for two days. Tephra was deposited on the shelf off north Iceland along a sector from Skagi in the west to Melrakkaslétta in the east. The axis of thickness crossed the Skagafjördur district and its continuation on the shelf probably lay some tens of kilometres west of Grímsey Island. The tephra layer was fairly large and had a calculated volume of 0.5 km^3.

The tephra sector crosses the rivers Y-Rangá and Thjórsá just north of Hekla. So much pumice was deposited into Y-Rangá that it was temporarily dammed and both rivers carried huge quantities of pumice to the sea off the south coast. The pumice rafts were still on the ocean between Iceland and the Faroes in the summer of 1301. It is possible that coastal currents transported the pumice rafts westwards and then clockwise along the coast to the shelf off north Iceland. Using the 1947 Hekla eruption as an analog, this is likely to have occurred within 3–4 months from the beginning of the eruption. It is possible that sea-transported grains of pumice and ash were deposited in marine sediments outside (and within) the areas affected by the tephra fall (based on Thorarinsson 1967; and references therein).

Katla 1262

Eruption site: Katla volcanic system, Katla caldera. The eruption and tephra fall is briefly mentioned in annals. The tephra has been identified in soil, mapped and geochemically fingerprinted.

The activity during this eruption is presumed to be typical for Katla eruption with a subglacial opening stage followed by explosive hydromagmatic activity. The only known product was basaltic tephra. The known tephra dispersal was towards the (north and) NE across NE Iceland. It is still uncertain whether the tephra fall reached the north coast but in that case deposition east of Grímsey Island is possible. The tephra volume is estimated at about 0.5 km^3 (based on Thorarinsson 1975; Larsen 2000).

Reykjanes 1226/27; 1211

Eruption sites: Reykjanes volcanic system. Locations were on and off Reykjanes.

Between 1240 and 1211 annals mention several eruptions on the Reykjanes volcanic system (1240, 1238, 1231, 1227, 1226, 1223, 1211, 1210), in most cases off Reykjanes but one annal mentions an eruption at Reykjanes, that is, onshore. 'Sandfall-winter' is reported 1231 and a minor tephra layer erupted off Reykjanes is attributed to this eruption. Another 'sandfall-winter' is reported 1226 and 1227 (probably winter 1226/1227); the tephra is now known as the 'Midaldalag 1226', and is thought to have erupted several kilometres off Reykjanes. Remnants of two tephra rings at the westernmost tip of Reykjanes are thought to have formed in an eruption in 1211.

Pumice rafting is likely in all these eruptions, but those of 1226/1227 and 1211 are known to have produced large quantities of pumice that could have resulted in extensive pumice rafts. Sea-transported pumice, found on salt-marshes in west Iceland, has been correlated to the younger of these two eruptions (based on Thoroddsen 1925; Thorarinsson 1965*b*; Sigurgeirsson 1995; and references therein; Gehrels *et al.* 2006).

Hekla 1158

Eruption site: Hekla volcanic system, Hekla central volcano. Eruption most likely began on 19 January 1158, although some annals place it in 1157. The duration is unknown. The tephra has been identified in soil, mapped and geochemically fingerprinted.

A strong Plinian opening stage emitted dacitic tephra that was carried towards NE and was deposited from Axarfjördur in the north to Nordfjördur in the east. Tephra was probably deposited on the shelf off NE Iceland, but the western limit was at least 50 km east of Grímsey Island. The tephra layer has a calculated volume of 0.3 km^3.

The axis of thickness crossed Tungnaá River 30 km NE of Hekla and also followed a 40 km stretch of Kaldakvísl River, the two main tributaries of Thjórsá River. Substantial quantities of pumice must therefore have been rafted into Thjórsá River and into the sea off the south shore. Although no reports mention pumice rafts off the coasts, the deposition of sea-transported grains of ash and pumice in marine sediments west and north of Iceland cannot be excluded (based on Thorarinsson 1967; Larsen *et al.* 1999; Sigurgeirsson 2000; and references in these papers).

Hekla 1104

Eruption site: Hekla volcanic system, Hekla central volcano. The eruption most likely took place in AD 1104, the year in which the metropolitan see in Denmark was established, but one annal places the tephra fall in 1105 as 'sandfall-winter'.

A violent Plinian opening stage emitted rhyolitic tephra that was carried towards the north and NE and was deposited from Steingrímsfjördur in the north to Axarfjördur in the east. Tephra was deposited on the shelf off north Iceland in a nearly 250 km-wide sector. The tephra layer was large and has a calculated volume of about 2 km^3.

The axis of thickness crossed the Y-Rangá and Thjórsá rivers 10–20 km north of Hekla. Substantial quantities of pumice must therefore have been rafted into the sea off the south shore. Although no reports mention pumice rafts off the coasts, the deposition of sea-transported grains of ash and pumice in marine sediments west and north of Iceland cannot be excluded (based on Thorarinsson 1967; Larsen *et al.* 1999; and references in these papers).

Eruptions in the eleventh century

Only four explosive or partly explosive eruptions are known during this century – an all-time low of the last 11 centuries – and no tephra fall is known to have occurred in north Iceland. One possible reason for this lull in volcanic activity could be that two major eruptions on 75 and 65 km-long fissures, the Eldgjá ~934 and Vatnaöldur ~870, took place in the preceding 130 years. This may have relaxed the 'system' temporarily. As a result, the beginning of the millennium is poorly defined by tephrochronology.

Summary: 1500–1001 and occurrences outside Iceland

Tephra from at least seven eruptions, possibly eight, was carried northwards across Iceland and onto the shelf off north Iceland in the period 1500–1001 (Fig. 8). The largest basaltic tephra layer is that of the ~1477 Veidivötn eruption, with an estimated volume of *c.* 10 km^3. The largest silicic tephra layer is that of the 1104 Hekla eruption, with an estimated volume of *c.* 2 km^3. It forms the most widespread marker horizon of the last millennium on the north Icelandic shelf and northern North Atlantic and it is also the oldest known north-bound tephra layer of the last 1000 years. No reports of pumice rafting onto the north Icelandic shelf are known, although jökulhlaup into Axarfjördur in 1477 is suspected but has not been verified.

Tephra from the G ~1500, V ~1477, V ~1410, H 1300 and H 1104 eruptions has been identified in marine sediments on the north Icelandic shelf (Table 1; Gudmundsdóttir *et al.* 2012*b*). The V ~1477 tephra has been described from western Ireland and Sweden, H 1158 has been described from northern Ireland and northern Norway and H 1104 has been found in Ireland and northern Norway (Wastegård & Davies 2009 and references therein).

A tephra layer assigned to the eruption off north Iceland in 1372, KOL 1372, has been described in several marine cores on the north Icelandic shelf (Table 1; Gudmundsdóttir *et al.* 2012*b*). This tephra layer is a local marker and unlikely to be found outside the Iceland shelf.

Tephra from the Öræfajökull 1362 eruption has hitherto not been identified in marine sediments north of Iceland but has been reported from Ireland and northern Norway and in glacier ice in Greenland and on Svalbard (Palais *et al.* 1991; Wastegård & Davies 2009 and references therein; Schupack *et al.* 2010).

Summary and conclusions

The number of known explosive and partly explosive eruptions in Iceland during the last millennium is close to 170, including those that are 'reported only'. The average number of tephra-producing eruptions per century is thus 17 eruptions and the average interval between eruptions is five to six years. The eruption frequency has, however, varied between three and 29 eruptions per century during this time-slice, and the intervals between eruptions vary from a couple of weeks to at least 30 years. The volumes of tephra generated in these eruptions has also varied greatly, from less than 0.01 km^3 to over 10 km^3.

Airborne tephra in 25–30 eruptions had the potential to reach the north Iceland shelf and the seas beyond. Among them are the G 2004, H 2000, G 1998, G 1996, H 1991, G 1934, G 1922, K 1918, A 1875 (basaltic), G 1873, V 1797, L 1783, H 1766, K 1721, V 1717, G 1681, G 1619, U 1598, G ~1500, V ~1477, K ~1440, V ~1410, H 1300, H 1158 and H 1104 eruptions. Many of these are small and their tephra would have limited dispersal on the shelf. The two largest are the V ~1477 and H 1104 eruptions with 10 and 2 km^3 of tephra (as freshly fallen), respectively.

More than 20 events that could have resulted in sea-transport of pumice into the shelf area off north Iceland are known, but reliable reports of pumice rafts off north Iceland are few. Pumice rafts were observed following the Hekla eruptions 1970, 1947 and 1766, and in the Askja eruption 1875 (29 March). The silicic Askja pumice was carried to the sea at the north coast by Jökulsá River. Pumice rafts reported off NE and north Iceland in 1965 and 1966 were assumed to be the product of the Surtsey eruption 1963–1967.

Tephra from 13 of the eruptions described here has been identified in marine sediments on the north Icelandic shelf. These are the H 1947, V 1797, H 1766, V 1766, V 1717, H 1693, V ~1511, G ~1500, V ~1477, V ~1410, KOL 1372, H 1300 and H 1104 eruptions (Table 1). Occurrence of the H 1947 and KOL 1372 tephra is most likely due to pumice rafting. Tephra from the K 1918 and A 1875 (silicic) eruptions has been detected in marine sediments of the eastern Norwegian Sea. No instances of pumice transport by volcanogenic jökulhlaups into the coring areas on the shelf have been verified.

Tephra from the H 1947, G 1903, A 1875 (silicic) L 1783, H 1510, V ~1477 Ö 1362, H 1158 and H 1104 eruptions has been identified in terrestrial sequences around the North Atlantic.

This study is a contribution to and partially funded by the European Union 6th framework project MILLENNIUM (Subgroup SG4, Marine proxies, Contract No EVK-CT-2006-017008). The contribution of Esther R. Gudmundsdóttir was additionally supported by a grant 100218021, TeMa, from RANNIS (The Icelandic Centre for Research), and the contributions of Gudrún Larsen and Jón Eiríksson were supported by the Institute of Earth Sciences, University of Iceland. The authors would like to thank T. Thordarson and an anonymous reviewer for constructive comments and suggestions on the manuscript.

References

Abbott, P. M. & Davies, S. 2012. Volcanism and the Greenland ice-cores: the tephra record. *Earth-Science Reviews*, **115**, 173–191.

Alloway, B. V., Larsen, G., Lowe, D. J., Shane, P. A. R. & Westgate, J. A. 2006. Tephrochronology. *Encyclopedia of Quaternary Science*, **4**, 2869–2898.

Andrews, J. T., Geirsdóttir, A. et al. 2002. Distribution, sediment mag- netism and geochemistry of the Saksunarvatn (10,180 ± 60 cal. yr BP) tephra in marine, lake and terrestrial sediments, northwest Iceland. *Journal of Quaternary Science*, **17**, 731–745.

Áskelsson, J. 1936. On the last eruptions in Vatnajökull. *Societas Scientiarum Islandica*, **XVIII**, 65.

Benjamínsson, J. 1982. Gjóskulag "a" á Nord-Austurlandi. *In*: Thórarinsdóttir, H., Óskarsson, Ó. H., Steinthórsson, S. & Einarsson, Th. (eds) *Eldur er í Nordri*. Sögufélag, Reykjavík, 181–185.

Carey, R. J., Houghton, B. F. & Thordarson, T. 2010. Tephra dispersal and eruption dynam- ics of wet and dry phases of the 1875 eruption of Askja Volcano, Iceland. *Bulletin of Volcanology*, **72**, 259–278.

Cunningham, L. K., Austin, W. E. N. et al. 2013. Reconstructions of surface ocean conditions from the northeast Atlantic and Nordic seas during the last millennium. *Holocene*, **23**, 921–935.

D'Andrea, W., Vaillencourt, D. A., Balascio, N. L., Werner, A., Roof, S. R., Retelle, M. & Bradley, R. S. 2012. Mild Little Ice Age and unprecedented recent warmth in an 1800 year lake sediment record from Svalbard. *Geology*, **40**, 1007–1010.

Davies, S. M., Larsen, G., Wastegård, S., Turney, C. S. M., Hall, V. A., Coyle, L. & Thordarson, T. 2010. Widespread dispersal of Icelandic tephra: How does the Eyjafjöll eruption of 2010 compare to past Icelandic events? *Journal of Quaternary Science*, **25**, 605–611.

Dugmore, A. J., Larsen, G. & Newton, A. J. 1995. Seven tephra isochrones in Scotland. *The Holocene*, **5**, 257–266.

Eggertsson, S. 1919. Ymislegt smávegis vidvíkjandi Kötlugosinu 1918. *Eimreidin*, **1919**, 212–222.

Einarsson, E. H., Larsen, G. & Thorarinsson, S. 1980. The Sólheimar tephra layer and the Katla eruption of 1357. *Acta Naturalia Islandica*, **28**, 1–24.

Eiríksson, J., Knudsen, K. L., Haflidason, H. & Heinemeier, J. 2000. Chronology of late Holocene climate events in the northern North Atlantic based on AMS 14C dates and tephra markers from the volcano Hekla, Iceland. *Journal of Quaternary Science*, **15**, 573–580.

Eiríksson, J., Larsen, G., Knudsen, K. L., Heinemeier, J. & Símonarson, L. A. 2004. Marine reservoir age variability and water mass distribution in the Iceland Sea. *Quaternary Science Reviews*, **23**, 2247–2268.

Eiríksson, J., Knudsen, K. L. et al. 2011. Coupling of palaeoceanographic shifts and changes in marine reservoir ages off North Iceland through the last millennium. *Palaeogeography, Palaeoclimatology, Palaeoecology*, **302**, 95–108, http://dx.doi.org/10.1016/j.palaeo.2010.06.002

Fiacco, R. J., Thordarson, T., Germani, M. S., Self, S., Palais, J. M., Whitlow, S. & Grootes, P. M. 1994. Atmospheric aerosol loading and transport due to the 1783–84 Laki eruption in Iceland, interpreted from ash particles and acidity in the GISP2 ice core. *Quaternary Research*, **42**, 231–240.

Gehrels, W. R., Marshall, W. A. et al. 2006. Rapid sea-level rise in the North Atlantic Ocean since the first half of the nineteenth century. *The Holocene*, **16**, 949–965.

Grönvold, K., Larsen, G., Einarsson, P., Thorarinsson, S. & Sæmundsson, K. 1983. The Hekla eruption 1980–1981. *Bulletin of Volcanology*, **46**, 349–363.

Grönvold, K., Óskarsson, N., Johnsen, S. J., Clausen, H. B., Hammer, C. U., Bond, G. & Bard, E. 1995. Ash layers from Iceland in the Greenland GRIP ice core correlated with oceanic and land based sediments. *Earth and Planetary Science Letters*, **135**, 149–155.

Gudmundsdóttir, E. R., Eiríksson, J. & Larsen, G. 2011. Identification and definition of primary and reworked tephra in Late Glacial and Holocene marine shelf sediments off North Iceland. *Journal of Quaternary Science*, **26**, 589–602.

Gudmundsdóttir, E. R., Larsen, G. & Eiríksson, J. 2012*a*. Tephra stratigraphy of core MD99-2275 on the North Icelandic shelf: extending tephrochronology into marine sediments off North Iceland. *Boreas*, **41**, 718–734.

Gudmundsdóttir, E. R., Larsen, G. & Eiríksson, J. 2012*b*. Holocene marine tephrochronology on the Icelandic shelf: an overview. *Jökull*, **62**, 53–72.

Gudmundsson, A., Oskarsson, N. et al. 1992. The 1991 eruption of Hekla, Iceland. *Bulletin of Volcanology*, **54**, 238–246.

Gudmundsson, F. & Ingólfsson, A. 1968. Helsingjanef á Surtseyjarvikri (Goose barnacles (*Lepas* spp.) on Surtsey pumice). *Náttúrufrædingurinn*, **37**, 222–235.

Gudmundsson, M. T., Sigmundsson, F. & Björnsson, H. 1997. Ice–volcano interaction of the 1996 Gjálp subglacial eruption, Vatnajökull, Iceland. *Nature*, **389**, 954–957.

Gudmundsson, M. T., Högnadóttir, Th., Pálsson, F. & Björnsson, H. 2000. *Grímsvötn: Eldgosið 1998 og breytingar á botni, rúmmáli og jarðhita 1996–1999*. RH-03-2000.

Gudmundsson, M. T., Sigmundsson, F., Björnsson, H. & Högnadóttir, Th. 2004. The 1996 eruption at Gjálp, Vatnajökull ice cap, Iceland: course of events, efficiency of heat transfer, ice deformation and subglacial water pressure. *Bulletin of Volcanology*, **66**, 46–65.

Gudmundsson, M. T., Thordarson, Th. et al. 2012. Ash generation and distribution from the April–May 2010 eruption of Eyjafjallajökull, Iceland. *Scientific Reports*, **2**, 572, http://dx.doi.org/10.1038/srep00572.

Gunnlaugsson, G., Gudbergsson, G. M., Thorarinsson, S., Rafnsson, S. & Einarsson, Th. 1984. *Skaftáreldar 1783–1784. Ritgerdir og heimildir*. Mál og menning, Reykjavík.

Haflidason, H., Eiríksson, J. & van Krefeld, S. 2000. The tephrochronology of Iceland and the North Atlantic region during the Middle and Late Quaternary: a review. *Journal of Quaternary Science*, **15**, 3–22.

Haraldsson, K. Ö., Árnason, S. G., Larsen, G. & Eiríksson, J. 2002. The Hekla eruption of 2000 – the tephra fall. *In*: *The 25th Nordic Geolocical Winter Meeting*, 6–9 January 2002, Reykjavík, Abstract Volume 71.

Hartley, M. E. & Thordarson, T. 2012. Formation of Öskjuvatn caldera at Askja, North Iceland: mechanism of caldera collapse and implications for the lateral flow hypothesis. *Journal of Volcanology and Geothermal Research*, **227–228**, 85–101.

Höskuldsson, Á., Óskarsson, N., Pedersen, R., Grönvold, K., Vogfjörd, K. & Ólafsdóttir, R. 2007. The millennium eruption of Hekla in February 2000. *Bulletin of Volcanology*, **70**, 169–182.

Isaksson, E., Divine, D., Kohler, J., Martma, T., Pohola, V., Motoyama, H. & Watanabe, O. 2005. Climate oscillations as recorded in Svalbard ice core d18O records between AD 1200 and 1997. *Geografiska Annaler*, **87A**, 203–214.

ÍSAFOLD II 1875. no. 9, 25 May, 67–68, B. Jónsson, Reykjavík.

Ísaksson, S. P. 1985. Stórhlaup í Jökulsá á Fjöllum á fyrri hluta 18. aldar. *Náttúrufrædingurinn*, **54**, 165–191.

Jakobsson, S. P. 1974. Eldgos við Eldeyjarboða. (Volcanic eruptions at Eldeyjarbodi, the Reykjanes Ridge.) *Náttúrufrædingurinn*, **44**, 22–40.

JAKOBSSON, S. P. 1979. Petrology of recent basalts of the Eastern Volcanic Zone, Iceland. *Acta Naturalia Islandia*, **26**, 1–103.

JAKOBSSON, S. P. & MOORE, J. G. 1982. The Surtsey research drilling project of 1979. *Surtsey Research Progress Report*, **9**, 76–93.

JAKOBSSON, S. J., JÓNASSON, K. & SIGURDSSON, I. A. 2008. The three igneous rock series of Iceland. *Jökull*, **58**, 117–138.

JÓHANNSDÓTTIR, G. 2007. *Mid Holocene to late glacial tephrochronology in West Iceland as revealed in three lacustrine environments*. MSc thesis, University of Iceland.

JÓHANNESSON, H. 1977. Thar var ei bærinn sem nú er borgin. *Náttúrufrædingurinn*, **47**, 129–141.

JÓHANNESSON, H. 1983. Gossaga Grímsvatna 1900–1983 í stuttu máli (A brief review of the volcanic activity of the Grímsvötn volcanic system 1900–1983). *Jökull*, **33**, 146–147.

JÓHANNESSON, H. 1984. Grímsvatnagos 1933 og fleira frá thví ári (Grímsvötn eruption 1933). *Jökull*, **34**, 151–158.

JÓHANNESSON, H. 1987. Heimildir um Grímsvatnagosin 1902–1910. *Fjölrit Náttúrufrædistofnunar*, **4**, 49.

JÓHANNESSON, H. & EINARSSON, S. 1988. Krísuvíkureldar I. Aldur Ögmundarhrauns og Midaldalagsins. *Jökull*, **38**, 71–87.

JÓHANNESSON, H. & SÆMUNDSSON, K. 1998. *Geological map of Iceland. Tectonics. 1:500.000*. Icelandic Institute of Natural History, Reykjavík.

JÓNSDÓTTIR, H. B. B. 2001. Late Holocene climatic changes on the North Iceland shelf. MSc thesis, University of Aarhus. 124 pp.

JÓNSSON, Ó. 1945. *Ódádahraun II*. Bókaútgáfan Nordri, Akureyri.

JÓNSSON, P. V. K., HÖSKULDSSON, Á., THORDARSON, Th. & EINARSSON, B. F. 2007. Eldgosid í Öræfajökuli 1362, framgangur og afleiðingar. *In*: *Geoscience Society of Iceland, Spring Meeting*.

JUDE-ETON, T. C., THORDARSON, T., GUDMUNDSSON, M. T. & ODDSSON, B. 2012. Dynamics, stratigraphy and proximal dispersal of supraglacial tephra during the ice-confined 2004 eruption at Grímsvötn Volcano, Iceland. *Bulletin of Volcanology*, **74**, 1057–1082.

KARMA, K., SELF, S., BLAKE, S., THORDARSON, Th. & LARSEN, G. 2008. The AD 1362 Öræfajökull eruption, S.E. Iceland: physical volcanology and volatile release. *Journal of Volcanology and Geothermal Research*, **178**, 719–739.

KEKONEN, T., MOORE, J., PERÄMÄKI, P. & MARTMA, T. 2005. The Icelandic Laki volcanic tephra layer in the Lomonosovfonna ice core, Svalbard. *Polar Research*, **24**, 33–40.

KNUDSEN, K. L., EIRÍKSSON, J. & BARTELS-JÓNSDÓTTIR, H. B. 2012. Oceanographic changes through the last millennium off North Iceland: temperature and salinity reconstructions based on foraminifera and stable isotopes. *Marine Micropaleontology*, **84–85**, 54–73.

KRISTJÁNSDÓTTIR, G. 1999. *Late Quaternary climatic and enciron- menta changes on the North Icelandic shelf, evidenc from sedimentological and geochemical analyses of marine sed- iment cores*. MSc thesis, University of Iceland.

KRISTJÁNSDÓTTIR, G. B., STONER, J. S., JENNINGS, A. E., ANDREWS, J. T. & GRÖNVOLD, K. 2007. Geochemistry of Holocene cryptotephras from the North Iceland Shelf (MD99-2269): intercalibration with radiocarbon and palaeomagnetic chronostratigraphies. *The Holocene*, **17**, 155–176.

LACASSE, C., KARLSDÓTTIR, S., LARSEN, G., SOOSALU, H., ROSE, W. L. & ERNST, G. G. J. 2004. Weather radar observations of the Hekla 2000 erutpion cloud, Iceland. *Bulletin of Volcanology*, **66**, 457–473.

LARSEN, G. 1982. Gjóskutímatal Jökuldals og nágrennis. *In*: THÓRARINSDÓTTIR, H., ÓSKARSSON, Ó. H., STEINTHÓRSSON, S. & EINARSSON, Th. (eds) *Eldur er í Norðri*. Sögufélag, Reykjavík, 51–65.

LARSEN, G. 1984. Recent volcanic history of the Veidivötn fissure swarm, southern Iceland – an approach to volcanic risk assessment. *Journal of Volcanology and Geothermal Research*, **22**, 33–58.

LARSEN, G. 2000. Holocene eruptions within the Katla volcanic system, south Iceland: characteristics and environmental impact. *Jökull*, **49**, 1–28.

LARSEN, G. 2010. Katla: Tephrochronology and eruption history. *In*: SCHOMACKER, A., KRÜGER, J. & KJÆR, K. (eds) *The Mýrdalsjökull Ice Cap, Iceland: Glacial Processes, Sediments and Landforms on an Active Volcano*. Elsevier, Amsterdam, Developments in Quaternary Science, **13**, 23–49.

LARSEN, G. & EIRÍKSSON, J. 2008. Late Quaternary terrestrial tephrochronology of Iceland – frequency of explosive eruptions, type and volume of tephra deposits. *Journal of Quaternary Science*, **23**, 109–120.

LARSEN, G. & THORARINSSON, S. 1977. H-4 and other acid Hekla tephra layers. *Jökull*, **27**, 28–46.

LARSEN, G., VILMUNDARDÓTTIR, E. G. & THORKELSSON, B. 1992. Heklugosid 1991: Gjóskufallid og gjóskulagid frá fyrsta degi gossins. (The Hekla eruption of 1991 – the tephra fall.) *Náttúrufrædingurinn*, **61**, 159–176.

LARSEN, G., GUDMUNDSSON, M. T. & BJÖRNSSON, H. 1998. Eight centuries of periodic volcanism at the center of the Iceland hotspot revealed by glacier tephrostratigraphy. *Geology*, **26**, 943–946.

LARSEN, G., DUGMORE, A. J. & NEWTON, A. J. 1999. Geochemistry of historical-age silicic tephras in Iceland. *The Holocene*, **9**, 463–471.

LARSEN, G., DUGMORE, A. J., NEWTON, A. J. & VILMUNDARDÓTTIR, E. G. 2001. Geochemistry, dispersal, volumes and chronology of Holocene silicic tephra layers from the Katla volcanic system, Iceland. *Journal of Quaternary Science*, **16**, 119–132.

LARSEN, G., EIRÍKSSON, J., KNUDSEN, K. L. & HEINEMEIER, J. 2002. Correlation of late Holocene terrestrial and marine tephra markers in North Iceland. Implications for reservoir age changes and linking land–sea chronologies in the northern North Atlantic. *Polar Research*, **21**, 283–290.

LOWE, D. J. 2011. Tephrochronology and its application: a review. *Quaternary Geochronology*, **6**, 107–153.

MAGNÚSDÓTTIR, S. & BRANDSDÓTTIR, B. 2011. Tectonics of the Theistareykir fissure swarm. *Jökull*, **61**, 65–79.

MORGUNBLADID 1918. 15 October, 1; 27 October, 1, Ísafoldarprentsmidja, Reykjavík.

MORGUNBLADID 1947. 14 June, 12, Árvakur, Reykjavík.

MORGUNBLADID 1965. 6 July, 2, Árvakur, Reykjavík.

MORGUNBLADID 1966. 10 September, 32, Árvakur, Reykjavík.

MOURITZEN, M. & NOE-NYGAARD, A. 1950. Volcanic ash from the Grímsvötn eruption in 1903. *Meddelser fra Dansk Geologisk Forening*, **71**, 583–584.

NEWTON, A. J. 1999. *Ocean-transported pumice in the North Atlantic*. PhD thesis, University of Edinburgh, Edinburgh.

NORDANFARI 7 1868. no. 3–4, 15 February, 8, B. Jónsson, Akureyri.

NORDANFARI 14 1875. no. 21–22, 22 April, 48, B. Jónsson, Akureyri.

ODDSSON, B. 2007. *The Grímsvötn eruption in 2004, Dispersal and total mass of tephra and comparison with plume transport models*. MS thesis, University of Iceland, Reykjavík.

ODDSSON, B., GUDMUNDSSON, M. T., LARSEN, G. & KARLSDÓTTIR, S. 2012. Monitoring of the plume from the basaltic phreatomagmatic 2004 Grímsvötn eruption – application of weather radar and comparison with plume models. *Bulletin of Volcanology*, **74**, 1395–1407.

ÓLADÓTTIR, B. A., LARSEN, G. & SIGMARSSON, O. 2011. Holocene volcanic activity at Grímsvötn, Bárdarbunga and Kverkfjöll subglacial centres beneath Vatnajökull, Iceland. *Bulletin of Volcanology*, **73**, 1187–1208.

ÓLADÓTTIR, B. A., LARSEN, G. & SIGMARSSON, O. 2012. Deciphering eruption history and magmatic processes from tephra in Iceland. *Jökull*, **62**, 21–38.

PALAIS, J. M., TAYLOR, K., MAYEWSKI, P. A. & GROOTES, P. 1991. Volcanic ash from the 1362 AD Öræfajökull eruption (Iceland) in the Greenland ice sheet. *Geophysical Research Letters*, **18**, 1241–1244.

PERSSON, C. 1971. Tephrochronological investigations of peat deposits in Scandinavia and on the Faroe Islands. *Sveriges Geologiska Undersökning Årsbok*, **65**, 1–34.

PILCHER, J. R., HALL, V. A. & MCCORMAC, F. G. 1996. An outline tephrochronology for the Holocene of the north of Ireland. *Journal of Quaternary Science*, **11**, 485–494.

RAN, L., JIANG, H., KNUDSEN, K. L. & EIRÍKSSON, J. 2011. Diatom-based reconstruction of palaeoceanographic changes on the North Icelandic shelf during the last millennium. *Palaeogeography, Palaeoclimatology, Palaeoecology*, **302**, 109–119, http://dx.doi.org/10.1016/j.palaeo.2010.02.001

SÆMUNDSSON, K. 1978. Fissure swarms and central volcanoes of the neovolcanic zones of Iceland. *Geological Journal Special Issue*, **10**, 415–432.

SÆMUNDSSON, K. 1979. Outline of the geology of Iceland. *Jökull*, **29**, 7–28.

SÆMUNDSSON, K. 1991. Jardfrædi Kröflukerfisins. *In*: GARDARSON, A. & EINARSSON, Á. (eds) *Náttúra Mývatns*. Hid íslenska náttúrufrædifélag, Reykjavík, 25–95.

SCHUPACK, B., MILLER, G. H., FROESE, D., KUEHN, S., HAFLIDASON, H. & WERNAR, A. 2010. Finding needles in a haystack: pushing the limits of tephrochronology in the Arctic. *In*: *The 40th International Arctic Workshop*, Winter Park, CO, Abstract.

SEJRUP, H. P., LEHMAN, S. J., HAFLIDASON, H., NOONE, D., MUSCHELER, R., BERSTAD, L. M. & ANDREWS, J. T. 2010. Response of Norvegian sea temperature to solar forcing since 1000 AD. *Journal of Geophysical Research*, **115**, C12034.

SEJRUP, H. P., HAFLIDASON, H. & ANDREWS, J. T. 2011. A Holocene North Atlantic SST record and regional climate variability. *Quaternary Science Reviews*, **30**, 3181–3195.

SICRE, M.-A., HALL, I. ET AL. 2011. Sea surface temperature variability in the subpolar Atlantic over the last millennium. *Paleoceanography*, **26**, PA4218, 1–10.

SIGMARSSON, O., KARLSSON, H. & LARSEN, G. 2000. The 1996 and 1998 subglacial eruptions beneath Vatnajökull ice sheet: contrasting geochemical and geophysical inferences on magma migration. *Bulletin of Volcanology*, **61**, 468–476.

SIGMUNDSSON, F. & GUDMUNDSSON, M. T. 2004. Gos í Grímsvötnum 1.–6. nóvember 2004. (Grímsvötn eruption, November 1–6, 2004.) *Jökull*, **54**, 139–142.

SIGURGEIRSSON, M. Á. 1995. Yngra–Stampagosid a Reykjanesi. (The Younger–Stampar eruption at Reykjanes, SW-Iceland.) *Náttúrufrædingurinn*, **64**, 211–230.

SIGURGEIRSSON, M. Á. 2000. *Gjóskulög á Austurlandi – Samantekt*. Greinargerd 00/04, Vedurstofa Íslands, Reykjavík.

SPARKS, R. S. J., WILSON, L. & SIGURDSSON, H. 1981. The pyroclastic deposits of the 1875 eruption of Askja, Iceland. *Philosophical Transactions of the Royal Society London, Series A*, **299**, 241–273.

STEINTHORSSON, S. 1978. Tephra layers in a drill core from the Vatnajökull Ice Cap. *Jökull*, **27**, 2–27.

STEINTHORSSON, S., HARDARSON, B. S., ELLAM, R. M. & LARSEN, G. 2000. Petrochemistry of the Gjálp-1996 subglacial eruption, Vatnajökull, SE Iceland. *Journal of Volcanology and Geothermal Research*, **98**, 79–90.

SVEINSSON, G. 1919. *Kötlugosid 1918 og afleidingar thess. (The Katla eruption of 1918 and its consequences.)* Prentsmidjan Gutenberg, Reykjavík.

SWINDLES, G. T., LAWSON, I. T., SAVOV, I. P., CONNOR, C. B. & PLUNKETT, G. 2011. A 7000 yr perspective on volcanic as clouds affecting northern Europe. *Geology*, **39**, 887–890.

THORARINSSON, S. 1944. Tefrokronologiska studier pa Island. *Geografiska Annaler*, **26**, 1–217.

THORARINSSON, S. 1950. Jökulhlaup og eldgos á jökulvatnasvædi Jökulsár á Fjöllum. *Náttúrufrædingurinn*, **20**, 113–133.

THORARINSSON, S. 1954. The tephra-fall from Hekla on March 29th 1947. *In*: EINARSSON, T., KJARTANSSON, G. & THORARINSSON, S. (eds) *The Eruption of Hekla 1947–48*. Societas Scientarum Islandica, Reykjavík, **II-3**, 1–68.

THORARINSSON, S. 1955. Öskufall svo sporrækt vard og Kötlugosið 1721. *Náttúrufrædingurinn*, **25**, 87–98.

THORARINSSON, S. 1958. The Öræfajökull eruption of 1362. *Acta Naturalia Islandica*, **2**, 1–100.

THORARINSSON, S. 1963. *Askja on Fire*. Almenna bókafélagið, Reykjavík.

THORARINSSON, S. 1965*a*. Sitt af hverju um Surtseyjargosid (Some facts about the Surtsey eruption). *Náttúrufrædingurinn*, **35**, 153–181.

THORARINSSON, S. 1965*b*. Nedansjávargos vid Ísland. (Submarine eruptions off the coasts of Iceland.) *Náttúrufrædingurinn*, **35**, 49–74.

THORARINSSON, S. 1966. *Surtsey – the New Island in the North Atlantic*. Almenna bókafélagid, Reykjavók.
THORARINSSON, S. 1967. The eruptions of Hekla in historical times. *In*: EINARSSON, T., KJARTANSSON, G. & THORARINSSON, S. (eds) *The Eruption of Hekla 1947–48*. Societas Scientarum Islandica, Reykjavík, **I**, 1–177.
THORARINSSON, S. 1968. Sídustu thættir Eyjaelda. (The last phases of the Surtsey eruption.) *Náttúrufrædingurinn*, **38**, 113–135.
THORARINSSON, S. 1970. *Hekla*. Almenna Bókafélagid, Reykjavík.
THORARINSSON, S. 1974. *Vötnin stríd. Saga Grímsvatnagosa og Skeidarárhlaupa*. Bókaútgáfa Menningarsjóds, Reykjavík.
THORARINSSON, S. 1975. Katla og annáll Kötlugosa. (Katla and its historical eruptions.) *Árbók Ferdafélags Íslands*, **1975**, 125–149.
THORARINSSON, S. 1981. Greetings from Iceland. *Geografiska Annaler*, **63A**, 109–118.
THORARINSSON, S. & SIGVALDASON, G. E. 1972. The Hekla eruption of 1970. *Bulletin Volcanologique*, **XXXVI-2**, 269–288.
THORDARSON, Th. 1991. *Skaftáreldar 1783–1785. Gjóskan og framvinda gossins*. Háskólaútgáfan, Reykjavík.
THORDARSON, Th. & HÖSKULDSSON, Á. 2008. Postglacial volcanism in Iceland. *Jökull*, **58**, 197–228.
THORDARSON, Th. & LARSEN, G. 2007. Volcanism in Iceland in historical time: volcano types, eruption styles and eruptive history. *Journal of Geodynamics*, **43**, 118–152.
THORDARSON, Th. & SELF, S. 1993. The Laki (Skaftár Fires) and Grímsvötn eruptions in 1783–1785. *Bulletin of Volcanology*, **55**, 233–263.
THORDARSON, T. & SELF, S. 2003. Atmospheric and environmental effects of the 1783–1784 Laki eruption, Iceland: a review and reassessment. *Journal of Geophysical Research*, **103**, 27411–27445, http://dx.doi.org/108(D1)10.1029/2001JD002042
THORDARSON, Th. & SIGMARSSON, O. 2009. Effusive activity in the 1963–1967 Sürstey eruption, Iceland: flow emplacement and growth of small lava shields. *In*: THORDARSON, Th., SELF, S., LARSEN, G., ROWLAND, S. K. & HÖSKULDSSON, Á. (eds) *Studies in Volcanology: The Legacy of George Walker*. Special Publications of the International Association of Volcanology and Chemistry of the Earth's Interior No. 2, The Geological Society, London, 53–84.
THORDARSON, Th., LARSEN, G., STEINTHORSSON, S. & SELF, S. 2003. The 1783–1785 A.D. Laki-Grímsvötn eruptions II: appraisal based on contemorary accounts. *Jökull*, **53**, 11–48.
THORKELSSON, Th. 1923. Eldgosin 1922 (Volcanic eruptions in Iceland 1922). *Tímarit V.F.Í.*, **1923**, 29–43.
THORODDSEN, Th. 1924. *Fjórar ritgjördir. Safn Frædafélagsins um Ísland og Íslendinga*. Copenhagen.
THORODDSEN, Th. 1925. *Die Geschichte der isländischen Vulkane. Det Kongelige danske Videnskabernes Selskabs Skrifter, 8. række. Naturvidenskabelig og mathematisk Afdeling B IX*. Høst og Søn, København.
VAN DEN BOGAARD, C. & SCHMINCKE, H. U. 2002. Linking the North Atlantic to central Europe: a high-resolution Holocene tephrochronological record from northern Germany. *Journal of Quaternary Science*, **17**, 3–20.
WANAMAKER, A. D., JR, BUTLER, P. G., SCOURSE, J. D., HEINEMEIER, J., EIRÍKSSON, J., KNUDSEN, K. L. & RICHARDSON, C. A. 2012. Surface changes in the North Atlantic meridional overturning circulation during the last millennium. *Nature Communications*, **3**, article no. 899, 1–7.
WASTEGÅRD, S. & DAVIES, S. 2009. An overview of distal tephrochronology in northern Europe during the last 1000 years. *Journal of Quaternary Science*, **24**, 500–512.
ZIELINSKI, G. A., GERMANI, M. S., LARSEN, G., BAILLE, M. G. L., WHITLOW, S., TWICKLER, M. S. & TAYLOR, K. 1995. Evidence of the Eldgjá (Iceland) eruption in the GISP2 Greenland ice core: relationship to eruption processes and climatic conditions in the tenth century. *The Holocene*, **5**, 129–140.

Iceberg-rafted tephra as a potential tool for the reconstruction of ice-sheet processes and ocean surface circulation in the glacial North Atlantic

MARION KUHS[1]*, WILLIAM E. N. AUSTIN[1,2], PETER M. ABBOTT[3] & DAVID A. HODELL[4]

[1]*School of Geography & Geosciences, University of St Andrews, North Street, St Andrews KY16 9LA, UK*

[2]*Scottish Association of Marine Sciences, Scottish Marine Institute, Oban, PA37 1QA, UK*

[3]*Department of Geography, College of Science, Swansea University, Singleton Park, Swansea SA2 8PP, UK*

[4]*Godwin Laboratory for Paleoclimate Research, Department of Earth Sciences, University of Cambridge, Downing Street, Cambridge CB2 3EQ, UK*

**Corresponding author (e-mail: mk578@st-andrews.ac.uk)*

Abstract: Ice-rafted tephra deposits, of Marine Isotope Stage 6 (MIS 6) age, from Site U 1304 on the Gardar Drift, North Atlantic were examined for their shard size distribution and major element composition. The heterogeneous composition, large shard sizes and association with ice-rafted debris (IRD) indicate that these late MIS 6 deposits were transported by iceberg-rafting from Iceland to Site U 1304. Comparison of individual shard geochemistry with the geochemistry of Holocene volcanic systems from Iceland allows the identification of different potential volcanic source regions. This detailed geochemical analysis, when combined with Icelandic Ice Sheet (IIS) flow models for the Last Glacial Maximum (LGM), suggests that the IIS had calving margins to both the north and south during the late MIS 6 and that icebergs could have been transported to the Site U 1304 by following surface ocean circulation patterns similar to those that prevailed during the LGM. We demonstrate that the descriptive concept of Icelandic glass in the characterization of tephra components within North Atlantic IRD can be significantly improved through quantitative characterization and that such data hold the potential to help constrain surface ocean circulation models, while also potentially yielding new information about the IIS during earlier glacial periods.

Supplementary material: Statistical tests, major element concentrations of analysed shards, primary and secondary standards are available at http://www.geolsoc.org.uk/SUP18716

In palaeoceanography, volcanic ash (i.e. tephra) not only serves as a tool in geochronology but also provides a potentially distinctive signature (i.e. 'fingerprint') of the source region of ice-rafted debris (IRD). Over the last few decades tephrochronology has increasingly become the focus of attention in palaeoclimate research for the correlation and synchronization of disparate records, whereas the application of tephra as an indicator for IRD provenances has received relatively little attention.

In the North Atlantic, Icelandic tephra shards have been used as IRD provenance tracers (Bond *et al.* 1997) in the same way that certain IRD lithologies, such as detrital carbonates, have been used as indicators for the potential IRD source regions, in turn giving an insight into the dynamics of the corresponding ice-sheet and/or hydrographic processes controlling IRD delivery (e.g. Hodell & Curtis 2008). However, the common geological evolution of the North Atlantic region often makes the explicit mapping of the provenance of IRD difficult. For example, hematite-coated grains, such as stained quartz, could originate from several locations: NE America (Bond & Lotti 1995), East Greenland (Bond *et al.* 1997), Fenno-Scandinavia (Habicht 1979) or the British Isles (Peck *et al.* 2007). In contrast, tephra shards in the North Atlantic are most likely to be derived from volcanoes on Iceland or Jan Mayen. The advantage of tephra shards over other petrological tracers is that, by means of relatively rapid and inexpensive geochemical analyses, the source region (volcanic system) may be identified. By comparing the composition of tephra shards with geochemical data from Holocene

From: Austin, W. E. N., Abbott, P. M., Davies, S. M., Pearce, N. J. G. & Wastegård, S. (eds) 2014. *Marine Tephrochronology*. Geological Society, London, Special Publications, **398**, 141–155.
First published online February 6, 2014, http://dx.doi.org/10.1144/SP398.8

volcanic systems, particular volcanic systems potentially can be identified as the most likely source region for these shards. Such a detailed analysis of shard geochemistry is a key element in tephrochronology, providing the basis by which isochronous horizons are identified and correlated between sequences (Davies *et al.* 2012). In past IRD provenance studies such detailed analyses of ice-rafted tephra are mostly missing. However, assigning these tephra to specific volcanic systems instead of generally labelling them as of 'Icelandic-origin' offers the potential to improve the reconstruction and the understanding of ice-sheet dynamics, iceberg pathways and surface ocean circulation pattern.

Aims and objectives

In this paper, we examine IRD-associated tephra deposits in marine sediments from the Integrated Ocean Drilling Program (IODP) Site U 1304 from the Gardar Drift, North Atlantic. The study interval occurs during a period of relatively stable cold surface water conditions prior to Heinrich Event 11 (HE 11) and the end of MIS 6 as defined by Termination II (Hodell *et al.* 2009). By means of geochemical and shard size analyses we determine potential volcanic source regions and the transport mechanism for these tephra shards. Instead of labelling the tephra as 'Icelandic-origin' we assign them to certain Icelandic volcanic systems and show how ice-rafted tephra can be used to obtain an insight into the dynamics of the Icelandic Ice Sheet (IIS) and surface ocean circulation during MIS 6 and potentially other glacial periods in the North Atlantic.

Material and methods

The 244 m-long U 1304 record was collected during Expedition 303 of the IODP in 2004 and was recovered from a water depth of 3082 m (Expedition 303 Scientists 2006*b*; Hodell *et al.* 2009) on the eastern flank of the Reykjanes Ridge on the southern Gardar Drift (latitude 53°03′N, longitude 33°32′W), within the subpolar North Atlantic (Fig. 1). Today, the North Atlantic Current passes the core site and transports warm surface water northeastwards to the Nordic Sea. Iceland Scotland Overflows mix with overlying Labrador Sea Water travelling southwestwards and represent the dominant water body at this site (Bianchi & McCave 2000; Hodell *et al.* 2009). More information about the site location, along with detailed description of the stratigraphy and isotope data, is published by Expedition 303 Scientists (2006*b*), Hodell *et al.* (2009) and Mazaud *et al.* (2009). An age-model for the interval 15.47 m composite depth (mcd) to 24.52 mcd of the U 1304 record was created by Hodell *et al.* (2009). Based on this age-model, the interval of high tephra concentration is older than 128 ka, and occurred prior to Termination II during MIS 6 (Fig. 2a). The analysed tephra were found between 23.32 and 23.17 mcd in an interval characterized by cold sea surface temperatures and high IRD concentrations, which followed a period of relative warm and fluctuating sea surface temperatures (Fig. 2b, d & e).

Quantitative analyses

The core material analysed from Site U 1304 was sampled at 5 cm intervals with a sample depth of 2 cm. A quantitative analysis (>150 μm sieved fraction) of tephra shards, lithic grains, *Neogloboquadrina pachyderma* sinistral (*Np* (s)) and the sum of all other planktonic foraminifera species was performed. With the aid of a microsplitter, the samples were halved *n* times until the residue evenly covered a 42-cell grided picking tray. Under a stereomicroscope the four components were counted until reaching ≥300 planktonic foraminifera specimens (i.e. combined values of *Np* (s) and others foraminifera species). The results were expressed as *Np* (s) percentages, grain and shard numbers per gram dry sediment and were recorded against depth (mcd). Lithic grains were categorized into five classes: quartz, hematite stained grains, detrital carbonate, basaltic and undifferentiated grains (labelled as nID). For four samples (23.32, 23.27, 23.22 and 23.17 mcd), tephra shards were initially classified by colour (brown or colourless) and overall morphology. The most commonly found morphologies are porous (shards with small bubbles) and conchoidal (blocky shards with large curvatures) shards. Fluted (shards with elongated striations) and wing-shaped (part of a bubble wall with some ridges) shards were only found in small numbers (Fig. 3). It should be noted that the morphology–terminology used here is based on an informal description of the main shard types encountered in the U 1304 samples and might deviate from the terminology used elsewhere, for example Heiken & Wohletz (1985).

Geochemical analyses

Forty-one tephra shards were analysed at the Tephrochronology Analytical Unit (TAU) at the Grant Institute, University of Edinburgh. By means of a wavelength dispersive spectrometer electron microprobe (Cameca SX100), major and minor elements of each single shard were determined. An accelerating voltage of 15 kV, a beam diameter of 5 μm, an initial beam current of 2 nA (to reduce Na loss) and later a beam current of

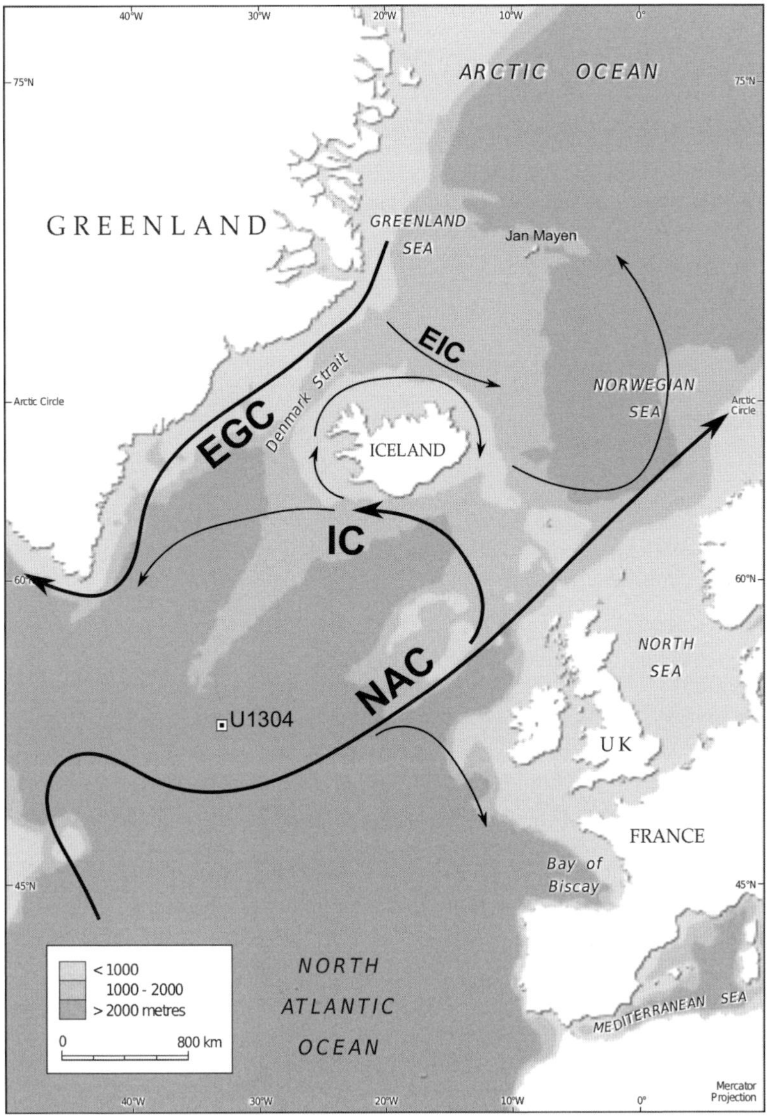

Fig. 1. Location of IODP Site U 1304 sediment core from the Gardar Drift, North Atlantic. Modern surface ocean circulation after Fratantoni (2001) and Ran *et al.* (2011). EGC, East Greenland Current; EIC, East Iceland Current; IC, Irminger Current; NAC, North Atlantic Current.

80 nA were applied (Hayward 2012). Further geochemical analyses of an additional 111 tephra shards were carried out at the Facility for Earth and Environmental Analysis (FEEA) at the University of St Andrews. The wavelength dispersive spectrometer electron microprobe (Jeol JXA 800) was operated with the following settings: accelerating voltage of 15 kV, a beam current of 10 nA and a beam size of 5 μm. For the primary calibrations a series of synthetic oxides, pure metals and minerals were used. Lipari and BHVO-2G were used as secondary standards for rhyolitic glass and basaltic glass, respectively. The results are expressed as oxide weight percentages after normalization to 100% and are presented on an anhydrous basis. To test the homogeneity of the tephra shards, up to five analyses per shard were carried out at FEEA. The mean of these different measurements, which was calculated after removing outliers and inclusions, was used for further analyses.

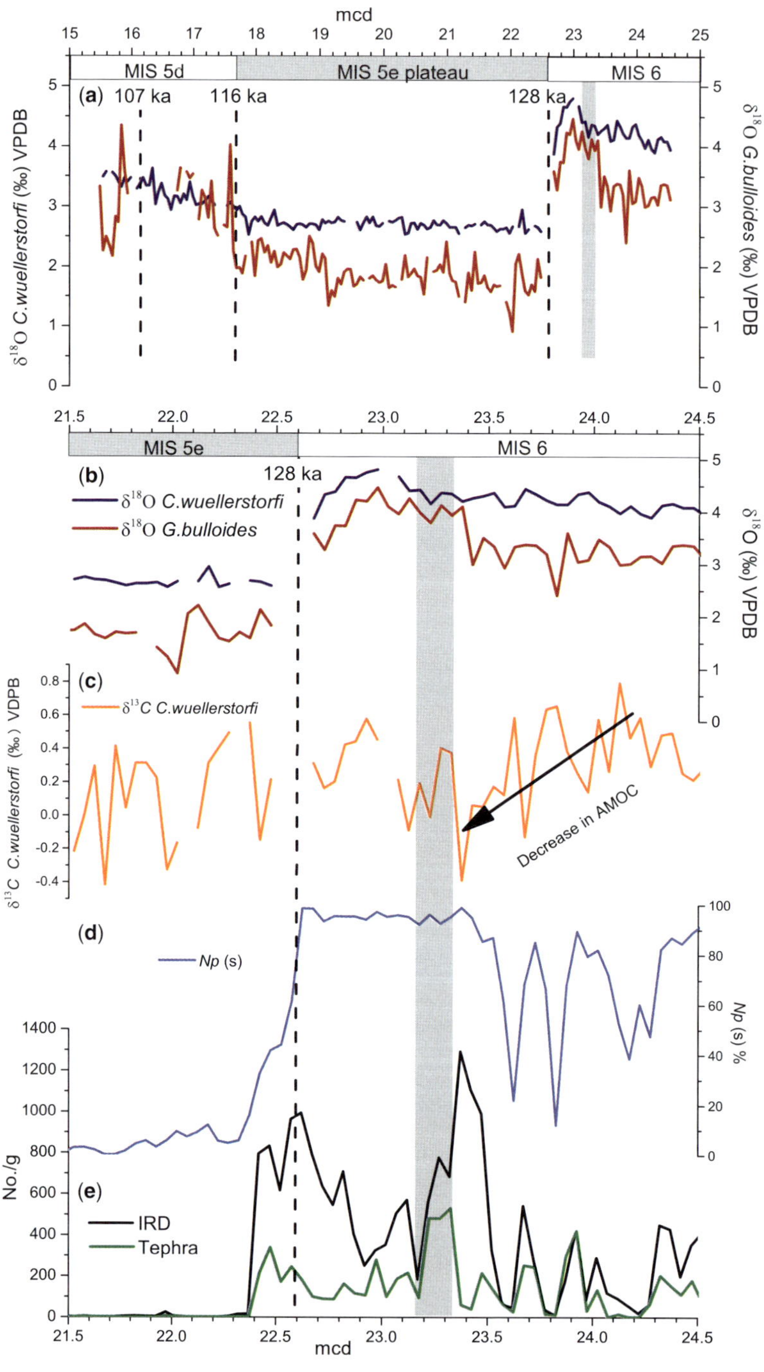

Fig. 2. Stratigraphical framework for IODP Site U 1304 based on Hodell *et al.* (2009). (**a**, **b**) Planktonic δ^{18}O, *G. bulloides* (red); benthic δ^{18}O, *C. wuellerstorfi* (blue); (**c**) bentic δ^{13}C, *C. wuellerstorfi* (orange); (**d**) percentage abundance of *Np* (s) (light blue; *this study*); (**e**) ice-rafted debris (IRD; black) and tephra (green) grains/gram dry sediment (>150 µm; *this study*). Studied interval: vertical grey shaded area.

Fig. 3. Different morphologies of tephra shards from IODP Site U 1304 based on a informal description of the main shard types encountered. Conchoidal (blocky shards with large curvatures), fluted (shards with elongated striations), porous (shards with small bubbles) and wing-shaped (part of a bubble wall with some ridges). Scale bar: 350 μm.

Shard size analyses

For each of the four samples analysed, the maximum grain size of at least 300 basaltic shards (>150 μm fraction) was randomly measured in order to gain an understanding of shard size distribution. However, it is important to highlight that shard size analyses can be influenced by the shape of the shard, that is, elongated shards can pass through a 150 μm sieve, even though their maximum diameter is >150 μm. The results are presented in frequency plots and the mean (Md_ϕ), mode, standard deviation (σ_ϕ) and skewness (α_ϕ) were calculated. In sedimentology it is common to quote the grain size in ϕ:

$$\phi = \log_2 r, \qquad (1)$$

where r is the grain size diameter in millimetres. For easier use of grain size equations and comparison with published data, the tephra shard sizes were transformed into ϕ values; σ_ϕ and α_ϕ were derived by the following equations (Sparks *et al.* 1981).

$$\sigma_\phi = \frac{1}{2}(\phi_{84} - \phi_{16}), \qquad (2)$$

$$\alpha_\phi = \frac{\frac{1}{2}(\phi_{84} + \phi_{16}) - \mathrm{Md}_\phi}{\sigma_\phi}. \qquad (3)$$

Results and discussion

Geochemical analyses

As expected from the inspection of the tephra shards under a light microscope, the majority of shards observed were of basaltic composition and only one rhyolitic shard was present among those analysed (Fig. 4). The total alkalis (Na_2O and K_2O) weight percentages (wt%) are relatively low (Fig. 4a), suggesting that the tephra shards are either tholeiitic or transitional-alkalic (Jakobsson 1979; Jakobsson *et al.* 2008). Basaltic tephra shards from tholeiitic and transitional-alkalic series tend to have large compositional overlaps (Fig. 4). In this study, seven shards can be clearly identified as

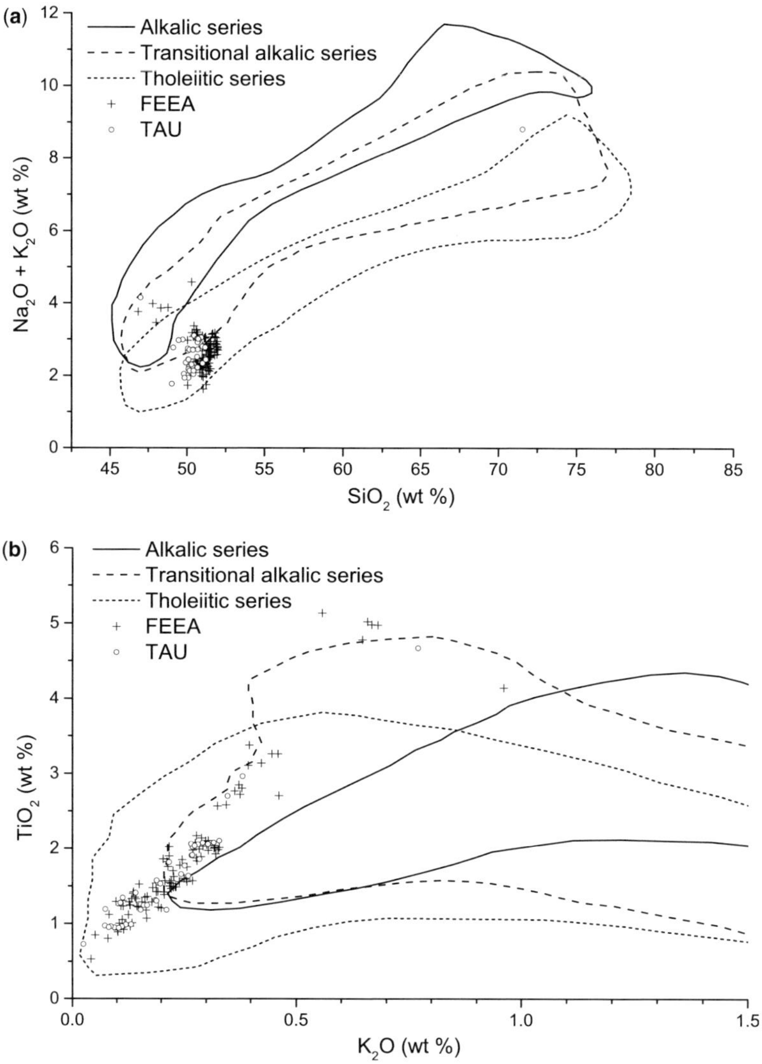

Fig. 4. (**a**) Total alkali-silica. (**b**) TiO_2 v. K_2O oxide bi-plots show the classification of the shards into the three different Icelandic rock suites: tholeiitic, transitional-alkalic and alkalic series. Envelopes are adapted from Jakobsson *et al.* (2008). FEEA, Facility for Earth and Environmental Analysis; TAU, Tephrochronology Analytical Unit.

transitional-alkalic and the remaining shards were assigned to the tholeiitic series because the majority of them can be plotted into the tholeiitic-envelope (Fig. 4) and all follow a similar linear trend (Fig. 4b). The high TiO_2 wt% and low K_2O wt% in the basaltic shards (Fig. 4b) are typical of Icelandic volcanic systems and distinguish the tephra from the Jan Mayen volcanic system (Jakobsson *et al.* 2008).

Figure 4 shows that the composition of the analysed tephra shards has a broad distribution, displaying greater heterogeneity than is normally observed for material from a single Icelandic eruption (e.g. Abbott *et al.* 2012). This suggests that the material was not derived from a single event but a series of eruptions, which is a key indicator of the transport and deposition of material by iceberg-rafting rather than sea ice-rafting or primary air-fall processes (Lackschewitz & Wallrabe-Adams 1997; Austin *et al.* 2004; Brendryen *et al.* 2010).

Tephra populations

To identify distinct populations, the geochemical data for each sample depth were plotted separately on various oxide/oxide bi-plots. By visual inspection for consistent ‘clusters’, more than one population can be discriminated at each depth. Ti_2O/MgO, Ti_2O/FeO, K_2O/CaO and K_2O/MgO ratios were the most useful binary plots for the discrimination of the different populations (Fig. 5). Single shards or clusters of two that could not be assigned to a population were classified as outliers. An assessment of these populations using the statistical tests of the similarity coefficient function and statistical distance demonstrates that they form distinct populations, supporting their definition using visual inspection.

To identify potential volcanic source regions for the U 1304 tephra deposits, the geochemistry of the tholeiitic populations was compared with available geochemical data from Icelandic Holocene volcanic systems using the compositional envelopes created by Jagan (2010) and Óladóttir *et al.* (2011). The Si_2O/CaO ratio v. Al_2O_3 (Fig. 6a) plot allows the separation of the Western volcanic zone (WVZ) and Reykjanes volcanic zone (RVZ) from the Eastern volcanic zone (EVZ, consisting of Grimsvòtn, Veidivötn–Bárdabunga, Askja and Krafla volocano systems; Jagan 2010). Even though the EVZ and WVZ envelopes overlap, the majority of populations were assigned to EVZ as they all follow the same slope as the EVZ envelope. Only two populations (2327 B and 2332 F) were classified as having WVZ-like geochemical signature. Owing to a tight envelope in the FeO/CaO v. Si_2O/Ti_2O plot, tephra shards with a signature like the Krafla volcano system (KVS) and Grimsvòtn volcano system (GVS) can be distinguished from shards with a geochemical affinity to the Veidivötn–Bárdabunga volcano system (VBVS) and Askja volcano system (AVS; Fig. 6b). Consistently high Ti_2O content of the GVS-like shards allow a clear separation from the three other volcanic systems (Fig. 6e). Shards with a KVS-like geochemistry have higher FeO and lower MgO contents than the other shards, whereas VBVS-like shards are characterized by low K_2O composition (Fig. 6c, d). Three populations, 2317 A, 2322 A and 2332 A, form a tight cluster in the KVS envelope (Fig. 6c–e). The statistical tests suggest that they could potentially have originated from the same eruption. However, owing to the ice-rafted nature of the shards, the use of this KVS-like deposit as an isochronous marker is rather limited (Brendryen *et al.* 2010).

It is important to note that the amount of material derived from each volcanic system changes over the depth interval studied (Fig. 7). While VBVS-type tephra shards dominate the three deepest samples, Grimsvòtn and Krafla-like tephra shards increase in number up-core until they are the most commonly found shards at 23.17 mcd. This stratigraphic pattern could be explained by changes in the activity of certain volcanic systems over time and/or transport mechanism (see the section Transport mechanism).

Shard size distribution

The shard size frequency plots (Fig. 8), as well as the calculated skewness, a measure of the symmetry in shard size distribution (Table 1), show that the distribution is nearly symmetrical to coarse/negative skewed (Folk & Ward 1957), which is probably caused by only including shards larger than >150 μm in this measurement. At all four sample depths the shard size distribution is similar. The relatively close mean, median and mode, as well as σ_ϕ values around 0.5 (Sparks *et al.* 1981), indicate a well-sorted shard size distribution (Fig. 8 and Table 1).

Shard size analyses of tephra deposits from known volcanic eruptions show that layers with a median shard diameter of 500 μm were not transported more than about 100 km by atmospheric processes (Fisher 1964; Kittleman 1973; Sparks *et al.* 1981). Such large distances between the source regions and the sites of deposition are typically only documented for highly explosive silica-rich eruptions. The U 1304 tephra deposits are of basaltic composition and their median shard diameter varies between 394 and 417 μm. Only if the magma responsible for these deposits had interacted with water could they have reached an intensity comparable to dry silica-rich eruptions (Thordarson & Larsen 2007), otherwise the erupted material would only have been transported tens of kilometres by airfall. At the end of MIS 6, Iceland was probably fully ice-covered, which would have allowed water–magma interaction to cause highly intensive basaltic eruptions and potentially deposit tephra with a radius of up to 100 km from the volcanic source. This somewhat restricted depositional range would have precluded tephra deposition on sea-surfaces. Thus, the majority of these shards are likely to have fallen originally onto the IIS and were subsequently more likely to have been iceberg-rafted rather than sea-ice-rafted to the core site. Therefore, if the well-sorted size distribution of the tephra is not associated with primary airfall and sea-ice rafting (Lackschewitz & Wallrabe-Adams 1997; Austin *et al.* 2004; Brendryen *et al.* 2010), we cannot exclude the possibility that it is the result of bottom current sorting. However, a detailed particle size analysis of different components of the sediment might help to confirm this hypothesis.

Transport mechanism

The MIS 6 tephra were deposited during an interval when the benthic foraminiferal $\delta^{13}C$ record at Site U 1304 (Hodell *et al.* 2009) suggests a reduction

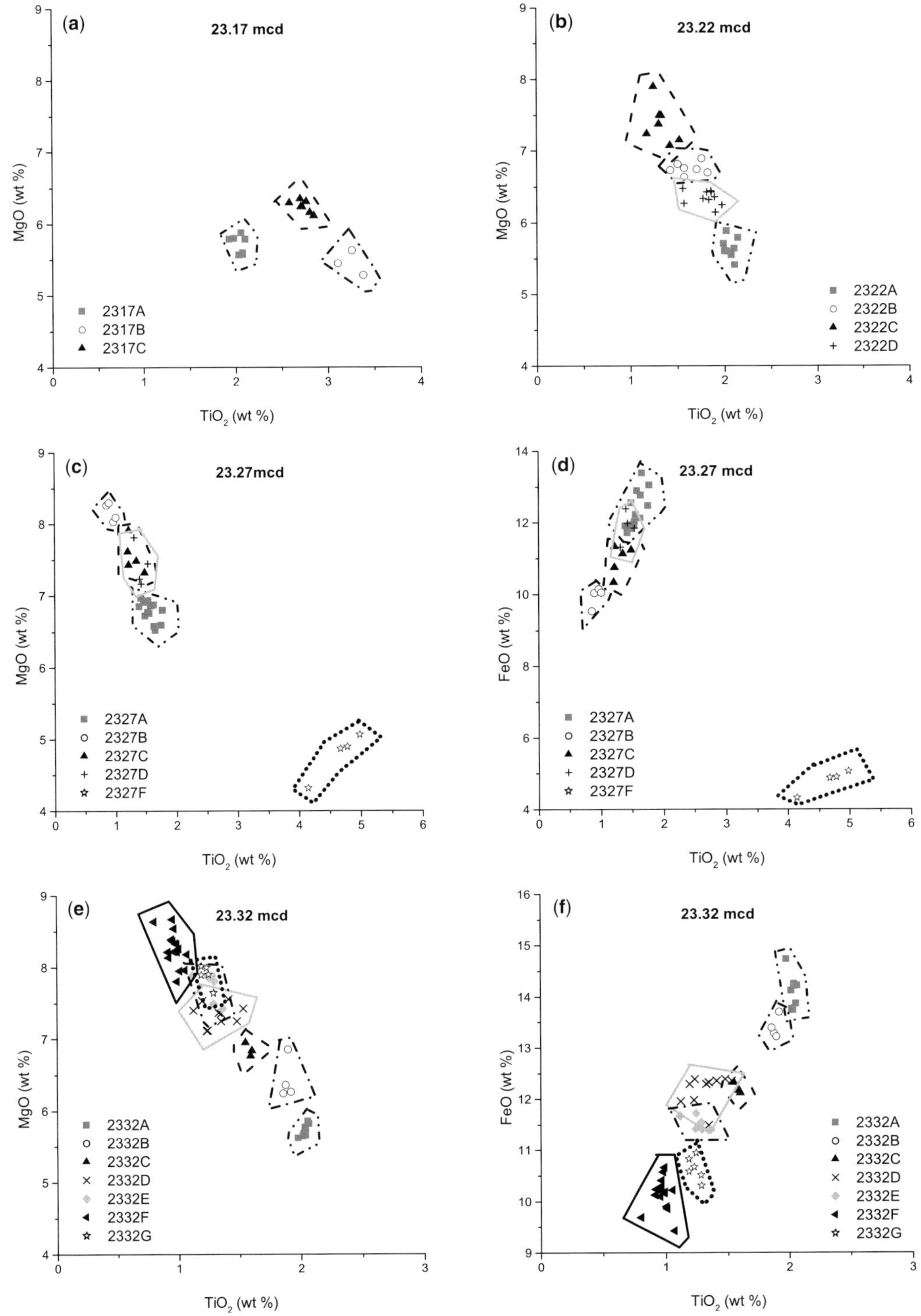

Fig. 5. Biplots of selected major oxides (wt %) of tephra shards from IODP Site U 1304. MgO v. TiO_2 and FeO v. TiO_2 show the different populations found in each sample depth. (**a**) 23.17 mcd (three populations); (**b**) 23.22 mcd (four populations); (**c**, **d**) 23.27 mcd (five populations); (**e**, **f**) 23.32 mcd (seven populations).

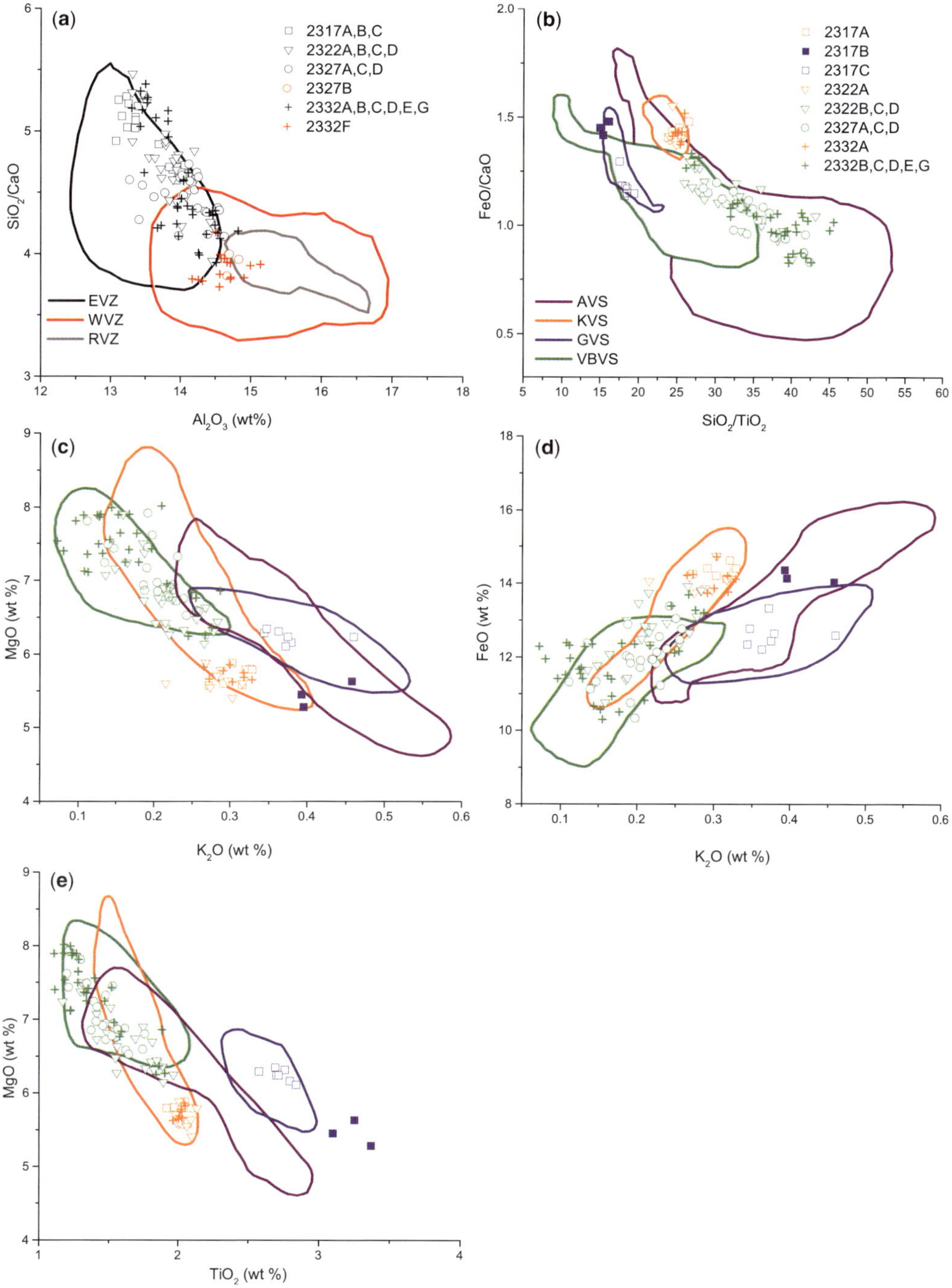

Fig. 6. Compositional envelopes from Jagan (2010) (**a**, **b**) and Óladóttir *et al.* (2011) (**c**–**e**). (a) SiO_2/CaO v. Al_2O_3 plot allows differentiation between tephra derived from Western volcanic zone (WVZ; red), Reykjanes volcanic zone (RVZ; grey) or Askja (AVS), Krafla (KVS), Veidivötn–Bárdabunga (VBVS) and Grimsvòtn volcano systems (GVS) summarized as Eastern volcanic zone EVZ (black). (b) FeO/CaO v. Si_2O/Ti_2O plot allows tephra shards with KVS- and GVS-like (orange and blue, respectively) signature to be distinguished from shards with a geochemical affinity to VBVS (green) and AVS (purple). (c) FeO v. K_2O; (d) MgO v. TiO_2; and (e) MgO v. K_2O separate shards with a geochemical signature similar to the VBVS (green) GVS (blue) and KVS (orange).

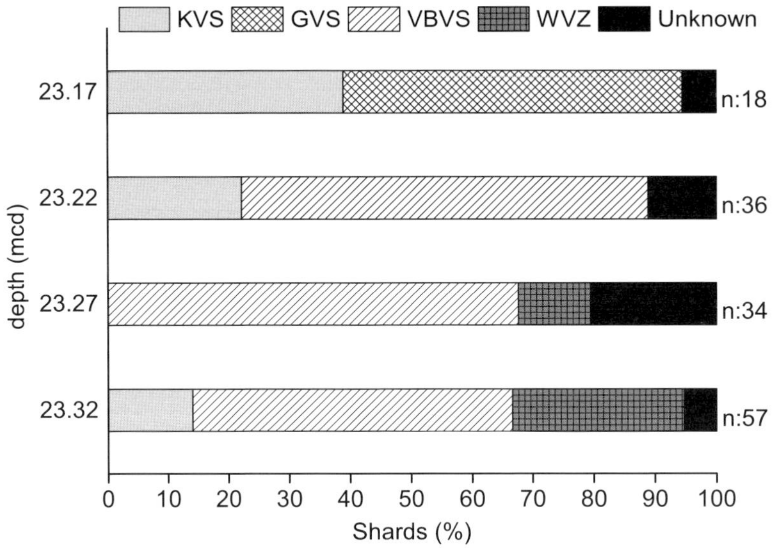

Fig. 7. The relative abundance of tephra shards from IODP Site U 1304 derived from the different Icelandic volcano systems changes over the four sample depths analysed. AVS, Askja; KVS, Krafla; GVS, Grimsvòtn; VBVS, Veidivötn–Bárdabunga volcano systems. WVZ, Western volcanic zone; and unknown: outliers and transitional-alkalic shards.

in ventilation of bottom waters and hence a weakening of the Atlantic Meridional Overturning Circulation (AMOC) (Fig. 2). The fact that the tephra are associated with an interval of enhanced IRD input and reduction in AMOC strongly points to hydrographic reorganization in the North Atlantic at this time, which further supports the argument that the tephra are iceberg-rafted (e.g. Small *et al.* 2013).

As Figure 7 shows, the proportions of tephra shards derived from different Icelandic volcanic systems vary over depth, which could have been caused by changes in the activity of certain volcanic

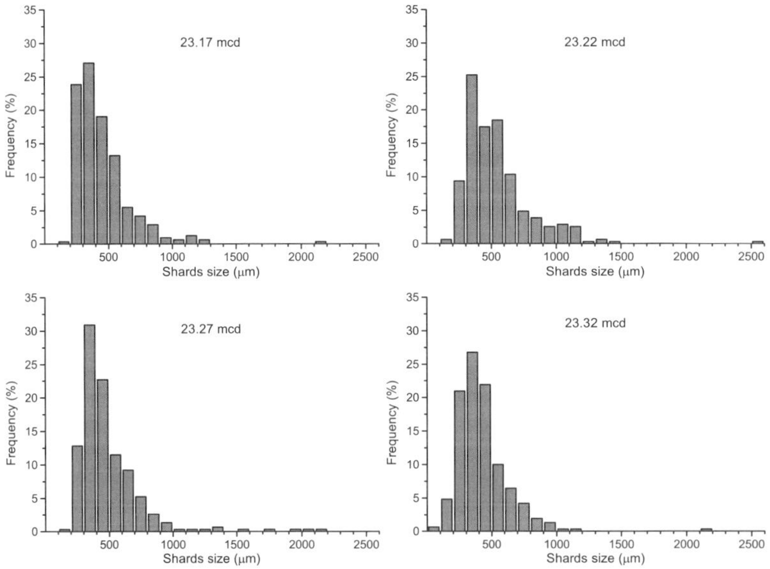

Fig. 8. Shard size frequency plots for each sample depth from IODP Site U 1304, North Atlantic. For each sample depth analysed, the maximum grain size of ≥300 shards was randomly measured. Note: samples only contain material >150 μm.

Table 1. *Mean, median, mode, standard distribution (sorting) and skewness (in units of [μm]/[ϕ]) for each sample depth based on the random sampling and measurement of maximum grain size of ≥300 basaltic glass shards from IODP Site U1304*

	23.17 mcd	23.22 mcd	23.27 mcd	23.32 mcd
Mean	453/1.27	546/1.00	490/1.17	425/1.37
Median	394/1.34	488/1.04	417/1.26	389/1.36
Mode	319/1.65	394/1.34	319/1.65	375/1.42
Sorting	−/0.57	−/0.56	−/0.58	−/0.57
Skewness	−/−0.15	−/−0.05	−/−0.23	−/0.02

Note: Samples only contain material >150 μm.

systems over time and/or transport mechanisms. Owing to the iceberg-rafted nature of the tephra, we suggest that the changes in geochemical dominance are most likely caused by changes in IIS dynamics. Very little is know about the IIS during the penultimate glacial period. Therefore, reconstructions and models of the IIS from the Last Glacial Maxiumum (LGM) have been used as a potential reference for the extent and behaviour of the IIS during MIS 6. During the LGM, Iceland and its shelf areas (Fig. 9) were almost completely ice-covered (Bourgeois *et al.* 2000; Hubbard *et al.* 2006). Bourgeois *et al.* (2000) identified six ice divides along which ice, including entrained detrital material such as tephra, was transported towards the marine calving margins (Fig. 10).

Estimating palaeo-drainage from tephra geochemistry in IRD deposits is problematic because tephra erupted from a given volcanic system can be transported by atmospheric processes in unknown directions and restricted distances from the source volcano depending on the shard size. Leadbetter & Hort (2011) simulated the transport, dispersion and deposition of an ash cloud from a short explosive Hekla eruption based on the meteorological conditions recorded for a six-year interval. The model shows that most of the time the majority of material would be transported eastwards owing to the prevailing westerly winds (Leadbetter & Hort 2011). Assuming that westerly winds also prevailed over Iceland during the LGM (Pollard & Thompson 1997) and that the analysed U 1304 tephra were probably only transported a relatively short distance from the source (see the section 'Shard size distribution'), it is likely that Grimsvòtn- and Veidivötn–Bárdabunga-like tephra were transported by the ice streams either to the north, south or SW. In contrast, WVZ- and Krafla-type tephra would have drained to the SW and north, respectively. Changes in ice-flow intensity along the different ice divides might therefore have caused the observed changes in abundances of tephra derived from different volcanic systems seen in Figure 7.

During the time of tephra deposition, surface circulation patterns must have existed which allowed the transport of icebergs from the northern as well as from the southern IIS margin to the core site at 53°N. Mapping IRD-deposition changes recorded in different marine cores from the North Atlantic region has allowed the reconstruction of iceberg trajectories during the late Quaternary (Ruddiman & Glover 1972; Ruddiman 1977; Bond & Lotti 1995; Bond *et al.* 1997). During the last interglacial–glacial cycle, significant changes in the IRD-deposition pattern are observed, indicating that icebergs travelled along different pathways between glacial and interglacial states (Ruddiman & Glover 1972; Ruddiman 1977; Bond & Lotti 1995; Bond *et al.* 1997). These changes in iceberg trajectories reflect changes in atmospheric and ocean circulation, in particular the surface circulation. Reconstructions (Robinson *et al.* 1995; Bond & Lotti 1995; Ruddiman 1977; Fig. 9) and models (Watkins *et al.* 2007) of surface circulation patterns for the last glacial show that surface circulation was able to transport icebergs from the northern as well as from the southern margin of the IIS; this may hold true for the penultimate glacial period if ocean circulation was the same or at least similar. Surface currents transported icebergs from the Fennoscandian Ice Sheet northwards before the current turned southward, entering the subpolar North Atlantic by passing eastern then southern Iceland as it travelled to the west (Ruddiman 1977; Robinson *et al.* 1995). Here, southern Icelandic icebergs were picked up by the South Iceland Current (Robinson *et al.* 1995) and were transported first to the west, before the current turned again to the south joining a counter-clockwise loop of surface circulation in the North Atlantic, the so-called subpolar cyclonic gyre (Ruddiman 1977; Robinson *et al.* 1995). The East Greenland Current transported icebergs originating from the northern IIS to the south into the North Atlantic, where the current merges into the subpolar gyre (Ruddiman 1977). This gyre circulation transported the Icelandic tephra-bearing icebergs southwards and then

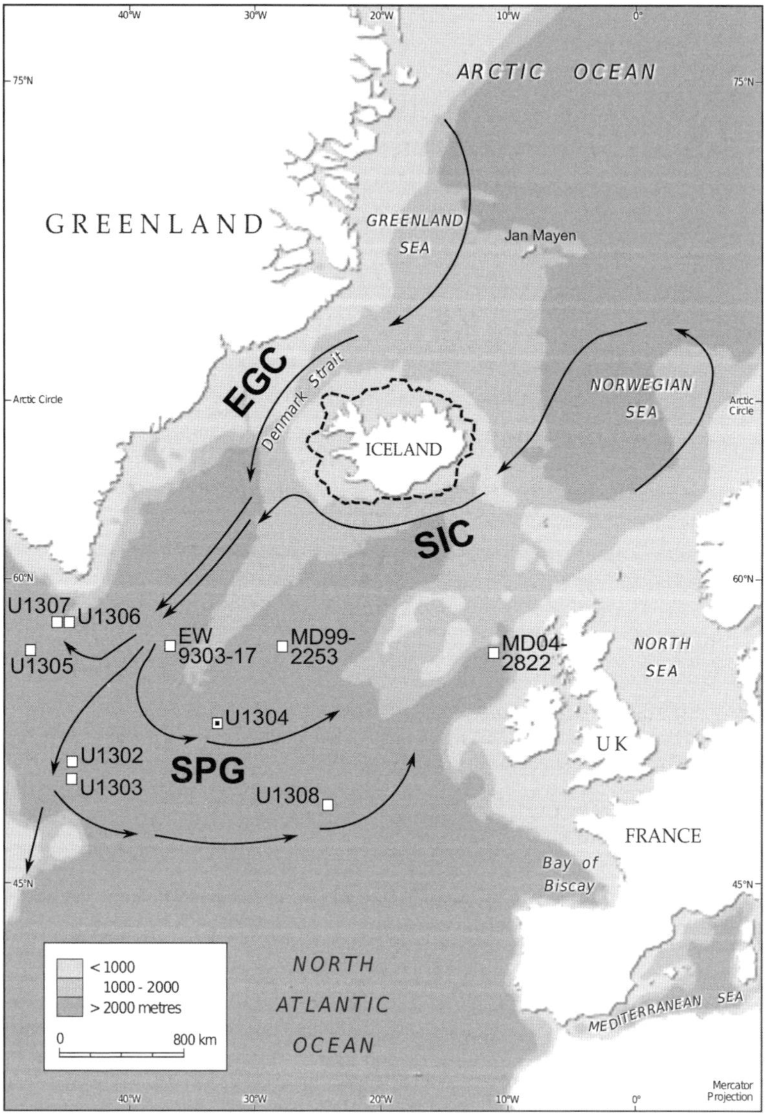

Fig. 9. Location of IODP Site U 1304 sediment core from the Gardar Drift, North Atlantic and additional core locations for potential further investigation. All cores U 1302 through U1308, see Expedition 303 Scientists (2006*a*); MD04-2822, see Hibbert *et al.* (2010); MD99-2253, see Farmer (2012) and Chapman (2012); and EW9303-17, see Bodén & Backman (1996). Dashed line indicates the Icelandic Ice Sheet (IIS) limits at the Last Glacial Maximum (LGM) after Geirsdóttir *et al.* (2007). Glacial surface ocean circulation after Ruddiman & Glover (1972), Ruddiman (1977) and Robinson *et al.* (1995). EGC, East Greenland Current; SIC, South Iceland Current; SPG, Subpolar Gyre.

eastwards to the U 1304 core site on the Gardar Drift. This subpolar cyclonic gyre appears to be a persistent feature of glacial and interglacial times. However, during interglacials this gyre is much more weakly developed and had a smaller spatial extent (Ruddiman & Glover 1972). During the last glacial period the gyre's southern extent reached as far south as 46°N, which led to maximum IRD deposition along a W–E trending zone around 46–50°N, that is, the so-called 'Ruddiman Belt' (Ruddiman 1977). This expansion to lower latitudes results from a southward-shift of the polar front, which allowed icebergs to travel in cold water further to the south before they melted. Evidence for a southward shifted polar front is also seen in the U 1304 surface proxies, that is, the high *Np* (s) percentage

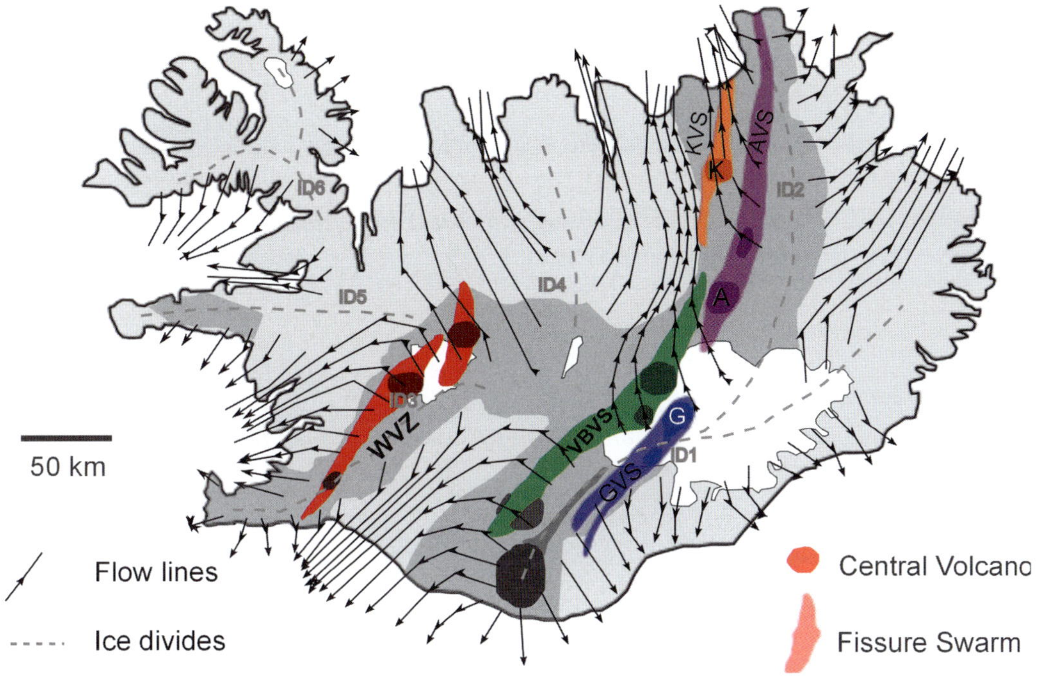

Fig. 10. Potential Icelandic source volcano systems and volcano zones for the U 1304 tephra. Flow lines (black arrowed lines) show the direction of ice flow patterns of the IIS during the LGM; after Bourgeois *et al.* (2000). Western volcanic zone (WVZ), Veidivötn–Bárdabunga volcano system (VBVS), Grimsvòtn volcano system (GVS), Krafla volcano system (KVS) and Askja volcano system (AVS). Light grey dashed lines: ice divides (ID) as identified by Bourgeois *et al.* (2000). The positions of major volcanoes and volcano systems are from Jónasson (2007) and Larsen & Eiríksson (2008).

as well as heavier planktonic $\delta^{18}O$ (Fig. 2) values, which indicate cold Arctic surface water at the core site.

Conclusions

The heterogeneous composition, large shard size and IRD association of the U 1304 tephra deposit indicate that it was transported by iceberg-rafting processes. These data suggest that the IIS had active marine calving margins during MIS 6, probably of similar extent to those of the LGM (Bourgeois *et al.* 2000; Hubbard *et al.* 2006).

Geochemical analyses confirm an Icelandic origin for all the tephra shards and comparison of this data with the published geochemistry of Holocene Icelandic volcanic systems allows shards to be assigned to discrete volcanic sources. The changes in tephra abundance derived from certain volcanic systems over depth/time are tentatively interpreted as resulting from fluctuations in the IIS ice flow pattern.

Icebergs are thought to follow well-defined pathways, which are dictated by prevailing wind and surface ocean circulation. During glacial periods, the surface circulation was dominated by a cyclonic gyre in the subpolar North Atlantic. Reconstruction and models of last glacial surface ocean circulation show that it was possible to transport tephra from Iceland to the U 1304 core site. However, in order to prove that these glacial surface circulation models are actually valid for the late MIS 6 period more data will be required from different core sites across the North Atlantic. Some of the potential core sites from this region are identified in Figure 9. Using geochemical analyses and shard size analyses of tephra from different core locations in the North Atlantic would not only allow us to determine the transport mode of the tephra and to identify the tephra source, but also to reconstruct the surface circulation, which transported the tephra to its depositional location.

The understanding of tephra transport mechanisms has a special importance for Quaternary research. During much of the Quaternary cold conditions prevailed and ice-rafting processes provide a challenge to (i.e. IRD-associated tephra) the detection of potential isochronous markers which might help to establish and to improve regional tephrochronologies. However, iceberg-rafted tephra are an equally important tool in palaeoclimate studies and their potential for the reconstruction of North Atlantic ice sheets and surface circulation has not been fully developed. Moving beyond the simple 'Icelandic glass' descriptive concept to a quantitative characterization assigning the Icelandic tephra to potential volcanic sources would allow us to test and improve established surface circulation models and would yield new information about the IIS during past glaciations.

We are grateful to D. Herd and C. Hayward for their assistance in the use of the electron microprobe facilities at the University of St Andrews and at the University of Edinburgh, respectively. Thanks to S. Pyne-ODonnell and an anonymous reviewer for their constructive comments on the manuscript. We acknowledge the support of the IODP and the Natural Environment Research Council (NERC); M. K. is funded through the NERC studentship (NE/G007373/1). W. E. N. A. and P. M. A. acknowledge the support of NERC through the Synchronising Marine and Ice-Core Records Using Tephrochronology (SMART) project (NE/F020600/1, NE/F02116X/1, NE/F021445/1).

References

Abbott, P. M., Davies, S. M. *et al.* 2012. A detailed framework of Marine Isotope Stages 4 and 5 volcanic events recorded in two Greenland ice-cores. *Quaternary Science Reviews*, **36**, 59–77.

Austin, W. E. N., Wilson, L. J. & Hunt, J. B. 2004. The age and chronostratigraphical significance of North Atlantic Ash Zone II. *Journal of Quaternary Science*, **19**, 137–146.

Bianchi, G. G. & McCave, I. N. 2000. Hydrography and sedimentation under the deep western boundary current on Björn and Gardar Drifts, Iceland Basin. *Marine Geology*, **165**, 137–169.

Bodén, P. & Backman, J. 1996. A laminated sediment sequence from the northern North Atlantic Ocean and its climatic record. *Geology*, **24**, 507–510.

Bond, G., Showers, W. *et al.* 1997. A pervasive millennial-Scale cycle in North Atlantic Holocene and glacial climates. *Science*, **278**, 1257–1266.

Bond, G. C. & Lotti, R. 1995. Iceberg discharges into the North Atlantic on millennial time scales during the last glaciation. *Science*, **267**, 1005–1010.

Bourgeois, O., Dauteuil, O. & van Vliet-Lanoë, B. 2000. Geothermal control on flow patterns in the last glacial maximum ice sheet of Iceland. *Earth Surface Processes and Landforms*, **25**, 59–76.

Brendryen, J., Haflidason, H. & Sejrup, H. P. 2010. Norwegian Sea tephrostratigraphy of marine isotope stages 4 and 5: prospects and problems for tephrochronology in the North Atlantic region. *Quaternary Science Reviews*, **29**, 847–864.

Chapman, M. 2012. Climate variability in the mid-latitude North Atlantic during the previous two interglacials. *Quaternary International*, **279280**, 84.

Davies, S. M., Abbott, P. M., Pearce, N. J. G., Wastegård, S. & Blockley, S. P. E. 2012. Integrating the INTIMATE records using tephrochronology: rising to the challenge. *Quaternary Science Reviews*, **36**, 11–27.

EXPEDITION 303 SCIENTISTS. 2006*a*. Expedition 303 summary. *In*: Channell, J. E. T., Kanamatsu, T., Sato, T., Stein, R., Alvarez Zarikian, C. A., Malone, M. J. & THE EXPEDITION 303/306 SCIENTISTS (eds) *Proceedings of the Integrated Ocean Drilling Program 303/306*. Integrated Ocean Drilling Program Management International Inc., College Station, TX.

EXPEDITION 303 SCIENTISTS. 2006*b*. Site U 1304. *In*: Channell, J. E. T., Kanamatsu, T., Sato, T., Stein, R., Alvarez Zarikian, C. A., Malone, M. J. & THE EXPEDITION 303/306 SCIENTISTS (eds) *Proceedings of the Integrated Ocean Drilling Program 303/306*. Integrated Ocean Drilling Program Management International Inc., College Station, TX.

Farmer, E. 2012. Climate and hydrography of the subpolar North Atlantic during the last interglacial. *Quaternary International*, **279280**, 137.

Fisher, R. V. 1964. Maximum size, median diameter, and sorting of tephra. *Journal of Geophysical Research*, **69**, 341–355.

Folk, R. L. & Ward, W. C. 1957. Brazos River bar (Texas); a study in the significance of grain size parameters. *Journal of Sedimentary Research*, **27**, 3–26.

Fratantoni, D. M. 2001. North Atlantic surface circulation during the 1990's observed with satellite-tracked drifters. *Journal of Geophysical Research: Oceans*, **106**, 22 067–22 093.

Geirsdóttir, A., Miller, G. H. & Andrews, J. T. 2007. Glaciation, erosion, and landscape evolution of Iceland. *Journal of Geodynamics*, **43**, 170–186.

Habicht, J. K. A. 1979. *Paleoclimate, Paleomagnetism and Continental Drift*. American Association of Petroleum Geologists, Tulsa, OK.

Hayward, C. 2012. High spatial resolution electron probe microanalysis of tephras and melt inclusions without beam-induced chemical modification. *The Holocene*, **22**, 119–125.

Heiken, G. A. & Wohletz, K. A. 1985. *Volcanic Ash. Los Alamos Series in Basic and Applied Sciences*. University of California Press, Berkeley, CA, 1–246.

Hibbert, F. D., Austin, W. E. N., Leng, M. J. & Gatliff, R. W. 2010. British Ice Sheet dynamics inferred from North Atlantic ice-rafted debris records spanning the last 175 000 years. *Journal of Quaternary Science*, **25**, 461–482.

Hodell, D. A. & Curtis, J. H. 2008. Oxygen and carbon isotopes of detrital carbonate in North Atlantic Heinrich Events. *Marine Geology*, **256**, 30–35.

Hodell, D. A., Minth, E. K., Curtis, J. H., McCave, I. N., Hall, I. R., Channell, J. E. T. & Xuan, C. 2009. Surface and deep-water hydrography on Gardar Drift (Iceland Basin) during the last interglacial period. *Earth and Planetary Science Letters*, **288**, 10–19.

Hubbard, A., Sugden, D., Dugmore, A., Norddahl, H. & Pétursson, H. G. 2006. A modelling insight into the Icelandic Last Glacial Maximum ice sheet. *Quaternary Science Reviews*, **25**, 2283–2296.

JAGAN, A. 2010. *Tephra stratigraphy and geochemistry from three Icelandic lake cores: a new method for determining source volcano of tepra layers*. MPhil, School of Geosciences, University of Edinburgh.

JAKOBSSON, S. P. 1979. Petrology of Recent basalts of the Eastern Volcanic System, Iceland. *Acta Naturalia Islandica*, **26**, 1–103.

JAKOBSSON, S. P., JÓNASSON, K. & SIGURDSSON, I. A. 2008. The three igneous rock series of Iceland. Jökull. *The Icelandic Journal of Earth Sciences*, **58**, 117–138.

JÓNASSON, K. 2007. Silicic volcanism in Iceland: composition and distribution within the active volcanic zones. *Journal of Geodynamics*, **43**, 101–117.

KITTLEMAN, L. R. 1973. Mineralogy, correlation, and grain-size distributions of Mazama tephra and other postglacial pyroclastic layers, Pacific Northwest. *Geological Society of America Bulletin*, **84**, 2957–2980.

LACKSCHEWITZ, K. S. & WALLRABE-ADAMS, H.-J. 1997. Composition and origin of volcanic ash zones in Late Quaternary sediments from the Reykjanes Ridge: evidence for ash fallout and ice-rafting. *Marine Geology*, **136**, 209–224.

LARSEN, G. & EIRÍKSSON, J. 2008. Late Quaternary terrestrial tephrochronology of Iceland-frequency of explosive eruptions, type and volume of tephra deposits. *Journal of Quaternary Science*, **23**, 109–120.

LEADBETTER, S. J. & HORT, M. C. 2011. Volcanic ash hazard climatology for an eruption of Hekla Volcano, Iceland. *Journal of Volcanology and Geothermal Research*, **199**, 230–241.

MAZAUD, A., CHANNELL, J. E. T., XUAN, C. & STONER, J. S. 2009. Upper and lower Jaramillo polarity transitions recorded in IODP expedition 303 North Atlantic sediments: implications for transitional field geometry. *Physics of the Earth and Planetary Interiors*, **172**, 131–140.

ÓLADÓTTIR, B. A., SIGMARSSON, O., LARSEN, G. & DEVIDAL, J.-L. 2011. Provenance of basaltic tephra from Vatnajökull subglacial volcanoes, Iceland, as determined by major- and trace-element analyses. *The Holocene*, **21**, 1037–1048.

PECK, V. L., HALL, I. R., ZAHN, R., GROUSSET, F., HEMMING, S. R. & SCOURSE, J. D. 2007. The relationship of Heinrich events and their European precursors over the past 60 ka BP: a multi-proxy ice-rafted debris provenance study in the North East Atlantic. *Quaternary Science Reviews*, **26**, 862–875.

POLLARD, D. & THOMPSON, S. L. 1997. Climate and ice-sheet mass balance at the last glacial maximum from the GENESIS version 2 global climate model. *Quaternary Science Reviews*, **16**, 841–863.

RAN, L., JIANG, H., KNUDSEN, K. L. & EIRÍKSSON, J. 2011. Diatom-based reconstruction of palaeoceanographic changes on the North Icelandic shelf during the last millennium. *Palaeogeography, Palaeoclimatology, Palaeoecology*, **302**, 109–119.

ROBINSON, S. G., MASLIN, M. A. & MCCAVE, I. N. 1995. Magnetic susceptibility variations in Upper Pleistocene deep-sea sediments of the NE Atlantic: implications for ice rafting and paleocirculation at the last glacial maximum. *Paleoceanography*, **10**, 221–250.

RUDDIMAN, W. F. 1977. Late Quaternary deposition of ice-rafted sand in the subpolar North Atlantic (lat 40° to 65°N). *Geological Society of America Bulletin*, **88**, 1813–1827.

RUDDIMAN, W. F. & GLOVER, L. K. 1972. Vertical mixing of ice-rafted volcanic ash in North Atlantic sediments. *Geological Society of America Bulletin*, **83**, 2817–2836.

SMALL, D., AUSTIN, W. E. N. & RINTERKNECHT, V. 2013. Freshwater influx, hydrographic reorganization and the dispersal of ice-rafted detritus in the sub-polar north atlantic ocean during the last deglaciation. *Journal of Quaternary Science*, **28**, 527–535.

SPARKS, R. S. J., WILSON, L. & SIGURDSSON, H. 1981. The pyroclastic deposits of the 1875 eruption of Askja, Iceland. *Philosophical Transactions of the Royal Society of London. Series A, Mathematical and Physical Sciences*, **299**, 241–273.

THORDARSON, T. & LARSEN, G. 2007. Volcanism in Iceland in historical time: volcano types, eruption styles and eruptive history. *Journal of Geodynamics*, **43**, 118–152.

WATKINS, S. J., MAHER, B. A. & BIGG, G. R. 2007. Ocean circulation at the Last Glacial Maximum: a combined modeling and magnetic proxy-based study. *Paleoceanography*, **22**, PA2204.

Holocene tephra from Iceland and Alaska in SE Greenland Shelf Sediments

ANNE JENNINGS[1]*, THORVALDUR THORDARSON[2,3], KATE ZALZAL[1,3], JOSEPH STONER[4], CHRISTOPHER HAYWARD[2], ÁSLAUG GEIRSDÓTTIR[3] & GIFFORD MILLER[1]

[1]*Institute of Arctic and Alpine Research and Department of Geological Sciences, University of Colorado, UCB 450, Boulder, CO 80309, USA*

[2]*School of GeoSciences, University of Edinburgh, Edinburgh EH9 3JW, UK*

[3]*Faculty of Earth Sciences, University of Iceland, Askja, Sturlugata 7, Reykjavík 101, Iceland*

[4]*College of Earth Ocean and Atmospheric Sciences, Oregon State University 104 COAS Admin Building Corvallis, OR 97331, USA*

**Corresponding author (e-mail: Anne.Jennings@Colorado.edu)*

Abstract: The record of Icelandic volcanic events in Holocene marine sediments off SE Greenland provides evidence for the frequency and timing of atmospheric tephra plume dispersal from Iceland towards Greenland. Geochemistry of tephra abundance peaks from two SE Greenland shelf cores: MD99-2322 and JM96-1215-2GC are compared with core MD99-2269, north Iceland shelf, to evaluate the dispersal direction of Icelandic eruptions. Glass shard counts (106–1000 μm) in MD99-2322 revealed 16 distinct cryptotephra peaks. Geochemical analyses of eight cryptotephra peaks in MD99-2322 and two in JM96-1215 indicate sources in the volcanic systems of Iceland and Alaska. A tephra layer matching in geochemistry and stratigraphy to the *c.* 3600 BP eruption of the Aniakchak Volcano in the Aleutian Islands was identified in JM96-1215/2GC. The Settlement Tephra (AD 871 $\pm$ 2) and Hekla B (H-B) were identified in MD99-2322. A new marker horizon, Katla EG-6.73, was found in both SE Greenland cores. Three basaltic peaks between 9.9 and 10.4 cal kyr BP, exhibit major-element geochemistry indistinguishable from the *c.* 10.2 kyr Saksunarvatn tephra. These layers represent 3 out of $\geq$ seven westward and northward-dispersed Grímsvötn layers found on the SE Greenland shelf and the north Iceland shelf between 9.9 and 10.4 cal kyr BP.

Supplementary material: a list of all analyses performed for this study is available at http://www.geolsoc.org.uk/SUP18715

Holocene explosive volcanic activity in the volcanic zones of Iceland has produced hundreds of tephra layers that are well preserved in soils, lake and marine sediments on and around Iceland. These numerous tephra layers have the potential to provide isochronous marker horizons to assist correlation among Icelandic archives (e.g. Thorarinsson 1944, 1967, 1981; Larsen 1984, 2002, Jóhannsdóttir 2007; Kristjánsdóttir *et al.* 2007; Larsen & Eiríksson 2008*a*; Thordarson & Höskuldsson 2008; Óladóttir *et al.* 2008, 2011*a*, *b*; Jagan 2010; Larsen *et al.* 2012; Meara 2012; Guðmundsdóttir *et al.* 2012). However, the high frequency of events from a few key central volcanoes, the tendency for these volcanoes to erupt magmas of near-identical compositions and the wind-induced atmospheric dispersal patterns all work together to complicate the establishment of marker horizons (Larsen *et al.* 1999; Larsen & Eiríksson 2008*b*). Tephra producing eruptions with tephra deposition well beyond the shores of Iceland (i.e. in the far-field) number far fewer than those with recorded tephra fall in Iceland (e.g. Thordarson & Höskuldsson 2008). However, hitherto the far-field identification is strongly biased towards tephra of felsic composition (>63 wt% SiO_2) because their identification has been more straightforward in the sediment archives in question (e.g. Dugmore *et al.* 1995; Hafliðason *et al.* 2000; Wastegård 2005). Over the last decade, high-resolution tephra studies of sediment cores from lakes within Iceland and on the north Iceland shelf have expanded the near- to far-field correlation of Holocene basalt to rhyolite tephra layers and extended the record to times predating the full deglaciation of Iceland (e.g. Larsen 2002; Eiríksson *et al.* 2004; Jóhannsdóttir 2007; Kristjánsdóttir *et al.* 2007; Guðmundsdóttir *et al.* 2012).

From: Austin, W. E. N., Abbott, P. M., Davies, S. M., Pearce, N. J. G. & Wastegård, S. (eds) 2014. *Marine Tephrochronology*. Geological Society, London, Special Publications, **398**, 157–193.
First published online February 27, 2014, http://dx.doi.org/10.1144/SP398.6

Explosive eruptions typically are short-lived events, with durations of hours to days (e.g. Francis & Oppenheimer 2004). The resulting ash plumes can be wide or narrow depending on the wind strength at the time of the eruption, height of tephra injection and variations therein during eruption as well as the duration of the eruption (e.g. Lacasse 2001). Therefore, the pattern of far-field tephra fallout can be complex, although the dispersal of the main ash plume commonly is strongly skewed in the downwind direction at the height of principal tephra injection. The overall pattern of tephra dispersal and far-field tephra deposition from the *c.* 20 km-high eruption plume of 2011 Grímsvötn event is a good example of the possible complexities (e.g. Stevenson *et al.* 2012). For these reasons a broad North Atlantic distribution for individual tephra time markers is unlikely. The dispersal pattern of tephra fall deposits in Iceland reveals the supremacy of eastward transport for volcanic plumes, consistent with net transport and dispersion within the east-flowing polar jet stream at heights at 7–15 km (e.g. Jónsson 1990). Recent tephrochronological studies in west Iceland (Jóhannsdóttir 2007; Thordarson & Höskuldsson 2008) indicate a sharp westward drop in the number of tephra layers in Holocene sediment archives within Iceland. These data imply very low tephra fall frequency – one every thousand years on average – for the offshore region west of Iceland. This information is consistent with the perception that the majority of explosive eruptions in Iceland feature 5–15 km-high eruption plumes or, in the case of higher plumes, main ash plume dispersal at or near the polar jet stream level (e.g. Guðmundsson *et al.* 1992; Lacasse *et al.* 2004; Höskuldsson *et al.* 2007; Oddsson *et al.* 2012; Guðmundsson *et al.* 2012). Stratospheric winds at heights >15 km above Iceland are seasonal, shifting from strong westerlies during the autumn and winter, to weaker easterlies in spring and summer (Lacasse 2001). Hence, it is only the most powerful explosive eruptions (e.g. Thorarinsson 1954, 1970; Carey *et al.* 2010), which correspond to the 1 in 200–500 year events (e.g. Thordarson & Höskuldsson 2008; Larsen & Eiríksson 2008*a*, *b*), that support >15 km-high eruption plumes and are capable of producing direct transport and deposition of tephra in Greenland and adjacent seas. Thus, utilization of Icelandic tephrochronology as a tool for synchronization of palaeoclimate proxy records across the North Atlantic requires knowledge of the tephra stratigraphy (including age), tephra composition and initial as well as principal plume dispersal direction in all areas circumscribing Iceland (e.g. Larsen 2002).

Over the last 60 ka, only six Icelandic eruptions have formed visible tephra layers in the Greenland ice cores. These are: (a) the 10.347 ka Grímsvötn tephra; *c.* 1 mm thick with grains up to 125 μm that has been linked to about 10 ka Saksunarvatn ash in the Faroe Islands; (b) the about 12 ka bimodal basalt–rhyolite Katla tephra; <1 mm thick with shards up to 70 μm in diameter that has been correlated with the Vedde Ash in Norway; (c) the basaltic (tholeiitic) *c.* 26.7 ka Fugloyarbanki tephra; (d) a mildly alkali basalt tephra at about 29.1 ka; (e) a tholeiite basalt tephra at about 38.1 ka; and (f) the *c.* 54 ka tephra present in several Greenland ice cores that has been correlated with the North Atlantic Ash Zone 2 and is thought to have been produced by an eruption within either the Torfajökull or Tindfjöll central volcano in Iceland (e.g. Grönvold *et al.* 1995; Zielinski *et al.* 1997; Mortensen *et al.* 2005; Davis *et al.* 2008; Svensson *et al.* 2008; Davies *et al.* 2010; Tomlinson *et al.* 2010, 2012). The relatively large grain size of these tephra horizons is consistent with direct deposition from westward or northwestward moving stratospheric plumes (Lacasse 2001). However, most of the Icelandic tephra horizons discovered so far in the Greenland ice cores, 45 in total spanning *c.* 60 ka (Grönvold *et al.* 1995; Mortensen *et al.* 2005; Abbott & Davies 2012) feature grains smaller than 25 μm and consistent with circumpolar transport within the polar vortex from Iceland to Greenland.

Little work has been done to develop the Holocene chronology for tephra in the marine sediment archives west of Iceland. In a previous study Jennings *et al.* (2002) focused on the tephrostratigraphy of the Kangerlussuaq Trough during the deglacial interval. They reported both Vedde Ash and Saksunarvatn ash on the SE Greenland shelf along with two younger, unidentified Holocene tephra horizons. Recent recognition of multiple Grímsvötn tephra layers in the 9.9–10.4 cal kyr BP interval in sediment archives in Icelandic lakes (Jóhannsdóttir *et al.* 2005; Jóhannsdóttir 2007) that have major element composition identical to Saksunarvatn ash in the Faroe Islands have exposed the uncertainty of ascribing any single tephra horizon to Saksunarvatn ash. Secondary transport on sea ice, which is thought to have contributed to the very wide dispersal of the late glacial Vedde Ash, is unlikely in warm periods such as the Holocene when sea ice in the Nordic Seas was much diminished (Koç & Jansen 1994). Hence, it is reasonable to assume that the marine sediment cores on East Greenland shelf contain a record of primary (or near-primary) tephra fall events. Therefore, there is a distinct advantage to working towards a complete Holocene tephrochronology for East Greenland shelf area because the identification of distinct tephra markers within the Holocene sediment sequence will enable robust correlation with other palaeoclimate proxy archives in the North Atlantic.

In this paper we expand on the tephrochronology of Jennings *et al.* (2002) and report on 22 basaltic to rhyolitic micro-tephra horizons identified within the Holocene section of two marine sediment cores (MD99-2322 and JM96-1215/2GC) from the Kangerlussuaq Trough, SE Greenland (Fig. 1a). We use established age models based on radiocarbon dating, in conjunction with geochemistry to assess the origin of 10 tephra horizons that were revealed by grain counting. Major element glass composition is used to assess the volcanic provenance and the source volcano and eruption when possible (Fig. 1b). We also compare and contrast the tephrochronology in our MD99-2322 core with that obtained from the MD99-2269 core from the North Iceland shelf (Kristjánsdóttir *et al.* 2007), because these two cores are correlated in detail on depth using palaeomagnetic secular variation (PSV) data and their radiocarbon ages were comingled to develop well-constrained age models that have been applied to both cores (Stoner *et al.* 2007; Jennings *et al.* 2011).

We discuss the origin, dispersal and potential utility of the Kangerlussuaq Trough tephra horizons as chronostratigraphic markers. A major motivation for this undertaking is to uncover marker horizons of known age that are useful for land sea correlation and for future assessment of the marine reservoir age. This will contribute to improved understanding of the marine environment through time as well as to more accurate and conforming correlative chronologies on regional scale (cf. Hafliðason *et al.* 2000; Eiríksson *et al.* 2004). Identification of specific tephra layers allows comparison to other archives (ice cores, lake, bog and marine records) to improve our understanding of the frequency and dispersal directions of Icelandic eruptive events through the Holocene.

Environmental setting

The Kangerlussuaq Trough is a deep cross-shelf bathymetric trough that terminates on the south side of the Denmark Strait (Fig. 1a). Cores JM96-1215/2GC (67°02.8′N, 30°51.6′W, water depth, 668 m, length 587 cm) and MD99-2322 (latitude 67°08.18′N, longitude 30°49.67′W, water depth 714 m, length 2617 cm) were collected from the deep middle basin of the trough in water depths well beyond the depths of iceberg scouring. The East Greenland Current sweeps cold, low-salinity Polar Water, sea ice and icebergs calved from the Greenland Ice Sheet southward along the shelf. A branch of the Irminger Current moves warm, salty Atlantic Water into Kangerlussuaq Trough beneath the surface Polar Water. The conjunction of these two currents forms the local marine Polar Front in the Denmark Strait. A belt of pack ice flowing with the East Greenland Current from the Arctic Ocean advances between late October and late December and persists for 6.5–8.5 months (Hastings 1960; Jennings *et al.* 2011). The offshore pack ice is important in this study because, were tephra to be deposited onto the sea-ice north of Kangerlussuaq Trough, it could be deposited in the trough during late summer as the sea-ice melts and the ice-edge retreats. Therefore we would expect the SE Greenland shelf to contain tephra from atmospheric dispersal of tephra to the NW and west of Iceland.

Materials and methods

Sediment core analyses and tephra counts

Tephrostratigraphic data are presented from two cores from the Kangerlussuaq Trough: MD99-2322 (shorthand = 2322) and JM96-1215/2GC (shorthand = 1215). The cores reveal very different sediment accumulation rates for the Holocene; 1215 was taken from the trough side where deglacial sediments were accessible with a 6 m core (Jennings *et al.* 2002) and the Holocene sequence is thin, whereas 2322 was raised from the thickest Holocene sediment pile in the basin centre (Labeyrie *et al.* 2003; Jennings *et al.* 2011). PSV data are used to evaluate the correlation of the tephra layers in the two Kangerlussuaq Trough cores. Core JM96-1215/2GC contains silicic tephra that has composition and age similar to that of the Vedde Ash as well as basaltic tephra that in age and chemical composition corresponds to the tephra sequence (that includes Saksunarvatn ash) formed by series of explosive eruptions at the Grímsvötn volcano around 10 ka BP (e.g. Jennings *et al.* 2002; Jóhannsdóttir 2007). Two tephra-abundance-peaks in the upper part of JM96-1215 previously identified by tephra grain counting (Jennings *et al.* 2002) are analysed for their chemical composition in this study. The multiple basaltic peaks in the *c.* 10 ka BP interval of this core (Jennings *et al.* 2002) are compared with basaltic peaks in MD99-2322 using the ages from the separate age models and comparison on the PSV.

Core MD99-2322 was sampled each centimeter from 0 to 450 cm. One-centimetre-wide samples were taken every 4–6 cm from 450 cm to the base of the core. Each sample was weighed wet, freeze-dried and re-weighed dry. Tephra counting was performed on the 106–1000 μm sieve fraction using samples prepared for foraminifera and isotope analyses (Jennings *et al.* 2011). Peaks in tephra grain abundance found by this technique are referred to as cryptotephras because they do not form visible or distinct layers and are only revealed by counting of shards (e.g. Dugmore *et al.* 1995; Turney *et al.*

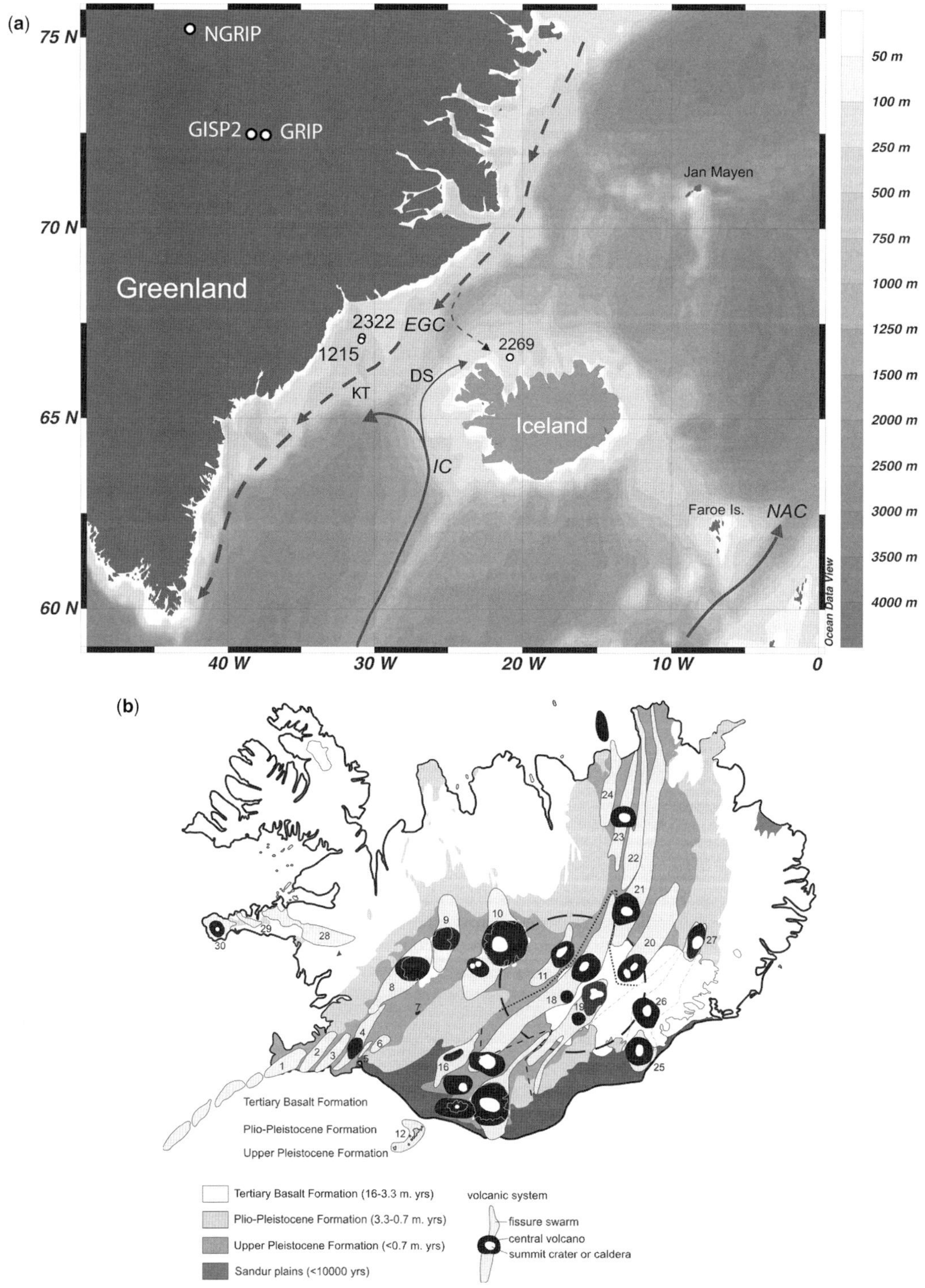

Fig. 1. (**a**) Bathymetric map of the Denmark Strait (DS) region showing location of cores MD99-2322, JM96-1215/2GC in the Kangerlussuaq Trough (KT) and MD99-2269 on the north Iceland shelf. General flowpaths of the East Greenland Current (EGC), Irminger Current (IC) and North Atlantic Current (NAC) are shown. Locations of Greenland

2004). Samples were split using a microsplitter to obtain a representative sample of tephra shards. In most cases, the entire sample was picked. Tephra grains were counted and categorized by morphology and colour. The total number of shards and shard types per gram of sediment in each sample were calculated by dividing the total number of shards in each sample by the dry bulk sediment weight. Tephra peaks were located by plotting the number shards per gram of dry bulk sediment. Tephra peaks of interest were selected for geochemical analysis in several stages as the counting in MD99-2322 progressed and distinct abundance peaks were defined. Available resources limited the number of peaks that could be geochemically analysed. It should be noted that the term 'pristine' is used here to refer to tephra grains that show no evidence of post-emplacement hydration (analytical sum >97.5 wt%) and preserve exceptionally delicate surface structures (i.e. not been exposed to abrasion) and thus have not been subjected to reworking or re-sedimentation.

The age model used for MD99-2322 is the ninth-order polynomial equation developed in Stoner *et al.* (2007) and applied by Jennings *et al.* (2011). For MD99-2269 the ages of tephra peaks presented by Kristjánsdóttir *et al.* (2007) were assigned using the linear age model of Stoner *et al.* (2007). Table 1 lists the ages and 1σ age errors of tephra peaks identified in MD99-2322 and -2269 using both the linear and the ninth-order polynomial age models. An age model for JM96-1215/2GC was made using the radiocarbon dates published for this core (Jennings *et al.* 2002). Calibration of the published ages was accomplished using using Calib 6.0 (Stuiver *et al.* 2010) with the Marine09 calibration curve (Reimer *et al.* 2009), assuming $\Delta R = 0$, and applying a zero-weighted fit between the calibrated ages. Ages of tephra layers defined in the marine cores are more likely to be slightly too old than too young because of the minimal marine reservoir correction applied.

Grain mounting

A selection of representative tephra shards in each sample were poured into 4.5 mm diameter holes drilled into a 1 inch epoxy resin block. A drop of Buehler Epo-Thin resin was added to the samples by running the resin down the side of the hole, taking care to avoid trapping air bubbles under the grains. After 10 min, more resin was added to fill the holes. The samples were cured at 2 bar pressure for 12 h and were fully cured after 72 h. The block surface was ground wet using 1200 then 2500 grit SiC paper until a sufficient number of grains were exposed. Then the grains were further polished using 0.3 μm aluminium oxide and distilled water on Kemet PSU-M polishing cloth for a minimum of 5–10 min. The presence of tephra grains with diameters as small as 50 μm in the geochemical analysis is related to breakage of grains during preparation, reduction in grain size during grinding or the two-dimensional size measurements made during geochemical analysis, which would miss a larger third dimension (Fig. 9).

Electron microprobe: analytical procedures

Electron probe microanalysis data were collected using the Cameca SX100 instrument at the School of Geosciences, University of Edinburgh. For analysis of groundmass glasses, two analytical set-ups were used, depending on the size of the areas of glass available. For the smallest areas, a beam diameter of 3 μm was used but for all other areas the beam diameter was 5 μm. For further details on analytical conditions see Hayward (2012), where it is also demonstrated that the use of small beam diameters in moderately hydrated glasses causes no mobilization of sodium if appropriately low beam currents (0.5 and/or 2 nA) are used. The glasses analysed for this study all fall well within the range of volatile contents at which such focused beams are appropriate. The accelerating voltage was 15 kV. Counting times for major elements was in all cases 20 s on peak and 10 s on each background position. Minor elements (P and Mn) were measured with counting times of 60 or 40 s at ≤80 nA. Analytical precision for individual elements is ≤5%.

The major element compositions as determined by the electron microprobe were used to determine

Fig. 1. (*Continued*) ice cores mentioned in the text are shown. (**b**) Distribution of active volcanic systems among volcanic zones and belts in Iceland as depicted by Jóhannesson & Sæmundsson (1998). Reykjanes Volcanic Zone: 1, Reykjanes–Svartsengi; 2, Krýsuvík; 3, Brennisteinsfjöll; Western Volcanic Zone: 4, Hengill; 5, Hróðmundartindur; 6, Grímsnes; 7, Geysir 8, Prestahnjúkur; 9, Kjölur; Mid-Iceland Belt; 10, Hofsjökull; 11, Tungnafellsjökull; Eastern Volcanic Zone: 12; Vestmannaeyjar; 13, Eyjafjallajökull; 14, Katla; 15, Tindfjöll; 16, Hekla–Vatnafjöll; 17, Torfajökull; 18; Veiðivötn–Dyngjuháls; 19, Grímsvötn; Northern Volcanic Zone; 20, Kverkfjöll; 21, Askja; 22, Fremrinámur; 23, Krafla; 24, Þeistareykir; Öræfajökul Volcanic Belt: 25, Öræfajökull; 26, Esjufjöll; 27, Snæfell; Snæfellsnes Volcanic Belt; 28, Ljósufjöll; 29, Helgrindur; 30, Snæfellsjökull. The large open circle indicates the approximate centre of the Iceland mantle plume/anomaly as depicted by Wolfe *et al.* (1997). The dotted line shows the northern limits of the East Volcanic Zone, whereas the hatched line indicates the boundary between the active and propagating rift segments of the zone. See key for further explanations.

Table 1. *Tephra layers identified in MD99-2322 and 2269 and their calibrated ages and 1σ age ranges based on published age models*

MD99-2269 Tephra name	Equivalent depth, 2322, cm	Void-corrected depth, 2269, cm	Ninth-order polynomial cal age yr BP, Jennings *et al.* (2011)	Linear age, cal yr BP, Stoner *et al.* (2007)	Maximum 1σ age range encompassing both age models
KOl 1372	84.86	156	725	664	608–762
Hekla 1104	141.78	243	1085	1077	1003–1171
Sn-1	227.46	412	1995	2028	1896–2102
Hekla 3	353.87	611	3176	3159	3074–3283
Hekla 4	497.44	937.5	4511	4472	4409–4572
TV5	792.26	1328	6798	6724	6649–6886
Suduroy	988.75	1508	8079	8140	7991–8217
G2 2269	1561.1	1888	9891	9912	9770–10051
G3 2269	1640.7	1958	10071	10043	9901–10191
Saks peak, 2269	1811.9	2086.5	10409	10373	10278–10497
base Saks, 2269	1815	2088	10416	10380	10283–10501
MD99-2322 Tephra name	**Tephra peak depth, 2322**	**Equivalent depth, 2269**	**Ninth-order polynomial age, Jennings *et al.* (2011)**	**Linear age, Stoner *et al.* (2007)**	**Maximum 1σ age range encompassing both age models**
2322/154.5	154.5	255	1134	1162	1056–1288
Settlement	158.5	262.03	1162	1221	1089–1325
H-B	336.5	588.34	3063	2984	2916–3128
2322/358.5; 'SILK MN?'	358.5*	618.39	3208	3217	3124–3329
2322/414.5	414.5	703.22	3663	3767	3545–3840
2322/446.5	446.5	799.55	3999	3918	3835–4118
2322/483.5	483.5	889.24	4359	4263	4194–4443
2322/555.5; 'Silk A1?'	555.5	1064.3	5089	5159	4918–5245
2322/663.5	663.5	1182.8	5802	5773	5698–5922
Katla EG- 6.73 ka	783.5	1318.2	6729	6636	6559–6822
2322/1209.5	1209.5	1651.5	8954	8947	8845–9120
2322/1605.5	1605.5	1933.5	10009	9996	9848–10130
Grims 2322-1617.5	1617.5	1942	10031	10013	9866–10151
Grims 2322-1719.5	1719.5	2003.5	10186	10151	10035–10302
2322/1749.5	1755.5	2026.7	10260	10218	10104–10372
Grims 2322-1797.5	1797.5	2083.1	10390	10365	10249–10489

1σ age ranges encompass uncertainties from both age models.
*Geochemistry at 334.5 cm.

the volcanic provenance of the grains from the analysed tephra horizons and when possible the source volcano and eruption (Fig. 1b) was determined via comparison with established reference compositions of tephra formations with known origin using the Iceland Reference Tephra Database compiled by T. Thordarson (unpublished data source 2013). The core of this database was constructed as part of the VAST project (Volcanism in the Arctic Systems) via analyses of representative near source tephra formations representing the spectrum of active volcanic systems in Iceland. The majority of these data (>3500 analyses) are still unpublished; they were acquired by the Iceland/Edinburgh VAST Tephra Study Group, including the dissertation work of Jóhannsdóttir (2007), Jagan (2010); Meara (2012), Hartley (2012) and unpublished data by T. Thordarson and C. Hayward (2012). This database has been supplemented by published data from the following sources: Grönvold & Makipaa (1978), Grönvold *et al.* (1983), Grönvold (1984), Jørgensen (1981), Steinþórsson (1978), Karhunen (1988), Trønnes (1990), Moore & Calk (1991), Sigurgeirsson (1992), Larsen *et al.* (1999, 2001), Moune *et al.* (2007), Óladóttir *et al.* (2008, 2011*b*), Thordarson *et al.*

(1996, 1998, 2001, 2003) and Tomlinson *et al.* (2010, 2012).

Results

Tephra abundance peaks located by grain counting

In this study, glass grains that exhibit pale brown, brown and black colour are mafic in composition. Microvesicular semi-equant grains (i.e. pumice) that are colourless, grey to pale yellowish white as well as colourless bubble-wall and bubble junction shards are demonstrably silicic in composition. Grains of intermediate composition range from brown to pale yellow brown in colour (Jóhannsdóttir 2007). Pale yellow brown vesicular grains and pale yellow brown pumice grains are included in the tally of dark tephra per gram (Fig. 2). Tephra-abundance peaks analysed for major element geochemistry are designated by an asterisk.

MD99-2322

Eleven mafic tephra abundance peaks (defined as >8 shards/g of dry bulk sediment) and six silicic tephra abundance peaks (defined as >5 shards/g of dry bulk sediment) were resolved by counting tephra grains in MD99-2322 (Fig. 2; Table 1). Mafic tephra peaks are most prominent in the interval from 1800 to 1600 cm, *c.* 10 400–10 000 cal years BP (five peaks), dropping to one peak in the interval 1600–790 cm (*c.* 10 000–6800 cal yr BP) before increasing again to four peaks from 790 to 0 cm (*c.* 6800–200 cal yr BP). Silicic peaks do not occur until 575 cm (*c.* 5200 cal yr BP) and all are present within the interval 575–330 cm (5200–3000 cal yr BP). Shards from eight of the defined peaks in 2322 were analysed for their chemical composition by electron microprobe.

154.5 cm (1144 cal yr BP). At 154.5 cm, a mafic to intermediate composition tephra peak of 11.4

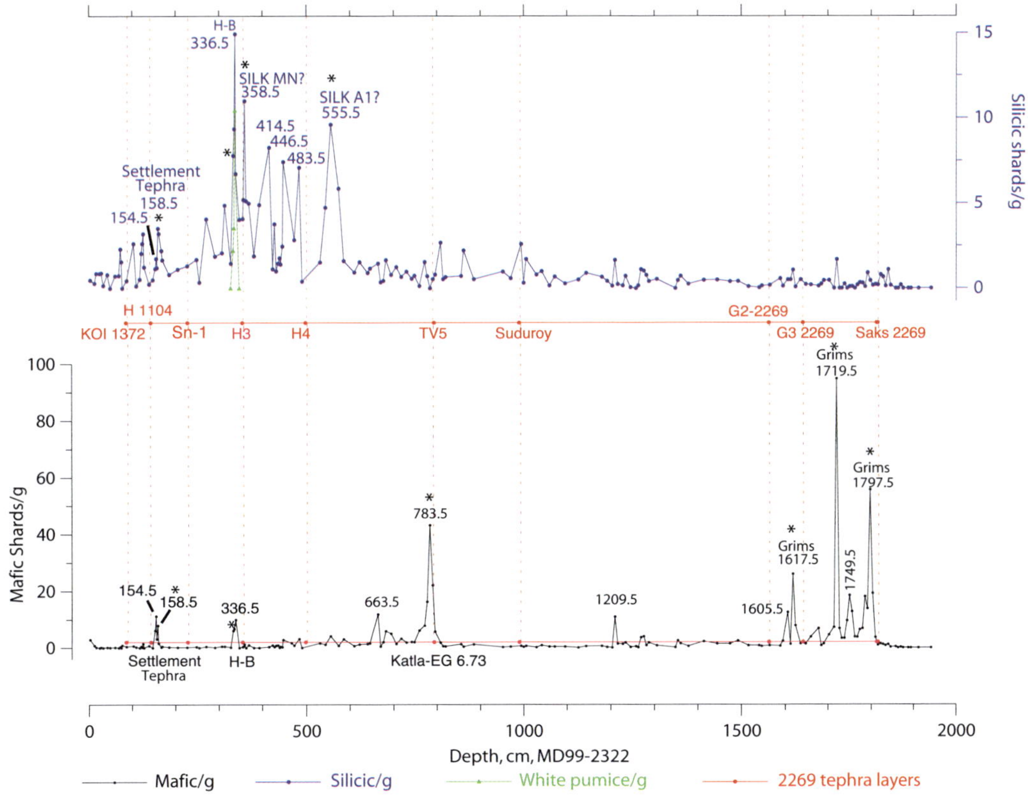

Fig. 2. Shards per gram of mafic (dark coloured; lower panel) and silicic (colourless and light coloured and white pumice; upper panel) tephra in MD99-2322. Asterisks denote levels with geochemical analyses presented in this paper. Red dots and labels are locations of MD99-2269 tephra layers mapped onto MD99-2322 depth based upon the PSV correlation of Stoner *et al.* (2007).

shards/g was found (Fig. 2). The shards are strongly vesicular light brown to pale yellow brown with a small component (1.7 shards/g) of colourless bubble-wall shards.

158.5 cm (1152 cal yr BP).* At 158.5 cm there is a two-component peak of strongly vesicular light brown to very pale yellow brown shards (8.0 shards/g) shards with a 3.5 shard/g peak of colourless bubble-wall and bubble-junction shards.

336.5 cm (3063 cal yr BP).* This peak extends from 332 to 337 cm with peak tephra abundance at 336.5 cm (Fig. 2). Geochemistry was determined on 334–335 cm. The peak contains both light and dark tephra grains. The light-coloured population of 15 grains/g comprises white to grey pumice (10.5 grains/g) and colourless platy bubble-wall shards (4.5 shards/g). The dark components comprise brown to pale yellow highly vesicular grains (8.4 mafic shards/g). Yellow vesicular grains are the most abundant.

358.5 cm (3208 cal yr BP).* This is a silicic peak (11.0 shards/g) of colourless platy bubble-wall and bubble-junction shards. Brown vesicular grains of 1.2 shards/g are present (Fig. 2).

414.5 cm (3663 cal yr BP). This is a silicic, 8.3 shards/g peak containing pristine colourless platy bubble-wall and -junction shards. Rare (*c.* 1 shard/g) brown vesicular mafic grains are also present at this level (Fig. 2).

446.5 cm (3999 cal yr BP). This is a silicic 7.4 shards/g peak of fresh-looking colourless platy bubble-wall and bubble-junction shards (Fig. 2). A minor amount of brown vesicular mafic shards are present (2.9 shards/g). The mafic shards are largely uniform in terms of size and colour, although a few feature rounded edges indicative of abrasion during transport and may represent a re-sedimented fraction.

483.5 cm (4359 cal yr BP). This is a silicic tephra-abundance peak of 7.1 shards/g comprising pristine-looking platy grains as well as bubble-wall shards (Fig. 2). A couple of the platy shards exhibit yellowish colour and have rounded edges, suggesting abrasion during re-sedimentation. Highly to moderately vesicular, light brown to dark brown mafic grains are present within this peak in minor amounts (3.2 shards/g).

555.5 cm (5089 cal yr BP).* This is a silicic peak of 9.6 shards/g typified by platy colourless bubble-wall shards (Fig. 2). The peak is associated with rare/minor (4.2 shards/g) small brown vesicular mafic grains.

663.5 cm (5802cal yr BP). This is an 11.9 shards/g peak in uniform brown vesicular mafic grains that includes rare (1.5 shards/g) colourless, platy bubble-wall shards (Fig. 2).

783.5 cm (6729 cal yr BP).* This is a 43.2 shard/g peak comprising only brown vesicular mafic grains (Fig. 2). Geochemical analyses were made on grains from depth of 795.5 cm representing the rising limb of the peak. Platy bubble-wall shards (4.2 shards/g) occur on the rising and falling limbs of the peak, but the core of the peak contains no rhyolitic shards.

1209.5 cm (8954 cal yr BP). This is a small peak in dark to light brown vesicular to strongly vesicular mafic grains defined by one data point of 11.1 shards/g (Fig. 2). A small peak of 1.7 shards/g of colourless platy bubble-wall shards coincides with the mafic peak.

1605.5 cm (10 009 cal yr BP). This is a 12.7 shard/g peak of brown vesicular mafic grains (Fig. 2). The shards are identical in appearance to the shards forming peaks at 1617.5, 1719.5, 1755.5 and 1797.5 cm. Very few colourless bubble-wall shards (0.5 shards/g) were found in the sample.

1617.5 cm (10 031 cal yr BP).* This peak is formed from 26.1 shards/g of brown vesicular mafic grains (Fig. 2). Very few colourless platy bubble-wall and bubble-junction rhyolitic shards (1.08 shards/g) were found in the sample.

1719.5 cm (10 186 cal yr BP).* This peak comprises brown vesicular mafic grains in concentration of 94.9 shards/g (Fig. 2). A few (1.7 shards/g) colourless bubble-wall shards are present.

1749.5 cm (10 260 cal yr BP). This 13 shards/g peak is uniformly formed from brown vesicular mafic grains identical in morphology and colour to the adjacent mafic peaks (1719.5 and 1797.5; Fig. 2). Colourless bubble-wall shards are essentially absent.

1797.5 cm (10 390 cal yr BP).* This peak is formed of 56.8 shards/g of brown vesicular mafic grains that are identical in composition to peaks at 1719.5 and 1617.5 cm.

JM96-1215/GC

Tephra abundance in JM96-1215/2GC published in Jennings *et al.* (2002) is shown against depth and age in Figure 3. We did not attempt a detailed tephrostratigraphy of this core. Instead we targeted previously identified tephra abundance peaks for geochemical analysis. Shards from tephra-abundance peaks at 18–26 and 50–52 cm were analysed for major element composition (Fig. 3). In addition, the previous counts from the early Holocene interval that showed multiple mafic peaks around

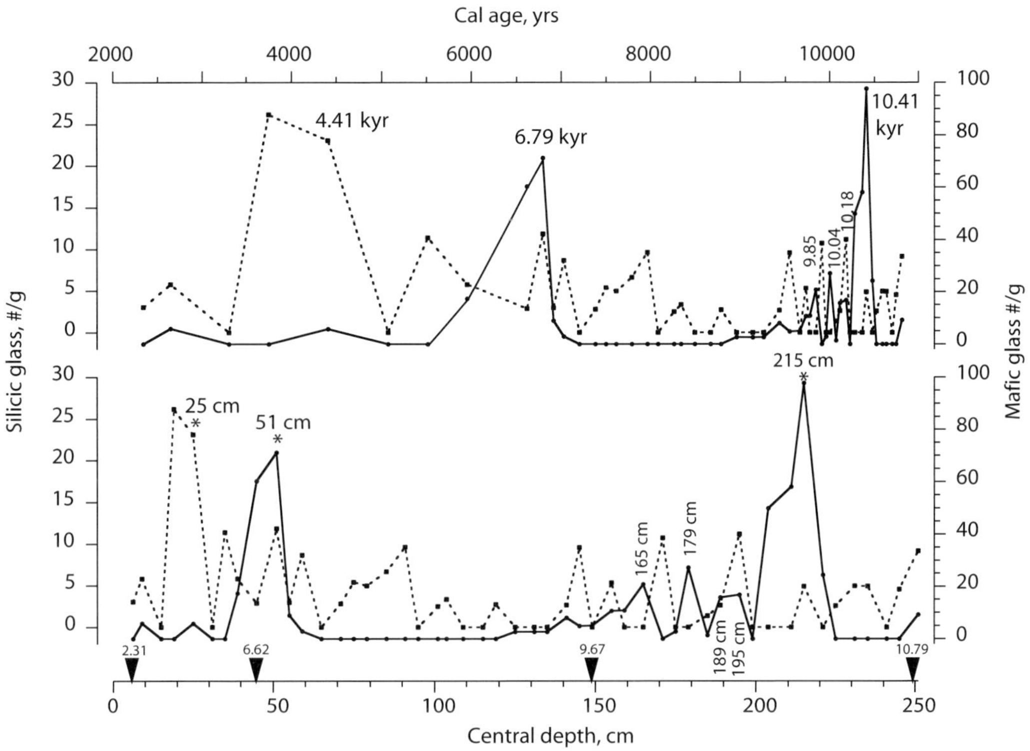

Fig. 3. JM96-1215/2GC shards per gram of mafic (solid line) and silicic (broken line) against calibrated age (upper) and depth in core, centimetres (lower). Positions of radiocarbon dates (details presented in Jennings *et al.* (2002, their table 2) shown by arrowheads; the calibrated ages are shown. One sigma age ranges for the calibrated ages are as follows: 5–7 cm, 2201–2422 yr; 44–46 cm, 6420–6815 yr; 148–150 cm, 9551–9779 yr; 248–250 cm, 10 656–10 932 yr. Asterisks denote tephra abundance peaks that have had geochemical analysis in this paper or in Jennings *et al.* (2002).

the 10 ka BP time period (Jennings *et al.* 2002) are re-evaluated in light of our findings in MD99-2322 (Fig. 3).

18–26 cm (3750–4410 cal yr BP).* This peak is defined by tephra grain counts of two samples: 18–20 cm (3750 cal yr BP) and 24–26 cm* (4410 cal yr BP). The ages of these samples are poorly constrained by two bounding dates in the core and the 2 cm width of the samples results in a broad age range for the peak (Fig. 3). At 18–20 cm there is a 26.1 shard/g peak in platy bubble-wall and bubble-junction shards of silicic glass with no mafic shards counted (Fig. 3). At 24–26 cm there are 23.1 colourless bubble-wall shards/g, and a small peak in brown vesicular mafic grains of 5.8 shards/g. This is the sample that was analysed for geochemistry.

50–52 cm (6793 cal yr BP).* This is a 71.1 shards/g peak of brown vesicular mafic grains (Fig. 3). Silicic bubble-wall and bubble junction shards estimated at 11.9 shards/g form this peak.

No basaltic shards were observed between 65 and 119 cm (7200–8790 cal yr BP; Fig. 3). Four peaks in brown vesicular mafic shards occur between 165 cm (9850 cal yr BP) and 221 cm (10 478 cal yr BP) (Fig. 3).

164–166 cm (9846 cal yr BP). This peak in brown vesicular mafic grains reaches 20.7 shards/g.

178–180 cm (10 004 cal yr BP). This brown vesicular basalt peak reaches 27.1 shards/g.

188–196 cm (10 117–10,184 cal yr BP). This peak is formed from two samples: 188–190 cm (10 117 cal yr BP) and 194–196 cm (10 184 cal yr BP). These two samples have similar concentrations of brown vesicular mafic grains: 15.7 and 16.8 shards/g, respectively, suggesting that a peak of higher concentrations may occur between them.

202–222 cm (10 283–10 478 cal yr BP). This is the largest of the brown vesicular mafic grain peaks with the highest shard counts per gram (97.7 shards/g) at 214–216 cm (10, 410 cal yr BPl Fig. 3). A small silicic bubble-wall shard content of 4.9 shards/g also occurred at this depth. Shards from 214–216 cm have composition indicating Grímsvötn as the source volcano and, based on its stratigraphic position, were identified as Saksunarvatn ash by Jennings *et al.* (2002).

Tephra chemistry

General aspects

We have measured the major element composition of glass grains from 10 of the tephra horizons identified in 2322 and 1215. In total, 141 grains were measured and the number from individual horizons ranges from 6 to 24, depending on the number of grains available for analyses in individual samples (Table 2). The majority of the measured grains (*c.* 95%) have totals in excess of 97% and thus can be considered to be unaffected by post-emplacement alteration or hydration (see Supplementary Material A). Only eight grains have totals below 97%, which most likely reflects post-emplacement hydration and thus dilution rather than modification of the original element concentrations. Overall, the chemical composition of the grains is distinctly bimodal, either basaltic or rhyolitic, with one exception – horizon MD99-2322 334–335 cm – which features four (out of 12) grains with intermediate (i.e. icelandite) composition (Table 2; Fig. 4).

The chemical composition indicates Icelandic provenance for the majority of the measured tephra grains (130 grains), which all fall on the Fe-rich trends of the Icelandic magma series. A smaller population, 11 grains of silicic composition, exhibits calc-alkaline provenance (Fig. 4) and, as will be demonstrated below, has composition that is compatible with silicic magmas from the Alaska–Cascade volcanic arcs.

The majority of the measured tephra grains (= 91) are of basaltic composition and all exhibit distinct Icelandic provenance as they follow the iron-enrichment trend that typifies the basaltic magmatism in Iceland (Fig. 4). Tholeiitic v. alkaline magmatism in Iceland exhibits a distinctive regional distribution. Production of tholeiite magma is confined to volcanic systems within the established axial rifts, namely the Reykjanes Volcanic Belt, Western and Northern Volcanic Zones and the northern half of the Eastern Volcanic Zone (Fig. 1b). On the other hand, eruption of alkaline magmas typifies the systems within the actively propagating sector (i.e. the southern half) of the Eastern Volcanic Zone and that of the intra-plate volcanic zones of Snæfellsnes and Öræfajökull (e.g. Thordarson & Höskuldsson 2008). This regional division of the magma series is particularly useful for basaltic compositions because it effectively limits the number of possible source volcanoes for alkali basalt tephra to three systems (Fig. 5). Six of the 10 tephra horizons subjected to chemical analysis are solely or predominantly basaltic in composition, while three contain a subsidiary fraction of basaltic grains (Fig. 5). The only tephra horizon for which no mafic shards were analysed is JM96-1215 24–26 cm (Table 2). Two of the basaltic horizons (JM96-1215 50–52 cm and MD99-2322 795–796 cm) are typified by mildly alkaline composition, while the remaining four horizons (MD99-2322 158–159, 1618.5, 1719.5 and 1797.5 cm) are distinctly tholeiitic (Fig. 5). Two of three tephra horizons with a subsidiary fraction of basaltic grains, MD99-2322 334–335 cm and 555–556 cm, feature equal number of grains with alkaline and tholeiite affinities, while in one, MD99-2322 358–359 cm, the basalt component is solely tholeiitic (Fig. 5).

A total of 46 grains are rhyolites and all exhibit mildly alkaline affinities (Fig. 6a). As indicated above, 11 of those are of calc-alkaline, while 35 of the grains have composition of Icelandic provenance (Fig. 4). The Icelandic v. calc-alkaline division of the rhyolite grains is further underpinned by $Na_2O + K_2O/FeO$, $Na_2O + K_2O/TiO_2$ and FeO/MgO values (e.g. Fig. 6b).

In terms of MgO, TiO_2, CaO and K_2O concentrations, the calc-alkaline grains are distinct from the rest of rhyolite tephra (Fig. 7; Table 2). Furthermore, nine of the calc-alkaline tephra grains show remarkable similarity in their composition and cluster very tightly on element-to-element plots, while two of the calc-alkaline grains are slightly different in composition (Fig. 7a). The majority of the silicic tephra grains of Icelandic provenance (33 out of 35) fall into two compositional groups defined by relatively high (/low) FeO, CaO, TiO_2 and low (/high) K_2O (Fig. 7b, c). Twenty-two of 35 grains have compositions similar to Katla rhyolite, while 11 are comparable to the rhyolites of the Tindfjöll volcano (Fig. 7). Shards of both compositional groups are present in all of the silicic peaks analysed in this study. However, it should be noted that many Hekla tephra layers feature, as a minor component, silicic grains with compositions similar to that of Torfajökull and Katla. Similarly, Katla tephra deposits often contain a minor component of silicic grains that resemble Tindfjöll rhyolites (Fig. 7d). Of the two remaining grains, the one from MD99-2322 158–159 cm has composition indicative of rhyolite tephra from Torfajökull,

Table 2. *Major element geochemical data from microprobe analyses of tephra peaks in MD99-2322 and JM96-1215/2-GC*

Tephra horizon	Grain	N	SiO_2	TiO_2	Al_2O_3	FeO	MnO	MgO	CaO	Na_2O	K_2O	P_2O_5	Total	Provenance – region	Compositional domain*	Likely source volcano
Jm96-1215 24–26 cm	1	4	71.71	0.51	14.95	2.24	0.14	0.50	1.73	4.94	3.06	0.08	99.84			
	3	3	70.47	0.51	14.83	2.36	0.15	0.52	1.70	5.16	2.97	0.07	98.75			
	4	2	71.01	0.50	14.71	2.37	0.16	0.48	1.83	5.16	3.00	0.08	99.30			
	5	1	72.22	0.53	15.08	2.36	0.17	0.55	1.84	4.99	3.04	0.09	100.88			
	7	1	72.12	0.50	15.11	2.33	0.15	0.52	1.86	4.78	3.09	0.09	100.57			
		11	**71.51**	**0.51**	**14.94**	**2.33**	**0.15**	**0.51**	**1.79**	**5.01**	**3.03**	**0.08**	**99.87**	**Alaska**	**Aniakchak**	**Aniakchak**
			0.75	**0.01**	**0.17**	**0.05**	**0.01**	**0.03**	**0.07**	**0.16**	**0.05**	**0.01**	**0.88**			
Jm96-1215 24–26 cm	2		71.61	0.30	13.46	3.73	0.15	0.19	1.36	4.94	3.33	0.05	99.12	Iceland	Katla	Katla
Jm96-1215 24–26 cm	6	2	71.21	0.10	12.49	1.47	0.05	0.05	0.54	3.13	4.70	0.01	93.75	Iceland	?	?
Jm96-1215 50–52 cm	1	2	47.92	4.40	13.12	14.36	0.21	4.93	9.66	2.93	0.85	0.50	98.87			
	2	2	47.64	4.47	13.03	14.67	0.22	5.01	9.80	2.90	0.76	0.46	98.96			
	3	2	47.20	4.43	13.20	14.23	0.22	4.92	9.70	3.07	0.82	0.47	98.25			
	4	2	47.98	4.45	13.15	14.41	0.22	4.86	9.69	3.10	0.80	0.47	99.13			
	5	2	47.39	4.45	13.29	14.51	0.23	4.86	9.50	3.15	0.81	0.48	98.67			
	6	1	48.07	4.41	13.31	14.03	0.22	4.80	9.47	3.28	0.89	0.47	98.94			
		11	**47.70**	**4.44**	**13.18**	**14.36**	**0.22**	**4.90**	**9.64**	**3.07**	**0.82**	**0.48**	**98.80**	**Iceland**	**Katla**	**Katla**
			0.35	**0.03**	**0.10**	**0.22**	**0.01**	**0.07**	**0.13**	**0.14**	**0.04**	**0.01**	**0.31**			
MD99-2322 158–159 cm	8		72.26	0.38	13.83	2.29	0.05	0.28	1.38	4.14	3.74	0.04	98.41	Alaska	Aniakchak	Aniakchak?
MD99-2322 158–159 cm	4		70.96	0.48	14.38	2.39	0.14	0.52	1.81	5.14	3.02	0.07	98.91	Alaska	Aniakchak	Aniakchak?
MD99-2322 158–159 cm	3	1	74.74	0.16	11.47	2.47	0.07	0.00	0.35	4.90	4.18	0.00	98.34			
	1	2	74.34	0.17	11.69	2.50	0.06	0.00	0.36	4.96	4.24	0.01	98.32			
	6	1	72.97	0.17	11.62	2.47	0.04	0.00	0.37	4.94	4.28	0.00	96.87			
		4	**74.01**	**0.17**	**11.59**	**2.48**	**0.06**	**0.00**	**0.36**	**4.93**	**4.24**	**0.00**	**97.84**	**Iceland**	**Tindfjöll**	**Hrafntinnuhraun?**
			0.93	**0.00**	**0.11**	**0.02**	**0.02**	**0.00**	**0.01**	**0.03**	**0.05**	**0.00**	**0.84**			
MD99-2322 158–159 cm	5	1	71.49	0.27	13.46	3.84	0.13	0.22	1.41	5.28	3.61	0.05	99.76			
	2	1	71.18	0.28	12.97	3.74	0.17	0.22	1.41	5.03	3.67	0.04	98.72			
		2	**71.34**	**0.27**	**13.22**	**3.79**	**0.15**	**0.22**	**1.41**	**5.16**	**3.64**	**0.05**	**99.24**	**Iceland**	**Katla**	**Hrafntinnuhraun?**
			0.22	**0.00**	**0.35**	**0.06**	**0.03**	**0.00**	**0.00**	**0.17**	**0.04**	**0.00**	**0.73**			
MD99-2322 158–159 cm	7	1	71.74	0.27	14.73	2.32	0.06	0.19	0.88	5.06	4.83	0.03	100.10	Iceland	Torfajökull	Hrafntinnuhraun
MD99-2322 158–159 cm	1		49.20	1.58	13.44	12.73	0.19	6.62	11.20	2.44	0.26	0.15	97.81			
	2		50.42	1.80	13.87	12.58	0.24	6.48	11.12	2.52	0.24	0.16	99.44			
	3		48.87	1.83	13.65	12.23	0.22	6.36	11.27	2.39	0.19	0.18	97.18			
	4		49.02	1.85	13.59	12.87	0.20	6.67	11.19	2.40	0.24	0.17	98.21			
	5		50.64	1.84	13.61	13.10	0.24	6.70	11.36	2.50	0.21	0.17	100.35			
	6		49.67	1.89	13.08	12.58	0.24	6.35	11.18	2.28	0.21	0.19	97.68			
	7		49.61	1.83	14.26	13.16	0.21	6.51	11.42	2.18	0.18	0.17	99.53			

(*Continued*)

Table 2. *Major element geochemical data from microprobe analyses of tephra peaks in MD99-2322 and JM96-1215/2-GC* (*Continued*)

Tephra horizon	Grain	N	SiO_2	TiO_2	Al_2O_3	FeO	MnO	MgO	CaO	Na_2O	K_2O	P_2O_5	Total	Provenance – region	Compositional domain*	Likely source volcano
	8		49.52	1.82	13.00	12.64	0.23	6.49	11.21	2.75	0.22	0.16	98.05			
	9	4	49.87	1.82	13.71	12.91	0.22	6.70	11.19	2.43	0.22	0.17	99.24			
	10		49.30	1.83	13.17	13.19	0.22	6.60	11.41	2.54	0.27	0.16	98.69			
	11		50.10	1.85	14.25	12.39	0.29	6.60	11.85	2.35	0.21	0.18	100.06			
	12		49.45	1.84	13.92	12.46	0.23	6.49	11.43	2.45	0.21	0.17	98.66			
		15	**49.64**	**1.81**	**13.63**	**12.74**	**0.23**	**6.55**	**11.32**	**2.44**	**0.22**	**0.17**	**98.74**	**Iceland**	**Bárðarbunga–Veiðivötn**	**Vatnaöldur**
			0.54	**0.08**	**0.41**	**0.31**	**0.02**	**0.12**	**0.20**	**0.14**	**0.03**	**0.01**	**1.00**			
MD99-2322 334–335cm	1	1	72.41	0.51	14.68	2.20	0.16	0.50	1.92	4.83	2.91	0.09	100.20			
MD99-2322 334–335cm	12	2	70.82	0.49	14.98	2.54	0.13	0.51	1.71	5.44	3.07	0.09	99.78			
		3	**71.61**	**0.50**	**14.83**	**2.37**	**0.15**	**0.50**	**1.82**	**5.13**	**2.99**	**0.09**	**99.99**	**Alaska**	**Aniakchak**	**Aniakchak?**
			1.12	**0.01**	**0.21**	**0.24**	**0.02**	**0.01**	**0.15**	**0.43**	**0.12**	**0.00**	**0.30**			
MD99-2322 334–335cm	4	1	71.21	0.31	13.03	3.74	0.15	0.25	1.32	4.73	3.50	0.04	98.29			
MD99-2322 334–335cm	2	2	71.64	0.29	13.43	3.81	0.17	0.20	1.38	4.94	3.60	0.04	99.49			
MD99-2322 334–335cm	5	2	71.12	0.28	13.36	3.70	0.14	0.20	1.29	4.65	3.44	0.04	98.23			
		5	**71.32**	**0.29**	**13.27**	**3.75**	**0.15**	**0.22**	**1.33**	**4.77**	**3.51**	**0.04**	**98.67**	**Iceland**	**Katla**	**Hekla?**
			0.28	**0.02**	**0.22**	**0.06**	**0.02**	**0.03**	**0.04**	**0.15**	**0.08**	**0.00**	**0.71**			
MD99-2322 334–335cm	3	1	75.52	0.18	12.30	2.60	0.09	0.03	0.35	4.64	4.15	0.01	99.85	Iceland	Tindfjöll	Hekla?
MD99-2322 334–335cm	13	1	62.03	1.12	15.27	8.39	0.20	1.59	4.54	4.73	1.62	0.37	99.85			
MD99-2322 334–335cm	15	3	62.02	1.03	16.33	7.43	0.19	1.45	5.13	4.73	1.42	0.37	100.11			
MD99-2322 334–335cm	6	2	62.09	1.09	15.46	7.90	0.22	1.58	4.82	4.42	1.56	0.38	99.51			
MD99-2322 334–335cm	10	1	62.04	1.06	15.89	7.52	0.21	1.72	4.98	4.40	1.61	0.39	99.82			
		7	**62.04**	**1.07**	**15.74**	**7.81**	**0.21**	**1.58**	**4.87**	**4.57**	**1.55**	**0.38**	**99.82**	**Iceland**	**Hekla**	**Hekla (H-B?)**
			0.03	**0.04**	**0.47**	**0.44**	**0.02**	**0.11**	**0.25**	**0.19**	**0.09**	**0.01**	**0.24**			
MD99-2322 334–335cm	7	2	47.25	4.50	13.08	14.87	0.24	5.16	9.76	3.07	0.78	0.47	99.18	Iceland	Katla	Katla
MD99-2322 334–335cm	8	2	49.62	3.11	13.21	14.06	0.25	5.42	9.68	2.78	0.50	0.31	98.93	Iceland	Grímsvötn	Grímsvötn
MD99-2322 358–359 cm	9		70.72	0.49	14.63	2.49	0.14	0.48	1.80	5.32	3.07	0.07	99.22			
MD99-2322 358–359 cm	14		71.74	0.51	14.86	2.47	0.15	0.57	1.85	5.24	3.05	0.09	100.53			
		2	**71.23**	**0.50**	**14.75**	**2.48**	**0.14**	**0.52**	**1.83**	**5.28**	**3.06**	**0.08**	**99.87**	**Alaska**	**Aniakchak**	**Aniakchak?**
			0.72	**0.01**	**0.17**	**0.01**	**0.01**	**0.06**	**0.04**	**0.05**	**0.02**	**0.01**	**0.93**			
MD99-2322 358–359 cm	10		72.89	0.17	11.36	2.56	0.07	0.04	0.38	4.94	4.02	0.00	96.41			
MD99-2322 358–359 cm	11		72.59	0.16	11.36	2.53	0.06	0.00	0.40	4.85	4.05	0.00	96.00			
		2	**72.74**	**0.16**	**11.36**	**2.55**	**0.06**	**0.02**	**0.39**	**4.89**	**4.04**	**0.00**	**96.21**	**Iceland**	**Tindfjöll**	**Katla?**
			0.21	**0.01**	**0.00**	**0.02**	**0.01**	**0.02**	**0.02**	**0.07**	**0.02**	**0.00**	**0.29**			
MD99-2322 358–359 cm	4		71.04	0.29	13.11	3.91	0.15	0.21	1.21	4.88	3.53	0.05	98.37			
MD99-2322 358–359 cm	12		71.63	0.28	13.66	3.90	0.14	0.21	1.31	5.23	3.51	0.04	99.91			

MD99-2322 358–359 cm	6		71.75	0.29	13.33	3.93	0.16	0.22	1.32	5.25	3.41	0.03	99.69			
MD99-2322 358–359 cm	2		71.26	0.29	13.32	3.96	0.15	0.22	1.33	5.13	3.57	0.05	99.28			
MD99-2322 358–359 cm	7		72.13	0.28	13.63	3.95	0.16	0.24	1.37	5.13	3.68	0.04	100.60			
MD99-2322 358–359 cm	8		70.99	0.29	13.72	3.96	0.16	0.18	1.38	5.15	3.56	0.04	99.41			
MD99-2322 358–359 cm	1		70.17	0.29	13.16	3.90	0.15	0.20	1.39	5.12	3.50	0.04	97.91			
MD99-2322 358–359 cm	3		70.93	0.28	13.07	3.87	0.15	0.21	1.39	5.12	3.51	0.04	98.60			
MD99-2322 358–359 cm	5		71.85	0.28	13.44	3.95	0.15	0.22	1.41	5.11	3.59	0.04	100.05			
		9	**71.31**	**0.29**	**13.38**	**3.93**	**0.15**	**0.21**	**1.35**	**5.13**	**3.54**	**0.04**	**99.31**	**Iceland**	**Katla**	**Katla**
			0.60	**0.00**	**0.25**	**0.03**	**0.01**	**0.02**	**0.06**	**0.10**	**0.08**	**0.00**	**0.87**			
MD99-2322 358–359 cm	17		50.30	2.90	12.74	13.93	0.25	5.57	9.63	2.68	0.44	0.31	98.74			
MD99-2322 358–359 cm	15		48.81	3.06	12.75	13.99	0.23	5.68	9.92	2.54	0.46	0.30	97.73			
		2	**49.55**	**2.98**	**12.75**	**13.96**	**0.24**	**5.63**	**9.77**	**2.61**	**0.45**	**0.31**	**98.23**	**Iceland**	**Grímsvötn**	**Grímsvötn**
			1.05	**0.11**	**0.01**	**0.04**	**0.01**	**0.08**	**0.21**	**0.10**	**0.01**	**0.01**	**0.72**			
MD99-2322 358–359 cm	16		49.25	2.28	13.55	12.01	0.22	7.08	11.51	2.66	0.24	0.20	99.01	Iceland	Grímsvötn	Grímsvötn
MD99-2322 358–359 cm	18		50.85	1.04	13.92	11.45	0.19	7.61	12.13	2.09	0.07	0.08	99.44	Iceland	Grímsvötn	Grímsvötn
MD99-2322 555–556 cm	38		72.52	0.16	11.38	2.42	0.06	0.00	0.38	5.03	3.99	0.01	95.95			
MD99-2322 555–556 cm	35		72.65	0.15	11.55	2.65	0.05	0.01	0.39	4.69	4.21	0.00	96.34			
MD99-2322 555–556 cm	42		69.76	0.17	11.34	2.25	0.09	0.00	0.39	4.97	4.26	0.00	93.22			
MD99-2322 555–556 cm	32		72.41	0.16	11.72	2.39	0.08	0.00	0.40	4.96	4.19	0.00	96.32			
MD99-2322 555–556 cm	33		74.49	0.17	10.99	2.58	0.07	0.00	0.40	5.03	4.31	0.00	98.04			
		5	**72.37**	**0.16**	**11.40**	**2.46**	**0.07**	**0.00**	**0.39**	**4.94**	**4.19**	**0.00**	**95.97**	**Iceland**	**Tindfjöll**	**Katla?**
			1.69	**0.01**	**0.27**	**0.16**	**0.01**	**0.00**	**0.01**	**0.14**	**0.12**	**0.00**	**1.74**			
MD99-2322 555–556 cm	41		71.21	0.28	13.13	3.82	0.18	0.17	1.20	5.35	3.57	0.04	98.95			
MD99-2322 555–556 cm	40		70.56	0.28	13.26	3.58	0.16	0.21	1.23	5.12	3.58	0.04	98.02			
MD99-2322 555–556 cm	36		69.91	0.28	13.11	3.84	0.15	0.24	1.27	5.21	3.41	0.03	97.44			
MD99-2322 555–556 cm	34		70.02	0.27	13.41	3.93	0.16	0.22	1.27	5.26	3.51	0.04	98.11			
MD99-2322 555–556 cm	37		69.73	0.28	13.71	3.73	0.16	0.18	1.34	4.99	3.35	0.04	97.50			
MD99-2322 555–556 cm	39		70.68	0.27	13.83	3.48	0.14	0.25	1.40	5.36	3.52	0.04	98.96			
MD99-2322 555–556 cm	43		69.64	0.28	13.43	3.57	0.12	0.22	1.45	5.43	3.46	0.03	97.62			
		7	**70.25**	**0.28**	**13.41**	**3.71**	**0.15**	**0.21**	**1.31**	**5.25**	**3.49**	**0.04**	**98.08**	**Iceland**	**Katla**	**Katla**
			0.58	**0.01**	**0.28**	**0.17**	**0.02**	**0.03**	**0.09**	**0.15**	**0.09**	**0.01**	**0.65**			
MD99-2322 555–556 cm	18		47.38	4.10	12.86	14.50	0.27	5.02	9.53	3.16	0.80	0.48	98.10			
MD99-2322 555–556 cm	21		47.99	4.09	13.21	14.86	0.22	5.12	9.83	3.16	0.80	0.47	99.75			
MD99-2322 555–556 cm	16		47.87	4.15	13.22	14.25	0.24	4.99	9.67	2.98	0.77	0.47	98.62			
		3	**47.75**	**4.12**	**13.10**	**14.54**	**0.25**	**5.04**	**9.68**	**3.10**	**0.79**	**0.47**	**98.83**	**Iceland**	**Katla**	**Katla**
			0.32	**0.03**	**0.21**	**0.30**	**0.02**	**0.07**	**0.15**	**0.11**	**0.02**	**0.01**	**0.84**			
MD99-2322 555–556 cm	4		48.76	3.04	12.96	14.45	0.25	5.47	9.60	2.74	0.52	0.33	98.10			
MD99-2322 555–556 cm	2		49.47	2.87	13.08	13.87	0.23	5.63	9.85	2.62	0.46	0.31	98.39			
MD99-2322 555–556 cm	7		49.69	2.89	13.30	14.30	0.22	5.49	10.01	2.63	0.42	0.31	99.25			
		3	**49.31**	**2.93**	**13.11**	**14.21**	**0.23**	**5.53**	**9.82**	**2.66**	**0.47**	**0.32**	**98.58**	**Iceland**	**Grímsvötn**	**Grímsvötn**

(*Continued*)

Table 2. *Major element geochemical data from microprobe analyses of tephra peaks in MD99-2322 and JM96-1215/2-GC* (*Continued*)

Tephra horizon	Grain	N	SiO_2	TiO_2	Al_2O_3	FeO	MnO	MgO	CaO	Na_2O	K_2O	P_2O_5	Total	Provenance – region	Compositional domain*	Likely source volcano
			0.49	**0.09**	**0.17**	**0.30**	**0.02**	**0.09**	**0.20**	**0.06**	**0.05**	**0.01**	**0.60**			
MD99-2322 555–556 cm	20		49.73	1.41	13.66	12.15	0.20	7.23	11.46	2.23	0.19	0.11	98.37	Iceland	Grímsvötn?	Grímsvötn?
MD99-2322 795–796 cm	1		48.08	4.41	13.58	14.46	0.21	4.82	9.75	3.09	0.85	0.44	99.71			
MD99-2322 795–796 cm	2		47.55	4.47	13.15	15.05	0.22	4.74	9.61	3.22	0.84	0.46	99.31			
MD99-2322 795–796 cm	3		47.39	4.36	13.46	14.84	0.21	5.02	9.87	2.88	0.78	0.45	99.28			
MD99-2322 795–796 cm	4		47.01	4.45	13.26	14.45	0.21	4.98	9.73	3.15	0.85	0.46	98.55			
MD99-2322 795–796 cm	6		47.07	4.38	13.11	15.07	0.22	5.09	9.76	2.60	0.78	0.45	98.52			
MD99-2322 795–796 cm	7		47.51	4.41	13.17	14.57	0.22	5.19	9.85	3.01	0.77	0.46	99.15			
MD99-2322 795–796 cm	8		47.17	4.41	12.79	14.30	0.22	5.11	9.52	2.98	0.77	0.45	97.72			
MD99-2322 795–796 cm	9		47.66	4.36	13.11	14.60	0.21	5.00	9.75	3.19	0.76	0.44	99.08			
		8	**47.43**	**4.41**	**13.20**	**14.67**	**0.22**	**4.99**	**9.73**	**3.02**	**0.80**	**0.45**	**98.91**	Iceland	Katla	Katla
			0.35	**0.04**	**0.24**	**0.29**	**0.01**	**0.15**	**0.11**	**0.20**	**0.04**	**0.01**	**0.62**			
MD99-2322 795–796 cm	5		48.90	3.00	13.11	13.61	0.24	5.65	9.95	2.69	0.40	0.28	97.83	Iceland	Grímsvötn	Grímsvötn
MD99-2322 1618.5 cm	19		49.37	3.10	12.94	14.54	0.23	5.11	9.48	2.96	0.52	0.29	98.55			
MD99-2322 1618.5 cm	2		49.70	3.23	12.60	14.97	0.25	5.15	9.51	2.37	0.46	0.30	98.54			
MD99-2322 1618.5 cm	9		49.36	3.18	12.97	14.67	0.24	5.20	9.81	2.78	0.49	0.30	99.00			
MD99-2322 1618.5 cm	6		49.75	3.10	13.19	14.44	0.25	5.25	9.67	2.73	0.50	0.26	99.14			
MD99-2322 1618.5 cm	16		49.40	3.22	12.90	14.50	0.26	5.29	9.73	2.56	0.45	0.30	98.60			
MD99-2322 1618.5 cm	4		50.07	3.24	12.95	14.33	0.24	5.31	9.59	2.65	0.54	0.28	99.20			
MD99-2322 1618.5 cm	18		50.43	3.34	12.65	14.31	0.24	5.32	9.81	2.73	0.50	0.30	99.63			
MD99-2322 1618.5 cm	11		49.56	3.13	13.13	14.06	0.23	5.33	9.63	2.66	0.50	0.28	98.52			
MD99-2322 1618.5 cm	17		50.52	3.16	12.79	14.32	0.24	5.34	9.58	2.79	0.52	0.29	99.54			
MD99-2322 1618.5 cm	14		50.25	3.20	13.09	14.35	0.24	5.36	9.82	2.88	0.48	0.30	99.97			
MD99-2322 1618.5 cm	10		49.35	3.27	13.18	14.17	0.23	5.36	9.52	2.73	0.52	0.32	98.65			
MD99-2322 1618.5 cm	5		49.79	3.00	13.42	13.79	0.22	5.46	9.82	2.63	0.45	0.28	98.84			
MD99-2322 1618.5 cm	15		49.54	3.20	12.84	14.63	0.25	5.46	9.64	2.68	0.48	0.28	99.01			
MD99-2322 1618.5 cm	7		50.11	2.94	13.17	13.93	0.25	5.48	9.92	2.55	0.45	0.27	99.07			
MD99-2322 1618.5 cm	3		50.25	3.10	13.05	14.08	0.24	5.55	9.86	2.72	0.50	0.27	99.62			
MD99-2322 1618.5 cm	12		49.97	3.05	12.74	13.93	0.25	5.58	9.93	2.82	0.47	0.28	99.02			
MD99-2322 1618.5 cm	8		49.74	2.96	13.26	13.52	0.23	5.58	10.08	2.56	0.42	0.25	98.60			
		17	**49.83**	**3.14**	**12.99**	**14.27**	**0.24**	**5.36**	**9.73**	**2.69**	**0.49**	**0.28**	**99.03**	**Iceland**	**Grímsvötn**	**Grímsvötn**
			0.39	**0.11**	**0.22**	**0.36**	**0.01**	**0.14**	**0.17**	**0.14**	**0.03**	**0.02**	**0.45**			

MD99-2322 1618.5 cm	1		49.98	2.75	12.82	13.47	0.24	5.89	10.05	2.72	0.42	0.24	98.57	Iceland	Grímsvötn	Grímsvötn
MD99-2322 1618.5 cm	13		50.52	1.20	14.16	10.91	0.20	7.53	12.18	1.95	0.16	0.10	98.91	Iceland	Grímsvötn?	Grímsvötn?
MD99-2322 1719.5 cm	1		49.09	2.86	12.87	13.88	0.22	5.80	10.22	2.47	0.39	0.30	98.10			
MD99-2322 1719.5 cm	2	2	49.87	2.84	13.38	13.64	0.23	5.72	9.98	2.67	0.41	0.30	99.03			
MD99-2322 1719.5 cm	3	2	49.66	2.84	13.14	14.02	0.24	5.31	10.03	2.71	0.39	0.31	98.66			
MD99-2322 1719.5 cm	4		49.05	3.05	12.85	14.11	0.22	5.49	9.59	2.81	0.35	0.32	97.84			
MD99-2322 1719.5 cm	5	2	48.07	2.81	13.05	13.84	0.24	5.82	9.95	2.61	0.40	0.30	97.10			
MD99-2322 1719.5 cm	6		48.51	3.02	13.46	14.56	0.24	5.61	9.77	2.60	0.49	0.33	98.59			
MD99-2322 1719.5 cm	8		49.49	3.12	12.95	14.24	0.24	5.23	9.56	2.57	0.45	0.35	98.20			
MD99-2322 1719.5 cm	9		49.16	3.17	12.80	14.69	0.24	5.19	9.44	2.61	0.42	0.36	98.08			
MD99-2322 1719.5 cm	10		49.08	3.03	12.95	14.21	0.23	5.32	9.68	2.68	0.47	0.37	98.02			
MD99-2322 1719.5 cm	12		49.35	3.11	13.28	14.19	0.26	4.78	9.83	2.64	0.41	0.32	98.19			
		14	**49.13**	**2.99**	**13.07**	**14.14**	**0.24**	**5.43**	**9.81**	**2.64**	**0.42**	**0.33**	**98.18**	**Iceland**	**Grímsvötn**	**Grímsvötn**
			0.53	**0.14**	**0.23**	**0.32**	**0.01**	**0.32**	**0.24**	**0.09**	**0.04**	**0.03**	**0.52**			
MD99-2322 1719.5 cm	7		50.72	3.00	13.93	12.46	0.23	5.17	10.08	3.05	0.51	0.33	99.48	Ice-Grimsvotn	Ice-Grimsvotn	Ice-Grimsvotn
MD99-2322 1797.5 cm	17		48.89	3.12	12.71	14.43	0.24	5.42	9.43	2.80	0.42	0.33	97.79			
MD99-2322 1797.5 cm	18		49.26	2.81	13.24	13.11	0.20	6.09	10.69	2.71	0.41	0.30	98.82			
MD99-2322 1797.5 cm	19		48.80	3.03	13.14	13.88	0.26	5.71	9.62	2.64	0.44	0.34	97.86			
MD99-2322 1797.5 cm	20		48.96	3.16	13.07	14.64	0.26	4.68	9.24	2.78	0.57	0.35	97.72			
MD99-2322 1797.5 cm	1		49.33	3.17	12.80	14.53	0.24	5.34	9.49	2.83	0.47	0.28	98.46			
MD99-2322 1797.5 cm	2		48.77	3.18	12.90	14.41	0.25	5.15	9.66	2.83	0.47	0.30	97.92			
MD99-2322 1797.5 cm	3		50.01	3.35	13.06	14.50	0.24	5.35	9.94	1.85	0.49	0.29	99.08			
MD99-2322 1797.5 cm	4		49.13	3.04	12.38	14.31	0.25	5.33	9.84	2.92	0.48	0.30	97.99			
MD99-2322 1797.5 cm	5		49.50	3.01	13.05	13.81	0.24	5.54	9.92	2.71	0.53	0.28	98.57			
MD99-2322 1797.5 cm	6	2	49.77	3.18	13.00	13.99	0.23	5.38	9.88	2.24	0.49	0.29	98.45			
MD99-2322 1797.5 cm	7		49.27	3.06	12.88	14.14	0.23	5.59	9.82	2.80	0.42	0.28	98.50			
MD99-2322 1797.5 cm	8		48.74	2.93	13.12	14.15	0.23	5.92	10.18	2.34	0.37	0.27	98.26			
MD99-2322 1797.5 cm	9		49.65	3.22	12.66	14.04	0.24	5.28	9.54	2.63	0.49	0.31	98.06			
MD99-2322 1797.5 cm	10		50.08	3.12	12.78	14.65	0.24	5.66	9.88	2.56	0.43	0.27	99.68			
MD99-2322 1797.5 cm	11	2	48.94	3.12	13.09	13.90	0.25	5.29	9.91	2.40	0.47	0.29	97.65			
MD99-2322 1797.5 cm	12		49.41	3.23	12.80	14.44	0.24	5.19	9.61	2.38	0.50	0.29	98.10			
MD99-2322 1797.5 cm	13		49.45	3.31	12.98	13.94	0.24	5.33	9.69	2.67	0.43	0.29	98.32			
MD99-2322 1797.5 cm	14		48.64	2.97	12.96	13.92	0.25	5.55	10.00	2.93	0.44	0.30	97.95			
MD99-2322 1797.5 cm	15		50.06	3.13	12.89	14.05	0.23	5.46	9.69	2.83	0.41	0.28	99.03			
MD99-2322 1797.5 cm	16		48.75	3.05	13.03	14.01	0.24	5.47	9.89	2.56	0.45	0.27	97.71			
		22	**49.27**	**3.11**	**12.93**	**14.14**	**0.24**	**5.44**	**9.80**	**2.62**	**0.46**	**0.30**	**98.30**	**Iceland**	**Grímsvötn**	**Grímsvötn**
			0.47	**0.13**	**0.20**	**0.36**	**0.01**	**0.30**	**0.30**	**0.27**	**0.05**	**0.02**	**0.54**			
MD99-2322 1797.5 cm	21		47.66	4.74	12.56	15.21	0.22	4.79	9.41	2.75	0.79	0.66	98.80	**Iceland**	**Hekla?**	**Hekla?**

*Compositional domain: indicates that the measured compositions of the grains in question resembles the magma compositions that are thought – on the basis of the exisiting reference database – to typify the volcanoes named in the column. However, it does not imply that they are the source (see text for further discussion).

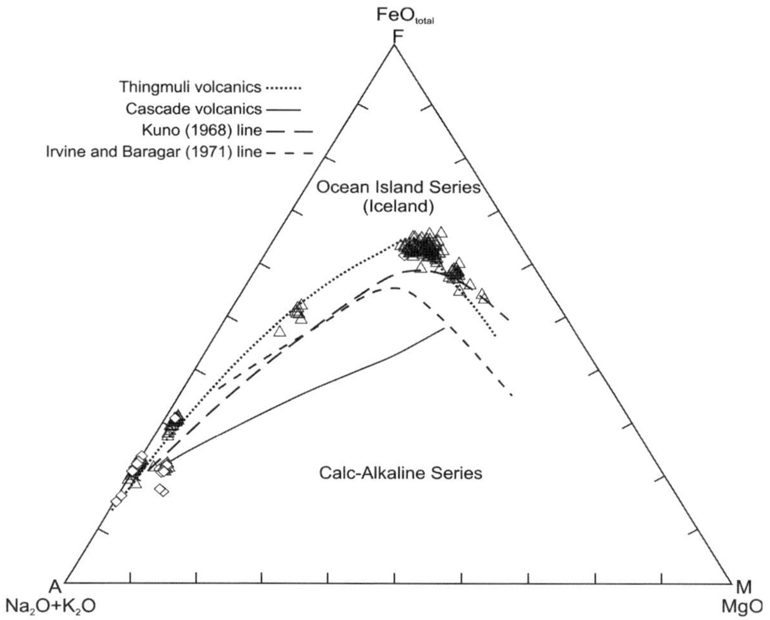

Fig. 4. AFM ternary diagram showing the magma series and volcanic environment affinities of the tephra analysed in this study. Tephra grains from MD99-2322, open triangles; grains from JM96-1215/GC, open diamonds. The broken line shows the Kuno (1968) division between the calc-alkaline and the tholeiitic fields, while the fine broken line shows the division of Irvine & Baragar (1971). Note, although majority of the data are consistent with Icelandic origin (i.e. follow the Thingmuli trend), a significant portion of the silicic tephra population exhibit calc-alkaline composition. This subpopulation is not sourced from Icelandic volcanoes.

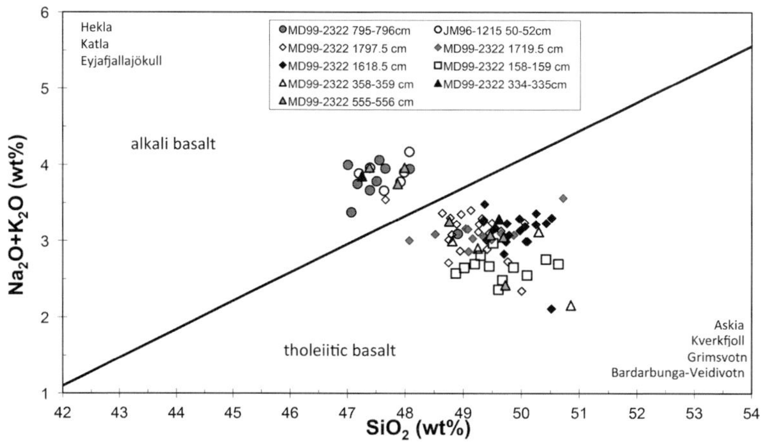

Fig. 5. Total alkalis v. SiO_2 plot showing the division of the basaltic grains into alkali and tholeiitic basalt. The black solid line separates the fields of alkali and tholeiitic basalt magma types (Macdonald & Katsura 1964). This distinction is important because one cannot be related to the other via simple fractionation paths owing to a thermal divide represented by the 'plane of critical silica undersaturation'. In other words these two magma types cannot come from the same source and almost exclusively are formed by separate eruptions. The likely source volcanoes for each magma type are given in the upper left corner for alkalic and lower right corner for tholeiitic volcanic systems (e.g. Thordarson & Höskuldsson 2008).

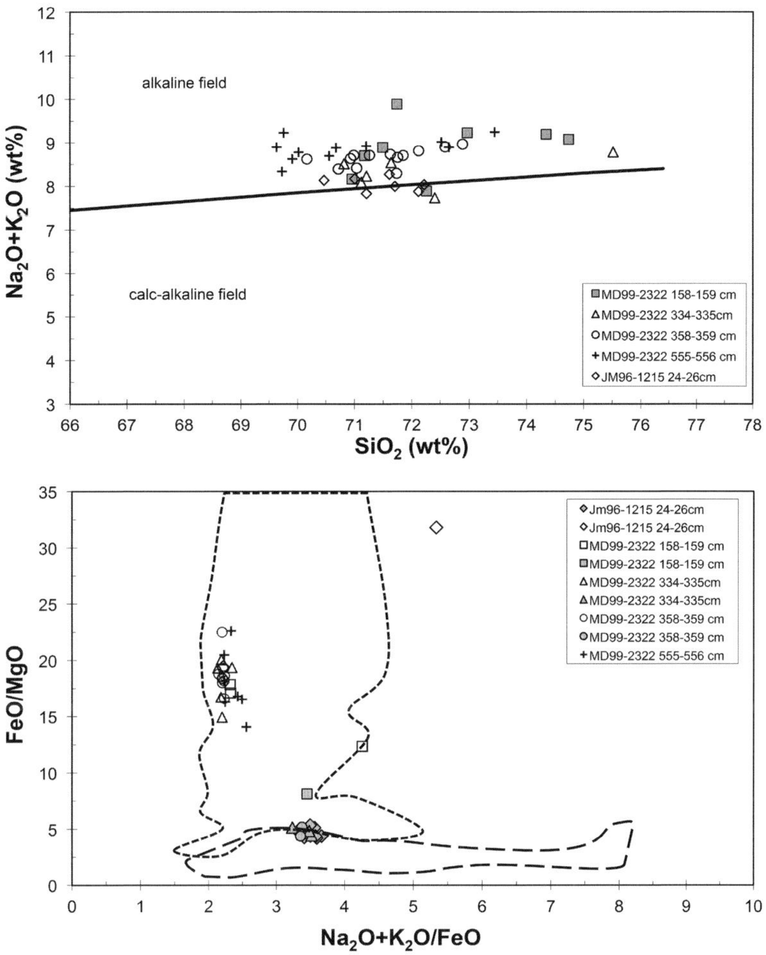

Fig. 6. (**a**) Total alkalis v. SiO_2 plot showing the mildly alkaline nature of the rhyolite grains of this study (see key and text for further details). (**b**) Total alkalis/FeO v. FeO/MgO plot illustrating the distinctive grouping of rhyolite tephra into grains with Icelandic v. calc-alkaline /volcanic arc provenance. The fine dashed line indicates the compositional field for Holocene rhyolite tephra in Iceland (data from T. Thordarson unpublished tephra database, 2012). The heavy dashed line delineates the compositional trend of rhyolite tephra from Alaska (data from Riehle *et al.* 1999).

while the other one, from JM96-1215 24–26 cm, does not exhibit affinities with any of the established compositional fields for Icelandic rhyolites (e.g. plots on Figs 6 & 7). The intermediate tephra in horizon MD99-2322 334–335 cm is clearly from the Hekla volcano (Fig. 8).

Individual tephra horizons

The geochemistry of each of the tephra horizons analysed is described and interpreted below. Each tephra horizon is defined by its depth in the core and the age estimate of the peak based on each marine sediment core's radiocarbon-based age model. Correlations to known volcanic eruptions are made based on the geochemistry and stratigraphy. The difference between the known age of a volcanic event and the age estimate of the tephra horizon provides useful information on the accuracy of the age models and the marine reservoir age used to calibrate the radiocarbon dates in the cores. For both SE Greenland cores we assumed $\Delta R = 0$ when calibrating the radiocarbon dates. However, ΔR may be as much as 150 years greater than 0 in this area because of the influence of Polar Waters in the East Greenland Current. Therefore when comparing the ages of known volcanic events with the age estimates of tephra horizons in the marine

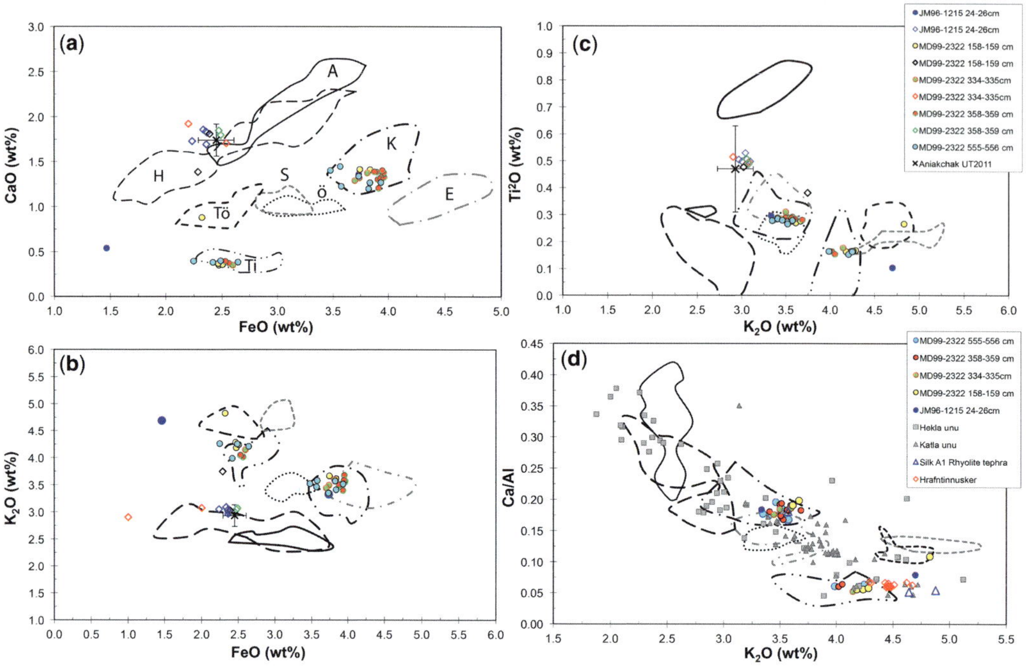

Fig. 7. (**a–c**) Element-element plots illustrating the compositional grouping of rhyolite grains (see key in c). The compositional fields shown (labelled on a) demarcate the provenance of individual volcanoes and are as follows: A, Askja (solid black line); H, Hekla (black coarse dashed line); K, Katla (black dashed–dotted line); To, Torfajökull (black fine dashed line); Ti, Tindfjöll (black dashed–double dotted line); E, Eyjafjallajökull (grey dashed–dotted line); Ö, Öræfajökull (grey fine dotted line); S, Snæfellsjökull (grey fine dashed line). Also shown is the averaged composition for the *c.* 3.6 ka BP Aniakchak tephra; the bars indicate two standard deviation (2σ) range of the dataset. (**d**) K_2O v. molecular Ca/Al plot demonstrating that tephra layers, in particular intermediate to silicic ones, from Iceland often contain grains with composition that is differs from the bulk grain population. These 'different' (or 'unusual') compositions are typically significantly more evolved than the host tephra, and are most commonly found near the base of a tephra layer (i.e. from the initial blast). The data feature clusters near the compositional fields typical of rhyolites from Tindfjöll, Katla and Öræfajökull volcanoes. The Katla-unu data points on this plot are from basaltic Katla tephra layers except for three points (blue triangles), which are from the dacitic SILK A1 tephra layer; some fall within the Katla compositional field while others exhibit compositions that are more in line with established compositions for the Torfajökull and Tindfjöll volcanoes. The Hekla-unu are from tephra layers that span the range basalt to rhyolite and, although spaning a wide compositional range, include ash grains with established Torfajökull and Tindfjöll compositions. Also shown is composition of the tephra produced by the $>$8 ka Hrafntinnusker eruption of the Torfajökull volcano (Martin & Sigmarsson 2007), which erupted from vents adjacent to source vents for the silicic component of the *c.* 870 AD Vatnaöldur (i.e. Settlement) tephra. The relevant MD99-2322 and JM96-1215 results are plotted for comparison (see key). Other data sources used in constructing these plots are Larsen *et al.* (2001), Óladóttir *et al.*(2008), Wastegård *et al.* (2003) and unpublished data by G. Larsen (2012) and T. Thordarson (2012). The compositional fields for individual volcanoes are constructed using the Iceland Reference Tephra Database (as described in the 'Electron microprobe' section of Materials and Methods).

cores, the ages of the known volcanic events are likely to be younger than the estimated ages of the tephra horizons in the marine cores.

Core JM96-1215/GC

Tephra horizon 18–26 cm (3750–4410 cal yr BP). Five out of seven analysed grains from 24 to 26 cm have a very uniform calc-alkaline rhyolite composition, while two have distinct Icelandic provenance (Figs 4 & 6b). The calc-alkaline tephra grains are surprisingly large (100 and 250 μm) and feature delicate protrusions as well as very thin bubble walls (Fig. 9a). When compared with the composition of mildly alkaline rhyolite tephra from Iceland, the five calc-alkaline grains are similar in terms of total alkalis at the lower end of the FeO spectrum, but have significantly higher CaO, TiO_2, MgO contents (e.g. Fig. 7). In fact, the relatively high TiO_2 and MgO contents of

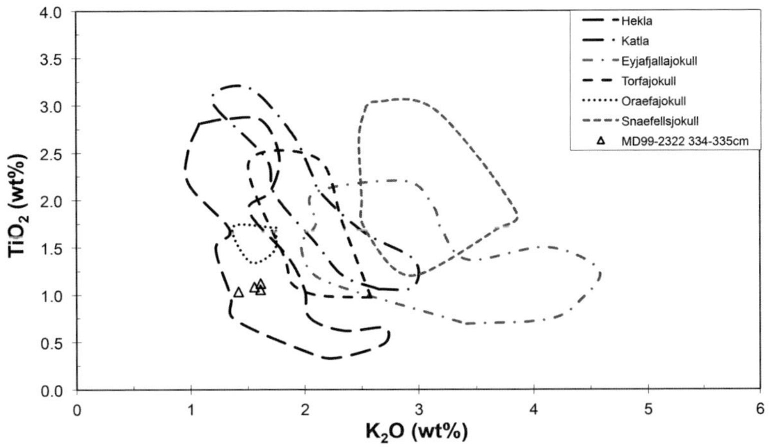

Fig. 8. K_2O–TiO_2 plots demonstrating the Hekla affinity of the intermediate grains from MD99-2322 334–335 cm (large open triangles). Also shown are compositional fields for Icelandic central volcanoes that have erupted intermediate magmas constructed using the Iceland Reference Tephra Database (as described in the 'Electron microprobe' section of Materials and Methods).

these tephra grains are very distinctive. On the strength of these differences in conjunction with preservation of delicate grain shapes and uniform chemical composition, we argue that these grains represent a distinct tephra horizon that was produced by an eruption at an arc volcano. When compared with available datasets on calc-alkaline tephra from North American arc volcanoes in Western USA and Alaska, the relatively high TiO_2 and MgO contents of this tephra are distinctive and compare favourably to the composition of the tephra from the 50–100 km^3 caldera forming eruption at Aniakchak volcano in Alaska at *c.* 3.6 ka BP (Fig. 7; Neal *et al.* 2001; Pearce *et al.* 2004; Dreher *et al.* 2005).

One of the two grains of Icelandic provenance falls within the compositional field of Katla rhyolites on all element-to-element plots (e.g. Fig. 7). The grain has pristine appearance (Fig. 9a) and has high totals (>99 wt%; Table 2). The source volcano for the other one cannot be determined at this stage, but its low totals (<94 wt%; Table 2) indicate hydration and hence it may represent a reworked grain. Rhyolite tephra layers corresponding to the composition of these two grains are not known in the Icelandic tephrochronology within the 3–4 ka time period and therefore it is not possible to identify their source eruptions at this stage with any degree of confidence.

Tephra horizon 1215 50–52 cm (6790 cal yr BP). In this tephra horizon we analysed six grains ranging in size from *c.* 100 to 200 μm (Table 2). This horizon is purely basaltic, comprising pristine looking tephra grains with a uniform chemical composition that is consistent with an origin from the Katla volcanic system (Figs 9b & 10a). Also, in terms of its composition it is indistinguishable from MD99-2322 795.5 cm, which is dated at 6729 cal yr BP (Fig. 10b). Furthermore, the composition of 1215 50–52 cm (and 2322 795–796 cm) falls well within the compositional field defined by the 24 Katla tephra layers identified in Iceland from the 6–7.1 ka BP period (e.g. Jóhannsdóttir 2007; Óladóttir *et al.* 2008). At close inspection, the composition of 1215 50–52 cm (and 2322 795–795 cm) is similar to that of the K-6750 cal yr BP Katla tephra layer identified in lake cores in western Iceland, but is distinctly different from the bracketing K-6630 and K-7060 cal yr BP layers (Fig. 10b). Therefore, we correlate this tephra horizon to the *c.* 6750 cal yr BP Katla eruption (e.g. Jóhannsdóttir 2007).

Core MD99-2322

Tephra horizon 2322 158–159 cm (1162 cal yr BP). Twenty tephra shards were analysed from this horizon and reveal a distinctive bimodality in terms of two magma types, basalt v. rhyolite (Table 2). The basaltic grains are the major constituent (12 out of 20 analysed grains) and comprise highly vesicular (bubble volume >75%) pumice-like grains featuring delicate pin-like protrusions (Figs 5 & 9c). Their composition is tholeiite (Fig. 5) and it indicates Bárðarbunga–Veiðivötn volcanic system provenance. The composition and grain morphology are consistent with the mafic part of the Settlement layer, which is also very widely dispersed (Figs 10a & 11; Larsen 1982,

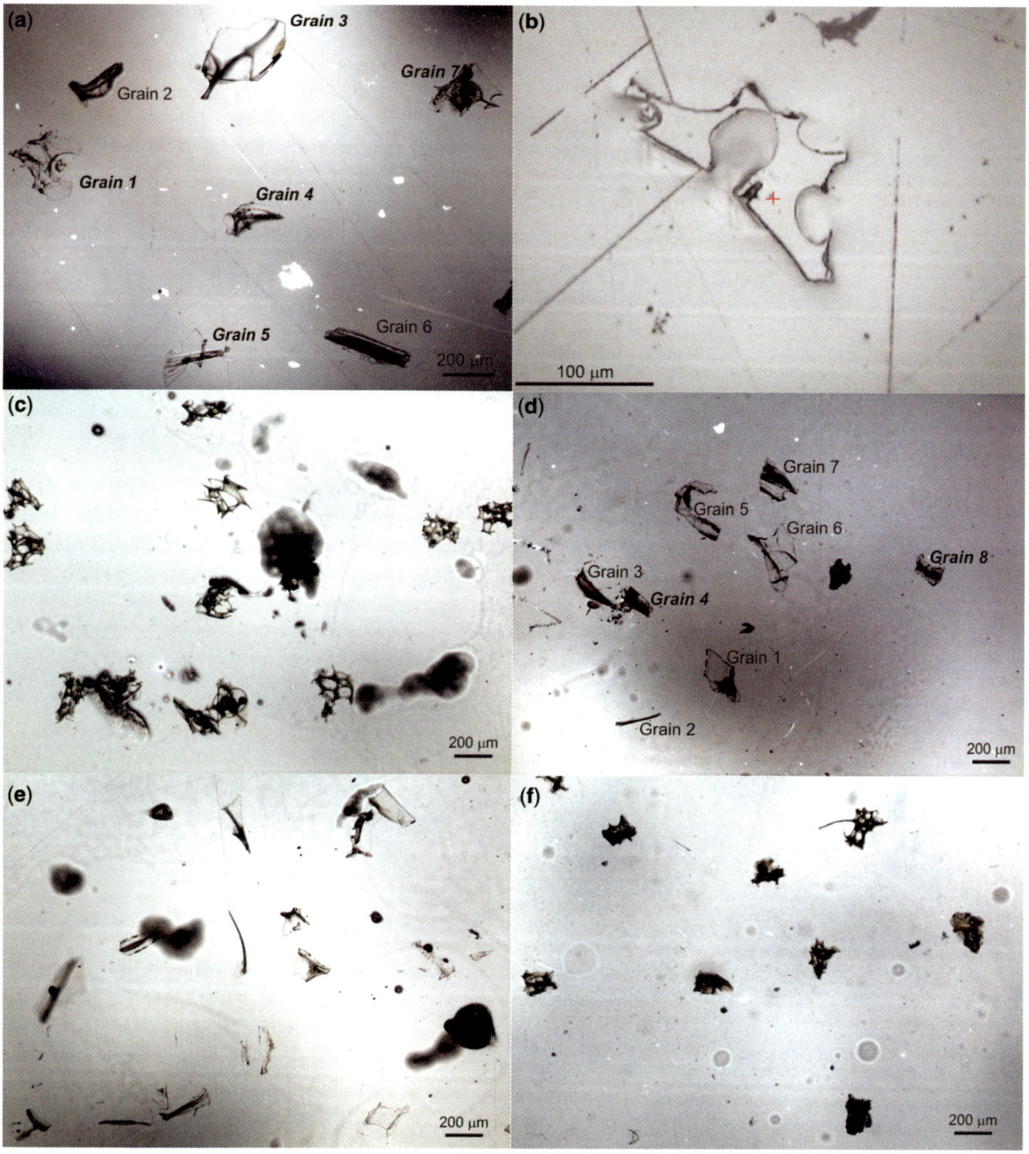

Fig. 9. Images showing examples of grains subjected to chemical analyses. All images are taken with transmitted light through a polished resin microprobe plug except image (b), which is an electron backscatter image. Scale bars are in the bottom right corner. The images are from the following tephra horizons: (**a**) JM96-1215 24–26 cm; (**b**) JM96-1215 50–52 cm; (**c**) MD99-2322 158–159 cm-basaltic grains (arrows); (**d**) MD99-2322 158–159 cm – rhyolitic grains; (**e**) MD99-2322 358–359 cm; and (**f**) MD99-2322 1719.5 cm.

1984). However, it should be noted that this chemical composition is indistinguishable from the other Bárðarbunga–Veiðivötn mafic layers that form the Settlement tephra layer series (cf. Larsen 2002; Wastegård *et al.* 2003).

The eight rhyolite grains range in size from *c.* 100 to 400 μm and exhibit pristine appearance as they show no evidence of abrasion or post-eruption alteration or hydration (Fig. 9d; Table 2). However, in terms of their composition, they reveal a more complex provenance than the basaltic component. Only one grain (i.e. grain 7 in Fig. 10d) is consistent with the established composition of the rhyolite part of the Settlement layer (Fig. 7a–c; Table 2), which originated from vents within the Torfajökull volcano in south Iceland and was dispersed to the west (Figs 1b & 11; Larsen 1984). Two grains (2 and 5 in Fig. 9d) have more evolved Katla-like composition, while three grains (1, 3, 6 in Fig. 9d) have compositions that resemble the highly-evolved rhyolites from the Tindfjöll volcanic system (Fig. 7a–c). The last two (4 and 8 in Fig. 9d) are of calc-alkaline provenance and their composition implies the same source volcano (i.e. Aniakchak) as the calc-alkaline grains in tephra horizon 1215 24–26 cm (Figs 6b & 7).

On the strength of the pristine and delicate nature of the grains along with uniform composition of the basaltic component, we identify this tephra horizon as time-equivalent to the Settlement layer tephra and thus the inferred age of the peak is AD 871($\pm$2), or 1079 BP (Grönvold *et al.* 1995), which is about 80 years younger than the age estimate from the ninth-order polynomial based age estimate (Table 1). This interpretation is supported by the presence of one rhyolite grain matching the silicic part of the Settlement layer. It is also conceivable that the grains with composition compatible to that of Tindfjöll and Katla rhyolite are from the Settlement layer eruption. Rhyolite tephra produced by Torfajökull in the past is known to feature such compositions, for example, the tephra produced by the >8 ka Hrafntinnusker event (Fig. 7d), whose source vents are situated adjacent to the Hrafntinnuhraun vents that produced the rhyolitic tephra in the Settlement layer (e.g. Larsen 1984; Martin & Sigmarsson 2007). Thus, the simplest explanation is that the initial phase of the rhyolitic event of the Settlement layer incorporated a portion of the older Hrafntinnusker tephra. The origin of the two Katla-like rhyolite grains remains enigmatic. The Aniakchak volcano featured at least two explosive eruptions around 900 AD (Neal *et al.* 2001), which may account for the presence of the two calc-alkaline grains, although this cannot be confirmed at this stage.

Tephra horizon 2322 334–335 cm (3063 cal yr BP). In total, 12 pristine grains were analysed from this tephra horizon ranging in size from 50 to 200 μm. The measurements reveal a mixed population in terms of their chemical composition: six grains are rhyolitic, four are intermediate and two are basaltic (Table 2). The rhyolite grains fall into three groups. Two grains (1 and 12; Table 2) are calc-alkaline and their composition indicates Aniakchak as the source volcano (Figs 6b & 7). Four grains have Icelandic provenance (Fig. 6b): three grains (2, 4, 5; Table 2) have Katla-like major element composition and one (grain 3) resembles the highly evolved Tindfjöll rhyolites (Fig. 7). The four grains of intermediate composition (i.e. 6, 10, 13, 15; Table 2) clearly have Hekla affinity (Fig. 8). The 2500–3000 cal yr Hekla H-A, -B, -C, -M, -N, -X, -Y, -Z two-coloured tephra series, which followed the eruption of H3 (Larsen & Eiríksson 2008*a*) have major and trace element composition that differ somewhat from intermediate tephra grains produced by other known post 4 ka BP Hekla eruptions, namely slightly lower K_2O and TiO_2, but higher MgO contents (Fig. 12), which on basis of our reference data appear to rule out the 3.1 cal kyr H3 and 3.8 cal kyr H-S as potential source events (Fig. 12; Meara 2012). The grains analysed in this study exhibit strongest affinity to the composition established for Hekla H-A, -B, -C, -M, -N, -X, -Y, -Z two-coloured tephra series (Meara 2012). An H3 link appears unlikely on basis of the established compositions; a notion supported by the fact that its tephra fallout did not extend beyond the western shores of Iceland (e.g. Larsen & Thorarinsson 1977). It is also noteworthy that Late Holocene Hekla tephra layers contain a minor population of highly evolved rhyolite grains that have major element compositions very similar to that obtained for the four rhyolite grains of Icelandic provenance mentioned above (e.g. Fig. 7d). The two basaltic tephra grains are unrelated; one is tholeiite (Grímsvötn-like composition), while the other is alkaline (Katla-like composition; Table 2, Figs 5 & 10a).

The highly variable composition and multi-source affinities of the tephra grains within this horizon make it very difficult to pinpoint its origin and age. The pristine nature of the grains – that is, unaltered composition (all totals are >98%; Table 2) and delicate grain shapes – indicates short residence time and minimal transport in surface environments and is suggestive of origin via primary tephra fallout. Furthermore, the intermediate tephra grains of the Hekla H-A to H-Z series provide important time constraints for this horizon. It cannot predate the oldest tephra layer (i.e. 3 ka) in the sequence and is unlikely to postdate the youngest tephra layer (2.5 ka). Considering the known dispersal direction for individual layers within the Hekla H-A to H-Z series, the eruption that produced layer H-B is the most likely source candidate. Its ^{14}C age of 2740 $\pm$ 20 years BP (Larsen & Eiríksson 2008*a*) calibrates to between 2779 and 2871 cal yr BP at 2σ, about 200 years younger than the age estimate of the 334–335 cm horizon in core 2322.

Tephra horizon 2322 358–359 cm (3208 cal yr BP). Seventeen grains were analysed from this tephra horizon ranging in size from 80 to 400 μm (Fig. 9e). This is a silicic tephra horizon with 13–17 grains having rhyolitic composition (Table 2, Fig. 6). The remaining four grains are basaltic; all are tholeiitic and their major element composition reveals Grímsvötn origin (Figs 5 & 10a). The

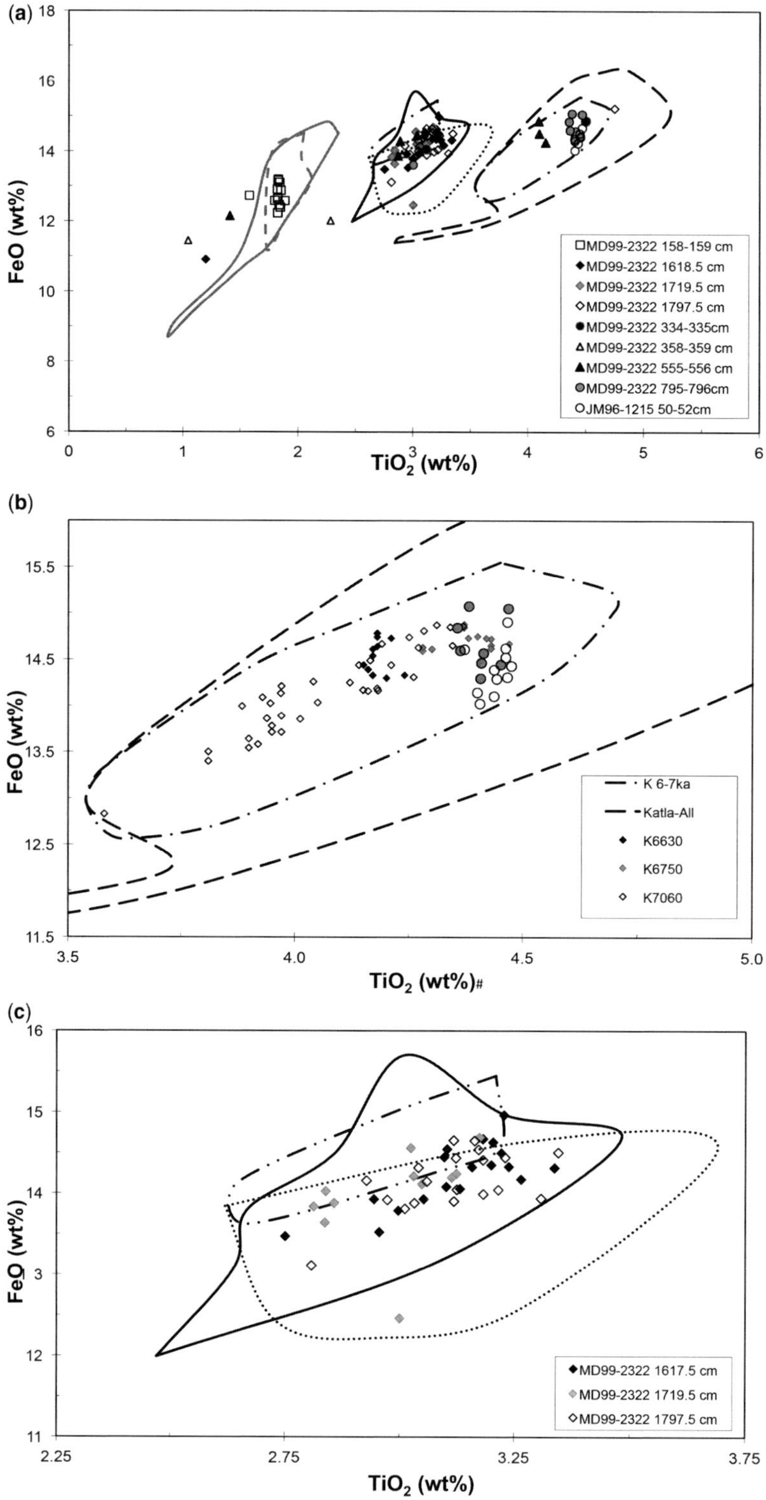
(a)
FeO (wt%)
TiO2 (wt%)
MD99-2322 158-159 cm
MD99-2322 1618.5 cm
MD99-2322 1719.5 cm
MD99-2322 1797.5 cm
MD99-2322 334-335cm
MD99-2322 358-359 cm
MD99-2322 555-556 cm
MD99-2322 795-796cm
JM96-1215 50-52cm
(b)
FeO (wt%)
TiO2 (wt%)
K 6-7ka
Katla-All
K6630
K6750
K7060
(c)
FeO (wt%)
TiO2 (wt%)
MD99-2322 1617.5 cm
MD99-2322 1719.5 cm
MD99-2322 1797.5 cm

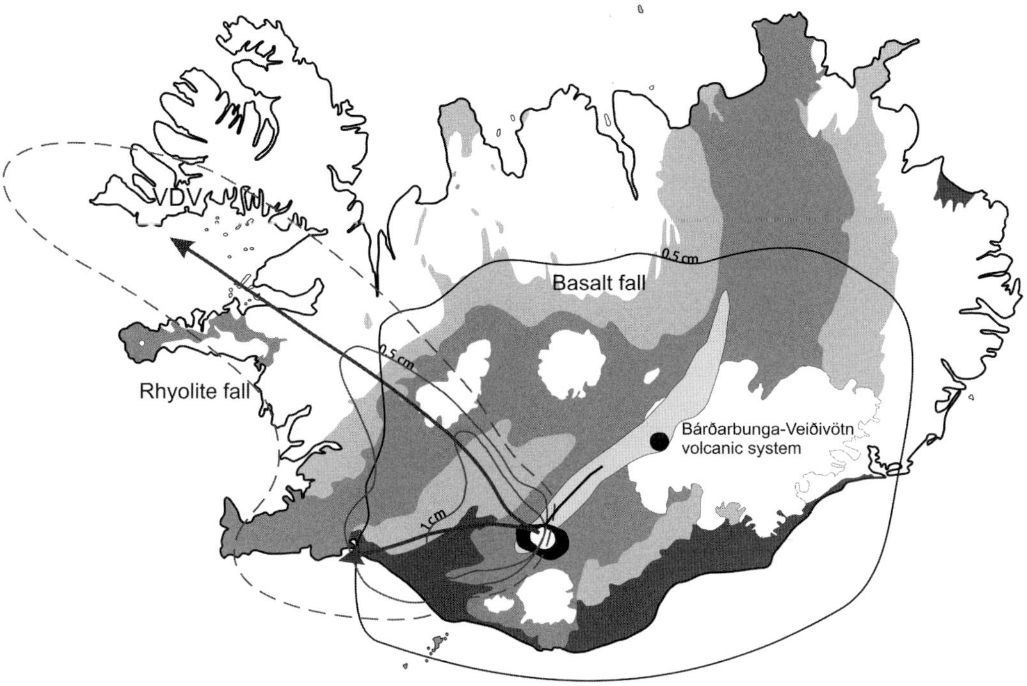

Fig. 11. Map outlining the distribution of the basaltic and rhyolitic components of the Settlement tephra layer produced by the Vatnaöldur eruption in Iceland in *c.* 870 AD (Larsen 1984; Grönvold *et al.* 1995). The map shown is a simplified version of the original isopach map by Larsen (1984); only the 0.5 cm isopach (heavy solid line) is shown for the basaltic component, illustrating that about half of Iceland was completely covered by tephra fallout. Three isopachs are shown for the rhyolite tephra fall component, the 1 and 0.5 cm isopachs (thin solid lines) and the *c.* 0.1 cm line (thin broken line), which has been extended on basis of on new data points shown from western Iceland (T. Thordarson, unpublished data, 2012). Also shown are the main dispersal directions of the rhyolite tephra (arrows), relevant volcanoes and volcanic system. For other geological structures see key in Figure 1b.

rhyolitic population is dominated by grains of Icelandic provenance (11 out of 13 grains; Figs 4 & 6b). Nine grains exhibit Katla-like composition and two are typified by Tindfjöll-like composition (Table 2; Fig. 7). Two grains (9 and 14; Table 2) are calc-alkaline and their composition is almost identical to the major component of JM96-1215 24–26 cm and thus suggestive of Aniakchak as the source volcano (Figs 6b & 7).

The multisource affinities of the tephra grains within this horizon make interpretations of its origin and age challenging. The pristine nature of the Katla-like and the Aniakchak rhyolite grains – that is, delicate form of the grains and

Fig. 10. TiO_2–FeO plots showing the source volcano provenance of basaltic tephra in MD99-2322 and JM96-1215/2GC. (**a**) All data on basaltic tephras from this study. (**b**) Composition of the mildly alkaline basalt grains from 1215 50–52 cm and 2322 795–796 cm tephra horizons compared with known Katla tephra layers from the 6.6–6.8 ka period in Western Iceland. (**c**) Composition of the tholeiitic grains from 2322 1618.5, 1719.5 and 1797.5 cm tephra horizons compared with established compositions for the *c.* 9.9–10.5 ka tephra series from the Grímsvötn volcano. The compositional field of the Saksunarvatn tephra in Faroe Islands is also shown for reference. Grey solid line, compositional field for basalt tephra from the Bárðarbunga–Veiðivötn system. Grey dashed line, compositional field for the basaltic component of the Settlement layer (Bárðarbunga–Veiðivötn system). The black line is the Grímsvötn compositional field. The dashed–double dotted line is the compositional field for all *c.* 9.9–10.5 ka Grímsvötn tephra layers in Western Iceland. The finely dotted line is the compositional field for the Saksunarvatn tephra from several localities in the Faroe Islands as defined by the data from Mangerud *et al.* (1986), Dugmore & Newton (1998), Wastegård *et al.* (2001) and Lind & Wastegård (2011). The black coarse dashed line shows the compositional field of basalt tephra from Katla volcanic systems, while the dashed–dotted line envelopes the compositional field for basalt tephra from Katla volcanic systems in the period 6–7 ka.

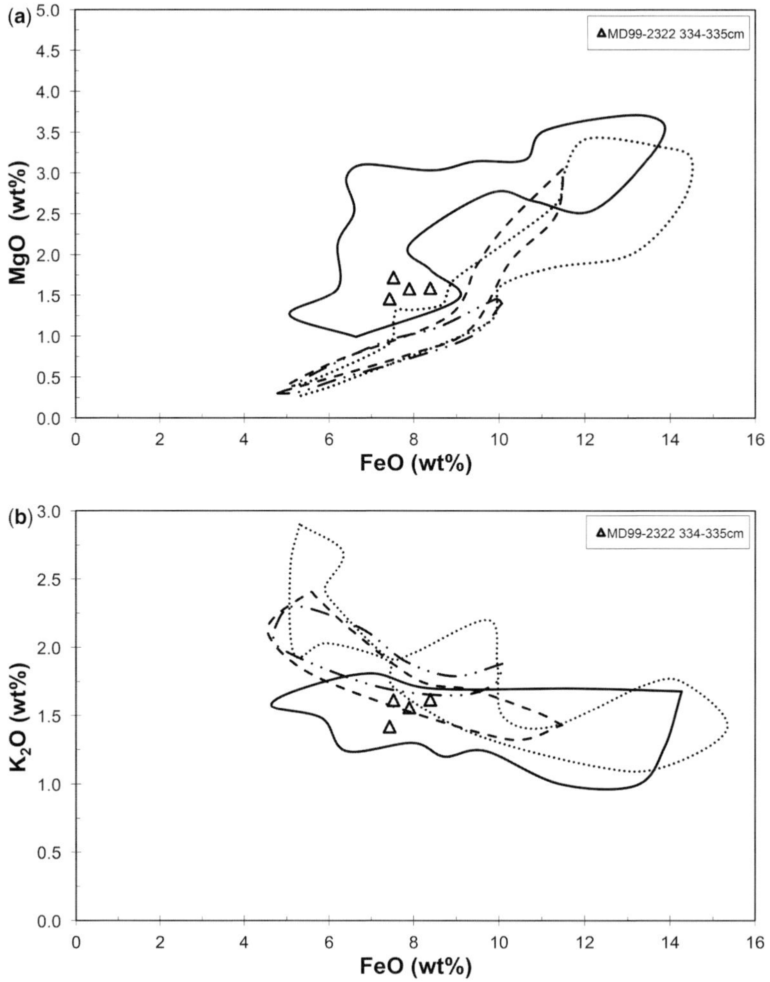

Fig. 12. FeO–MgO (**a**) and FeO–K_2O (**b**) plots demonstrating the compositional affiliation of the intermediate grains from MD99-2322 334–335 cm with the 2.5–2.9 ka H-A to H-Z tephra series from the Hekla volcano. Solid black lines outline the compositional field of H-A to H-Z tephra deposits; the fine dotted line shows the field of historic basaltic icelandite to icelandite tephra from Hekla; fine dashed–dotted and the fine dashed lines show the fields for the intermediate components of the 3.1 ka H3 and *c.* 3.8 ka H-S tephra deposits from Hekla. The compositional fields are constructed using the Iceland Reference Tephra Data Base (T. Thordarson, unpublished data, 2012).

totals are >97.9 wt% (Table 2; Fig. 9e) – suggests origin via primary tephra fallout. However, the relatively low totals (96.0 and 96.4 wt%) for the two Tindfjöll-like rhyolite grains imply that they were subjected to post-emplacement hydration and may have had extended residence in surface environments such as in the Greenland Ice Sheet (cf. Jennings *et al.* 2002) or in surficial sediments of the SE Greenland shelf, before they were deposited at the core site. However, as stated above, the majority of the rhyolite grains exhibit Katla affinities and the pristine nature of all of these grains implies origin via primary fallout, rather than reworking of pre-existing rhyolite Katla tephra (e.g. the Vedde ash). However, the Holocene tephra succession in Iceland does not feature a distinctively rhyolitic Katla layer in the time period (i.e. around 3.2 ka) indicated by the established 2322 age model. A possible explanation for the Katla rhyolite component is that it represents the very initial phase of a Katla SILK eruption. Katla SILK events are known to emit a small amount of rhyolite (e.g. Larsen *et al.* 2001; Fig. 7d). This rhyolite magma would have been sitting at the top of the magma body and thus the first thing to be erupted. As part of the very initial phase, it may have been ejected to the greatest height in the atmosphere – as observed in Hekla 1947 (Thorarinsson 1954) and Grímsvötn 2011 (T. Thordarson, unpublished data, 2012) – and thus decoupled from the main tephra plume. We propose that MD99-2322 358–359 cm is the early rhyolite component of tephra layer SILK MN (e.g. Larsen *et al.* 2001). This layer has a southwards dispersal from the Katla volcano, but if the early rhyolite component was decoupled it may have been dispersed to the west towards the SE Greenland shelf. An added support to this suggestion is that the composition of the few rhyolite grains that have been measured in a SILK-tephra features both the Katla-like and Tindfjöll-like rhyolite compositions, as shown by the blue triangles on Figure 7d, which could explain the presence of the two Tindfjöll-like rhyolite grains in this tephra horizon. SILK MN has a ^{14}C age of 2975 ± 12 years BP (Larsen *et al.* 2001) and calibrates to between 3101 and 3214 years BP at 2σ, providing a reasonable match with the age of 3208 cal yr BP calculated for the 2322 358–359 cm horizon. The presence of two pristine calc-alkaline Aniakchak rhyolite grains confirms that this tephra peak is no older than *c.* 3600 years BP (Pearce *et al.* 2004).

Tephra horizon 2322 555–556 cm (5089 cal yr BP). In this tephra horizon we analysed 19 grains ranging in size from *c.* 100 to 250 μm, where 11 grains are rhyolitic and seven grains are basaltic, all of Icelandic provenance (Figs 4–6). The measurements show that the rhyolite population comprises seven grains with Katla-like composition and four of Tindfjöll-like composition (Fig. 7). The latter group of grains exhibit distinctly lower totals (93.2–98.0 wt%) than the others (Table 2). Of the seven basaltic grains, four are tholeiitic with uniform composition of Grímsvötn affinities, while three mildly alkaline grains have typical Katla composition (Figs 5 & 10a).

As in the case of tephra horizon 2322 358–359 cm, the multisource affinities of the tephra grains within this horizon make interpretations of its origin and age challenging. The pristine nature of the Katla-like rhyolite and the basaltic grains are suggestive of origin via primary tephra fallout. However, the relatively low totals for the four Tindfjöll-like rhyolite grains can be taken as an indication of post-emplacement hydration and thus extended residence in surface environments. For example, it is possible that the four Tindfjöll-like rhyolite grains could have been stored in the Greenland Ice Sheet (cf. Jennings *et al.* 2002) and released to the core site by Greenland icebergs. However, we do not consider this likely because the tephra horizon at 555.5 cm lies within an interval of MD99-2322 that lacks IRD (Jennings *et al.* 2011). Reworking of these grains from Greenland shelf surface sediments is not considered to be likely because of the pristine nature of the grains. It is more likely that these shards were produced by the same eruption as the Katla rhyolite grains, because tephra layers from the Katla volcano are known to feature this composition as a minor component (Fig. 7d). By the same token, the basaltic Katla grains may also be from the same event, because silicic tephra layers from the Katla volcano are commonly bimodal, indicating simultaneous eruption of rhyolite and basalt magmas (e.g. Tomlinson *et al.* 2012), while the basaltic tephra grains of Grímsvötn affinity must represent a near-simultaneous eruption at the Grímsvötn volcano.

The Holocene tephra succession in Iceland does not feature a Katla rhyolite layer in the time period (i.e. between 5 and 6 ka) suggested by the established 2322 age model. With reference to the arguments given above for horizon 2322 358–359 cm, we propose that 2322 555–556 cm is the rhyolitic component of SILK-A1 tephra layer. This layer has been identified in soil profiles near the Katla volcano, but the exact dispersal direction is unknown (Larsen *et al.* 2001). However, the SILK-A1 tephra layer does contain a minor population of rhyolitic grains which have compositions that match favourably to the rhyolite grains measured in 2322 555–556 cm (Fig. 7d), although the grains analysed by Larsen *et al.* (2001) are too few to allow for statistical validation of this proposed correlation. The SILK A1 tephra layer has an estimated age of 5000 ^{14}C yr BP, which is obtained by assuming constant soil accumulation rate between radiocarbon dated tephra layers H4 (Hekla, 3826 ± 12 ^{14}C yr BP) and SILK A8 (Katla, 6400 ± 80 ^{14}C yr BP (Larsen *et al.* 2001). In calibrated years the age estimate of 5000 ^{14}C yr BP gives 5589–5661 cal yr BP or substantially older than the modelled age obtained here for the 2322 555–556 cm horizon (5089 cal yr BP; Table 1). However, given that the age of this tephra layer is obtained by extrapolated soil accumulation rates over *c.* 2.5 ka, the age discrepancy is understandable.

Tephra horizons 2322 795–796 cm (6729 cal yr BP). In this tephra horizon we analysed nine grains ranging in size from *c.* 100 to 250 μm (Table 2). The results show that it is purely basaltic, comprising pristine-looking tephra grains. Eight grains exhibit a uniform chemical composition that is consistent with an origin from the Katla volcanic system, while one grain shows Grímsvötn affinity (Fig. 10a). Also, it is indistinguishable from JM96-1215 50–52 cm and the composition of the eight

Katla grains falls well within the compositional field defined by the 6–7.1 ka basaltic tephra layers from Katla (e.g. Jóhannsdóttir 2007; Óladóttir *et al.* 2008). As in the case of the 1215 50–52 cm tephra horizon (see above), 2322 795–795 cm is similar to the K-6750 cal yr BP Katla tephra layer identified in lake cores in western Iceland (Fig. 10b). Therefore, we correlate it to the *c.* 6750 cal yr BP Katla eruption.

Tephra horizons 2322 1617.5 cm (10 031 cal yr BP), 1719.5 cm (10 186 cal yr BP), 1797.5 cm (10 390 cal yr BP). These three tephra horizons are described collectively as they all exhibit the same physical and chemical characteristics. We have analysed 19 (1617.5 cm), 11 (1719.5 cm) and 21 (1797.5 cm) grains from these horizons (Table 2). The grains range in diameter from *c.* 150 to 250 μm and the results show that they comprise pristine-looking basaltic tephra grains with uniform chemical composition, verifying origin at the Grímsvötn volcano (Figs 9f & 10a, c). In terms of their chemical composition, the tephra grains from these horizons are similar to the Saksunarvatn tephra in the Faroe Islands (Mangerud *et al.* 1986; Dugmore & Newton 1998; Wastegård *et al.* 2001), but the distribution of data points is more in line with the compositional field defined by the 9.9–10.4 ka Grímsvötn tephra layer series in Western Iceland (Fig. 10a, c; Jóhannsdóttir *et al.* 2005; Jóhannsdóttir 2007). In this context, it is interesting that multiple tephra layers of Grímsvötn composition have now been identified in the Faroe Islands (e.g. Kylander *et al.* 2012).

The overall uniform composition and delicate nature of the grains, which are highly vesicular and feature needle-thin protrusions (e.g. Fig. 10f), strongly suggest that they represent pristine tephra fallout (i.e. not reworked or re-sedimented). Furthermore, the presence of multiple but closely spaced tephra horizons of Grímsvötn affinities at this level in the core (i.e. within the age bracket 10–10.4 ka) is consistent with findings in west Iceland, where up to five Grímsvötn tephra layers have been identified in this time period (Jóhannsdóttir *et al.* 2005; Jóhannsdóttir 2007; T. Thordarson unpublished data 2012). In addition, the basaltic peaks at 1755.5 and 1605.5 cm comprise tephra shards identical in appearance to the shards from these confirmed Grímsvötn horizons. Although these were not analysed in this study, we suspect, but do not confirm, that these basaltic horizons represent additional Grímsvötn eruptions. Thus, the MD99-2322 contains evidence of three, and possibly five, Grímsvötn tephra horizons in early Holocene times, supporting the notion put forth by Jóhannsdóttir *et al.* (2005) and Jóhannsdóttir (2007) that the basaltic tephra marker layer in the North Atlantic referred to as the Saksunarvatn tephra was formed by multiple eruptions at the Grímsvötn volcano and spanning *c.* 500 years (9.9–10.4 ka).

Records of volcanic events and palaeomagnetic secular variations

Kangerlussuaq Trough cores. The inclination of PSV and basaltic tephra peaks of JM96-1215/2GC and MD99-2322 were compared against age based on the independent age models of each core (Fig. 13). The early to middle Holocene parts of the cores compare surprisingly well given their very different resolutions, but the PSV correlation is quite poor by 6500 cal yr BP, when the sediment accumulation rate in JM96-1215 declines. The major basaltic peaks in these cores line up very well in terms of both the number of peaks and their ages. The Katla 6.73 kyr BP (2322) and Katla 6.79 kyr BP (1215) peaks could not be correlated on the basis of the PSV because of the low resolution of 1215 (Fig. 13). However, on the basis of geochemistry and age, they most likely represent the same volcanic event from Katla, that is, K6750 (Fig. 10b).

Both cores have several basaltic tephra horizons between 10.4 and 9.9 cal kyr BP. The three main horizons in JM96-1215 (10.41, 10.18, 10.04 cal kyr BP) compare closely in age with the three most prominent horizons in MD99-2322 (10.39, 10.19 and 10.03 cal kyr BP). This interval matches the overall pattern of the PSV inclination, although the relatively low resolution of JM96-1215 does not allow any specific inclination features to be correlated (Fig. 13).

SE Greenland v. northern Iceland shelves. Correlation in depth of the PSV records from MD99-2269 on the north Iceland shelf and MD99-2322 on the SE Greenland shelf (Stoner *et al.* 2007) allows tephra layers from both cores to be put onto the same depth scale (Figs 2 & 14). The combined stratigraphy provides information on the north v. west dispersal of important tephra horizons and shows the stratigraphic positions of tephra layers relative to one another (Figs 2 & 14). Based on the PSV depth correlation between these two cores, only two tephra markers show a strong correspondence on depth. These are the first Grímsvötn layer in each core and the Katla basaltic peaks at 6.73 and 6.72 cal kyr, respectively, in 2322 and 2269. The first Grímsvötn layer in each core was identified in previous work as Saksunarvatn tephra (Kristjánsdóttir *et al.* 2007; Stoner *et al.* 2007). However, given the presence of multiple early Holocene Grímsvötn layers in the two cores we cannot say which, if any, of the layers corresponds to the

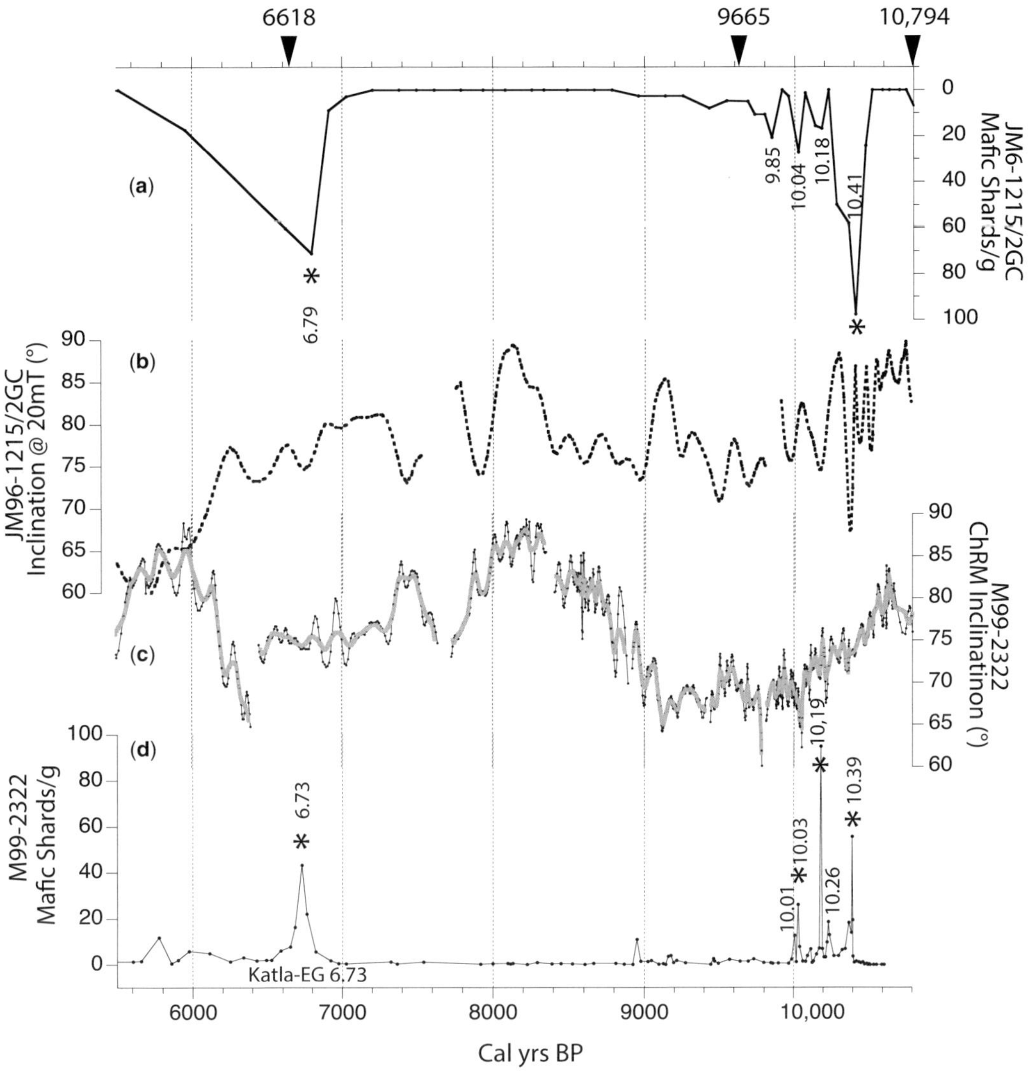

Fig. 13. Comparison between basaltic shards/g and palaeomagnetic secular variation (inclination) for two cores on their own independent radiocarbon based age models. (**a**) JM96-1215 basaltic tephra peaks and (**b**) inclination against *c*. MD99-2322 basaltic tephra peaks and (**d**) inclination on independent age scales. PSV is derived from u-channel palaeomagnetic measurements. Inclination is defined based on a single step for core 1215, whereas inclination for MD99-2322 is derived from component magnetization calculated by principal component analysis. See Stoner *et al.* (2007) for details. The smooth grey line reflects 100 year running mean. Peak ages are given above peaks. Asterisks denote samples for which geochemical data are available.

Saksunarvatn tephra recorded in the Faroe Islands by Mangerud *et al.* (1986). Based on the depth correlation, the first Grímsvötn layer in each core is the only one that potentially is correlative from core to core and thus has potential to be a regional marker horizon. The clear lack of correlation on depth among the other early Holocene Grímsvötn layers (i.e. G2 and G3 in 2269, and the four other Grímsvötn layers in 2322) indicates that these peaks reflect relatively discrete volcanic events with largely westward (2322) or northward (2269) dispersal of ash plumes. The 1σ age envelopes of the first Grímsvötn peak in 2322 and 2269 are comparable, although in either case there is an overlap with the ages obtained for the Grímsvötn peaks immediately above (Table 1).

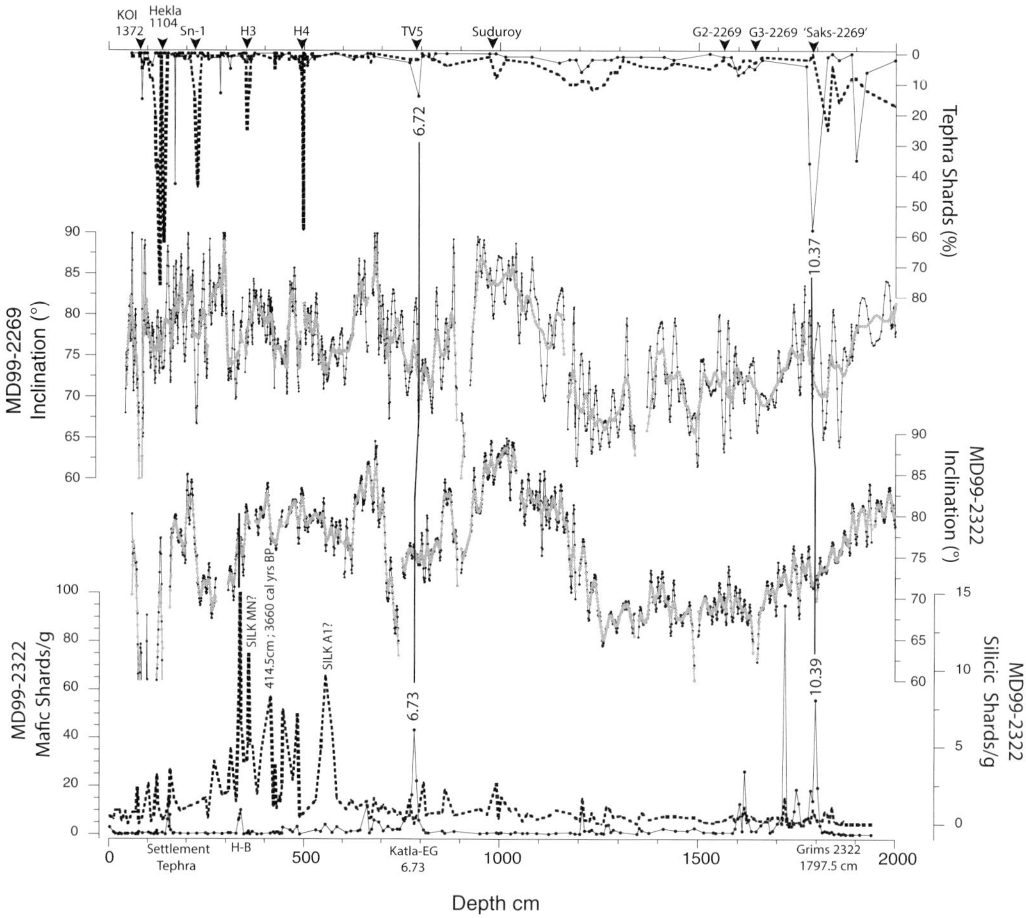

Fig. 14. Inter-comparison of PSV (inclination) and tephra for cores MD99-2322 and MD99-2269 on a common depth scale derived by correlation of the PSV records (Stoner *et al.* 2007). Characteristic remanent magnetization (ChRM) inclination defined through principal component analysis of the demagnetization data (see Stoner *et al.* 2007 for details). The smooth line reflects 20 cm running mean. Note that the tephra peaks in MD99-2269 are defined as percentage peaks based on counts of types of tephra shards in the >150 μm fraction (Kristjánsdóttir *et al.* 2007) rather than as counts per gram as shown for MD99-2322. Vertical lines join the two possible correlative peaks in the two cores.

Although there is a close correspondence in depth, age and chemical composition of the 783.5 cm Katla basaltic layer in 2322, Katla EG-6.73 (= Katla 6750 cal yr BP) and the layer identified as TV5 by Kristjánsdóttir *et al.* (2007), the current data do not allow direct correlation of the two horizons (see discussion below).

In terms of stratigraphy, it is a testament to the PSV correlation that H3 (only in 2269) underlies H-B (only in 2322) on the PSV-correlation-based depth scale (Figs 2 & 14). This stratigraphic sequence is what would be expected based upon the stratigraphy and timing of these events in Icelandic soil sections (Larsen & Eiríksson 2008*a*, *b*).

Discussion

Marker tephra horizons

Of the 10 tephra horizons analysed in this study, eight can be assigned to a source volcano and an eruption with a satisfactory degree of confidence. These are the horizons 1215 24–26 and 50–52 cm in core JM96-1215/2GC and 2322 158–159, 334–335, 795–796, 1618.5, 1719.5 and 1797.5 cm in core MD99-2322.

The calc-alkaline nature, distinct composition and stratigraphic position of the 1215 24–26 cm horizon strongly suggest that it represents tephra fallout associated with the caldera-forming event at the Aniakchak volcano at 3595 ± 4 yr BP based on its stratigraphic position in the GRIP ice core (Pearce *et al.* 2004).

The grain morphology and uniform chemical composition of the basaltic component in 2322 158–159 cm horizon leaves little doubt about its origin and age; it corresponds to the widespread *c.* 870 AD Settlement layer marker tephra in Iceland (e.g. Figs 10a & 11).

The presence of intermediate tephra grains with composition corresponding to the Hekla H-A to H-Z tephra series pins the age of the 2322 334–335 cm tephra horizon to the interval 2.5–3 cal kyr BP. The most likely source candidate for these intermediate grains is the eruption that produced the layer H-B, which has a calibrated age of *c.* 2850 cal yr BP (Larsen & Eiríksson 2008*a*). This age is 100–200 years younger than the modelled age for the 334–334 cm horizon.

The tephra grains in the 1215 50–52 cm and 2322 795–796 cm horizons are identical in terms of the physical appearance and chemical composition and undoubtedly represent the same fallout event from the Katla volcano. Furthermore, based on chemical composition, they correlate well with the K-6750 tephra that is an established marker horizon in western Iceland (Fig. 10b; Jóhannsdóttir 2007).

The presence of three, and possibly five, Grímsvötn tephra horizons in the time period 10–10.4 cal kyr BP in the MD99-2322 core corresponds well with recent findings in western Iceland, where five Grímsvötn tephra layers are identified in the early Holocene record (e.g. Jóhannsdóttir *et al.* 2005; Jóhannsdóttir 2007).

Multiple Grímsvötn tephra events between 9.9 and 10.4 cal kyr BP

The so-called Saksunarvatn ash, an important marker horizon in early Holocene marine and continental archives of the North Atlantic region, has on grounds of major element composition been correlated with the Grímsvötn volcano in the Eastern Volcanic Zone of Iceland (Pétursson & Larsen 1992; Grönvold *et al.* 1995). It has been considered an important marker horizon in early Holocene marine and continental archives of the northern North Atlantic. The Saksunarvatn tephra was first identified in the Faroe Islands, where it occurs as a distinct layer in lacustrine sediments and has been radiocarbon dated to 9000–9100 ^{14}C yr BP (Mangerud *et al.* 1986) or *c.* 10 200 cal yr BP. Originally, its observed distribution appeared to be mainly to the north and east of Iceland (Birks *et al.* 1996), but identification of basaltic tephra horizon with Grímsvötn (Saksunarvatn-like) composition at marine sites north and west of Iceland and in the Greenland Ice Sheet led to correlations with the Saksunarvatn ash and the assumption that these tephra horizons were the product of a single eruption (e.g. Grönvold *et al.* 1995; Zielinski *et al.* 1997; Andrews *et al.* 2002; Mortensen *et al.* 2005). Grímsvötn ash of Saksunarvatn-like composition was located in both the GRIP (Grönvold *et al.* 1995) and GISP2 ice cores (Zielinski *et al.* 1997), giving an age of 10 347 ± 45 yr B2 K (GICC05) and 10 297 ± 45 cal yr BP, but with maximum counting errors of ±89 years (Rasmussen *et al.* 2007). In lake sediments at Kråkenes, Western Norway, the Saksunarvatn tephra layer was identified and well dated to 10 210 ± 30 cal yr BP (Gulliksen *et al.* 1998), which falls within the maximum counting errors of the age of westward-dispersed tephra identified as Saksunarvatn ash in the Greenland ice cores (Rasmussen *et al.* 2007). Saksunarvatn ash has been thought to represent a single volcanic event with a very wide dispersal. Its presence in the Greenland ice cores, marine cores and well dated terrestrial sediment archives promoted its use as an isochron and a marker horizon for correlating marine and terrestrial sedimentary sequences (Ólafsdóttir 2010) and for examining early Holocene marine reservoir corrections (Hafliðason *et al.* 2000; Jennings *et al.* 2002; Eiríksson *et al.* 2004).

However, recent observations from Icelandic lake cores show five early Holocene tephra layers with Grímsvötn origin deposited between 9.9 and 10.4 cal kyr BP (Jóhannsdóttir *et al.* 2005; Jóhannsdóttir 2007), with chemical composition similar to that of Saksunarvatn ash as determined in the Faroe Islands (e.g. Figs 10a & 15). Kristjánsdóttir *et al.* (2007) noted several Grímsvötn tephra events in MD99-2269 on the North Iceland shelf. The oldest of these horizons is a visible tephra layer that they identified as the Saksunarvatn ash with an estimated age of 10 380 cal yr BP (10 284–10 501 cal yr BP 1σ) based on the calibrated radiocarbon ages and linear age model presented in Stoner *et al.* (2007) (Table 1). Both MD99-2322 and JM96-1215 have multiple tephra horizons of brown vesicular basalt of Grímsvötn affinities and major element concentrations that are more compatible with the composition of early Holocene Grímsvötn tephra series in western Iceland as determined by Jóhannsdóttir (2007) rather than that of the Saksunarvatn ash in the Faroe Islands (Figs 10a, c & 15). On the PSV correlation and the age models for these two cores, the early Holocene Grímsvötn horizons in the Kangerlussuaq Trough are correlative, but the higher resolution analyses of 2322 allow the definition of five peaks (Fig. 13). The brown vesicular basalt peaks at 10.24 and 10.01 in MD99-2322 were not

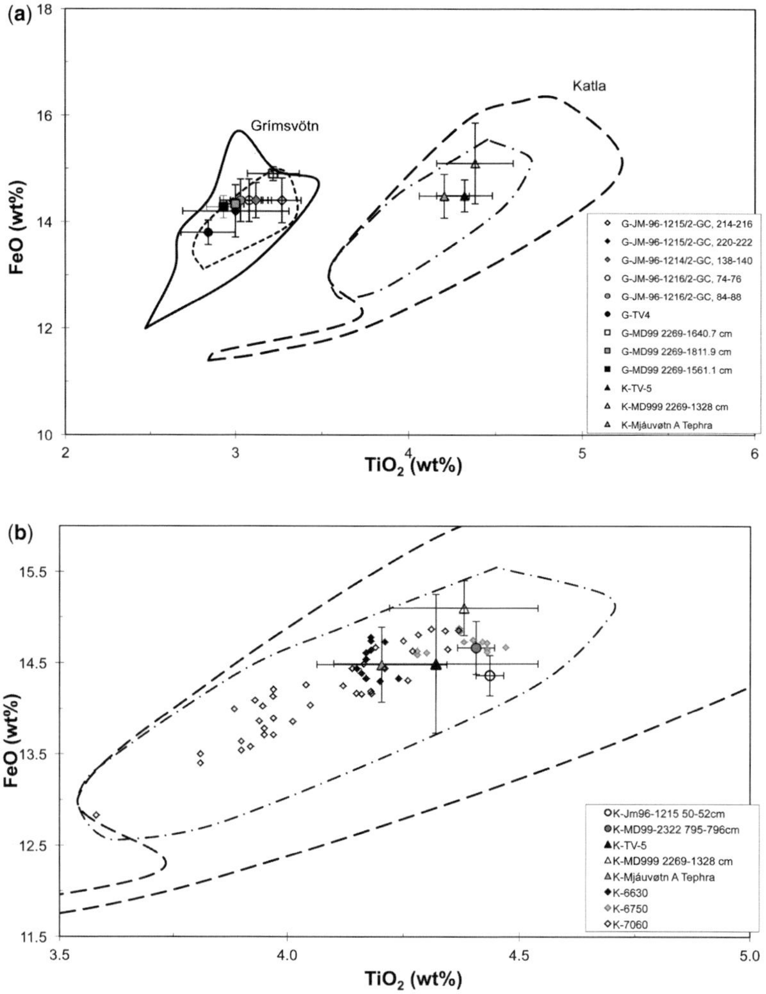

Fig. 15. (**a**) TiO_2–FeO plot showing the source volcano provenance of mid to early Holocene Grímsvötn and Katla basaltic tephra by previous studies in marine sediment cores west and north of Iceland and in sediment core from the Mjauvøtn lake in the Faroe Islands (Björck *et al.* 1992; Wastegård *et al.* 2001; Jennings *et al.* 2002; Kristjánsdóttir *et al.* 2007; see key and text for details). Abbreviations in the key are G, Grímsvötn (identified as Saksunarvatn in original studies) and K, Katla and indicate representative source volcanoes for these tephras. The solid line outlines the compositional field of Grímsvötn tephra and the fine dashed line shows the compositional range of Grímsvötn tephra analysed in this study. The compositional field of Holocene Katla tephra is indicated by the coarse dashed line, while the dashed–dotted line shows the range of Katla tephra deposits in the 6–7 ka period. The compositional fields are constructed using the Iceland Reference Tephra Data Base (T. Thordarson, unpublished data 2012). (**b**) Close up of the Katla field in (a), comparing the averaged compositions of 6–7 ka Katla tephra horizons found in marine cores west and north of Iceland and in the Mjáuvøtn (Mjauvotn in legend) lake core in the Faroe Islands. The bars indicate one standard deviation (1σ) of individual datasets.

analysed, but the morphology and colour of the grains suggest that they also represent volcanic events dispersed to the west or NW from Grímsvötn. An additional younger peak of basaltic shards occurs in JM96-1215/2GC. This peak does not have a correlative in MD99-2322.

Comparison of the PSV inclination records of MD99-2322 and MD99-2269 shows that the first of the Grímsvötn basalt tephra horizons in each core is correlative in terms of both depth and age (Figs 2 & 14; Table 1). The basal Grímsvötn peak in 2322 at 1797.5 cm overlaps in age only with the lowest Grímsvötn peak in 2269 (Table 1; Fig. 14). The full counting errors on the age of the Grímsvötn (inferred Saksunarvatn) tephra in the ice cores (10 252–10 342 yr BP; Rasmussen *et al.* 2006) shows that this peak could correspond to any of the three lowest Grímsvötn basalt tephra horizons in MD99-2322, but only to the lowest peak in MD99-2269 (Table 1). In terms of age, the 2322-1797.5 cm tephra peak has the best match to the Grímsvötn (inferred Saksunarvatn) tephra in the ice cores. Furthermore, the lowest Grímsvötn tephra peak in MD99-2269 and -2322 (Stoner *et al.* 2007; Kristjánsdóttir *et al.* 2007) and in JM96-1215/2GC (Jennings *et al.* 2002) have previously been interpreted as the correlative to the Grímsvötn (inferred Saksunarvatn) ash in the ice cores. Although such a correlation is likely, it cannot be confirmed by our data. Collectively, the data from MD99-2269, MD99-2322 and JM96-1215 clearly indicate that, in the interval between 9.9 and 10.4 cal kyr BP, there were at least five and possibly as many as seven tephra-producing (i.e. explosive) Grímsvötn eruptions that dispersed tephra to the north and west and that these eruptions were large enough to produce tephra horizons to be preserved in archives beyond the shores of Iceland (e.g. Table 2). Recent observations of as many as four Grímsvötn tephra layers on the Faroe Islands (Kylander *et al.* 2012), where Saksunarvatn ash was originally defined as a single event, support the concept of multiple explosive eruptive events from Grímsvötn in the early Holocene. Whether any of the eastward-dispersed events discovered in the Faroe Islands is correlative with any of the northward- and westward-dispersed events found in MD99-2322, -2269, and in the Greenland Ice Cores, is not yet determined.

Katla EG-6.73 ka; a new tephra marker for the SE Greenland shelf

Well-defined abundance peaks of brown vesicular basaltic tephra were found in the two SE Greenland shelf cores (1215 50–52 cm and 2322 795–796 cm; Fig. 13). In terms of composition they are statistically indistinguishable, their ages match closely and the peaks in the two cores provide a correlation point on the PSV inclination curves (Fig. 13). Their composition indicates mildly alkali (Fe–Ti) basalt erupted from Katla (Figs 2, 5 & 10a). Katla is a glaciated central volcano of the Katla Volcanic system of the Eastern Volcanic Zone, where over 300 phreatomagmatic eruptions have produced numerous basaltic tephra layers throughout the last 8400 years, at a frequency of about two to three per century (Óladóttir *et al.* 2005, 2008). Their dispersal is poorly known in prehistoric times, but volcanic plumes and tephra fallout sectors from historic Katla eruptions tend to have strongly unidirectional dispersal (Larsen 2010). The geochemistry of tephra peaks, 1215 50–52 cm and 2322 795–796 cm, falls within the compositional field of Katla basaltic eruptions between 6 and 7 ka. Based on their age and geochemistry, they are tentatively linked to the *c.* 6.75 ka tephra layer identified in lake sediment archives in western Iceland (Fig. 10b). However, it is clear from the frequency in basaltic tephra peaks in the SE Greenland cores that this is the only Katla basalt event that was recorded with a westward dispersal beyond Iceland (Figs 2, 13 & 14). Thus, this tephra layer can be used as a marker horizon for the SE Greenland shelf.

Several distal Katla basalt tephra layers have been reported within the timeframe of concern here, which is not a surprise because at that time the volcano was producing two to three eruptions per century (Óladóttir *et al.* 2005, 2008). All Holocene basaltic tephra layers from Katla have a similar major element composition and define a tight compositional field on all element-to-element plots (Fig. 15). However, despite the overall compositional uniformity among Katla tephra layers, small but marked differences are observed in the minor element concentrations as well as in distribution pattern of data points (e.g. Óladóttir *et al.* 2008). The Mjáuvøtn A was described from the Faroe Islands (Wastegård *et al.* 2001) with an age of 6658–6533 cal yr BP (Olsen *et al.* 2010). Although its composition is consistent with that of tephra layers produced by Katla in a 6–7 ka period, the composition (at least on 1σ level) and data point spread differ markedly from the SE Greenland marker (Fig. 15). This difference along with its eastward dispersal makes it an unlikely correlative with the SE Greenland event. TV-5 was described from Lake Torfadalsvatn on northern Iceland (Björck *et al.* 1992). Kristjánsdóttir *et al.* (2007) correlated a basaltic tephra peak in MD99-2269 with a 1σ age estimate of 6646–6848 cal yr BP to TV-5 (Fig. 14; Table 1). This correlation to TV-5 is sensible given the northward dispersal (Fig. 15). PSV controlled depth correlation of 2322 and 2269 inclination shows that the Katla tephra in 2269 matches within 9 cm (Figs 2 & 14 and Table 1) or within 10–70 years depending on the choice of age model (Table 1). Even though these peaks occur at similar times and exhibit similar geochemistry, given the high frequency of Katla basaltic eruptions in this time period and the different dispersal directions (i.e. west v. north), it is not likely that these peaks represent the same eruption and a correlation is not advised. However, given the apparent rarity of Katla basaltic eruptions in the marine environment beyond the Iceland landmass, and their very close ages and geochemistry, these peaks could be used as an approximate isochron.

Aniakchak eruption identified on the SE Greenland shelf

The age of the Aniakchak Eruption is 3595 ± 4 yr BP in the GRIP ice core (Pearce *et al.* 2004). Its identification in GRIP is still debated owing to its previous attribution to the Minoan eruption of Santorini by Hammer *et al.* (2003). The age estimate of 4410 cal yr BP of the Aniakchak horizon in JM96-1215 is based upon interpolation between two calibrated ages of 6620 yr BP (6420–6815 yr 1σ range) and 2310 yr BP (2201–2432 yr 1σ range; Fig. 13). The large discrepancy between the known age of the Aniakchak eruption and the age estimate of the tephra in JM96-1215 demonstrates that the 1215 age model is poorly constrained in the upper slow sedimentation part of the core. The tephra abundance peak at 414.5 cm (3660 cal yr BP) in MD99-2322 has the greatest likelihood of having been produced by the Aniakchak eruption, but this peak has not yet been analysed for its geochemical composition. The Aniakchak tephra was carried to the NE across Alaska (Begét *et al.* 1992; Pearce *et al.* 2004). Very small shards (10 μm diameter) were deposited on the Greenland Ice Sheet (Hammer *et al.* 2003; Pearce *et al.* 2004). In JM96-1215 the Aniakchak shards are much larger (200 μm), which can be taken to suggest that they were deposited on sea ice in the Arctic Ocean and carried by surface circulation to SE Greenland, where the shards melted out at the sea-ice edge. Direct transit to the site by air fall cannot be ruled out.

Settlement tephra in MD99-2322 154.5 cm (1144 cal yr BP) and 2322 158.5 (1152 cm) double peak, an historical tephra layer on the SE Greenland shelf

Based on the age, the presence of both basaltic/intermediate and rhyolitic shards, and the geochemistry, one or both of these peaks represents the Settlement or Landnám tephra (Thorarinsson 1944). The two-coloured Landnám tephra, dated in GRIP and GISP2 ice cores to AD 871 ± 2 (Grönvold *et al.* 1995; Zielinski *et al.* 1997), or 1079 yr BP, resulted from simultaneous eruptions of magma chambers in the Veidivötn volcanic system and the Torfajökull caldera volcano (Larsen 1984; Larsen *et al.* 1999) and provide an important historical reference marking the beginning of human settlement in Iceland and the Faroe Islands. Both the widespread basaltic component and the more directed plus smaller volume of rhyolitic tephra (e.g. Fig. 11) have been identified in Holocene soil and sediment archives in Iceland to the west of the Torfajökull volcanic system (Larsen 1984). Both components have been found in Greenland ice cores (Grönvold *et al.* 1995; Zielinski *et al.* 1997) and the basaltic component has been found in Faroe Islands (Hannon & Bradshaw 2000; Wastegård *et al.* 2001) and, most recently, in Scottish fjord sediment cores (Cage *et al.* 2011). The Settlement tephra was not found in MD99-2269. On the PSV record of MD99-2322 it lies at a recognizable shift in declination that should make it relatively easy to locate in other high-deposition-rate cores from the region (Fig. 14). Radiocarbon dates on foraminifers from this depth in MD99-2322 can be used in future to evaluate the marine reservoir correction along the SE Greenland shelf (cf. Eiríksson *et al.* 2004).

Influences on Tephra distribution

Tephra layers are not evenly distributed through the Holocene on the SE Greenland shelf. There are some lengthy time intervals with no tephra deposition, and intervening periods of more frequent events (Fig. 16). In terms of basaltic eruptions, after the high frequency basaltic tephra delivery from Grímsvötn between 9.9 and 10.4 cal kyr BP, there is a hiatus between 8.9 and 6.8 cal kyr BP during which we find no tephra events that reached the SE Greenland shelf. This hiatus is most likely related to pervasive westerly wind direction and/or relatively small eruptions that did not disperse tephra beyond Iceland. There are frequent eruptions throughout the Holocene (Larsen *et al.* 2001; Óladóttir *et al.* 2005) but they either were not large enough to be exported outside of Iceland or they were dispersed away from SE Greenland. The period between 6.8 and 3.0 cal kyr BP records more frequent west and NW dispersed events, beginning with Katla-EG 6.73, the basaltic layer that is recorded in both MD99-2322 and JM96-1215.

Intermediate and rhyolitic tephra layers increase in frequency after 5.1 cal kyr BP. Microprobe analyses show that both Katla and Hekla were source volcanoes for west/NW-dispersed tephra. Larsen *et al.* (2001) describe eight SILK layers between *c.* 7500 and 3400 cal yr BP and only two of these had a westward dispersal. The SILK layers with the best match geochemically to the SE Greenland tephra peaks are not known to have dispersed westward, but it is possible that separation of the early rhyolitic component of these eruptions was dispersed westward onto the SE Greenland shelf, in a different direction from the rest of the tephra. Hekla A, B and C are three westward dispersed events that form part of a seven-part series of Hekla eruptions with indistinguishable

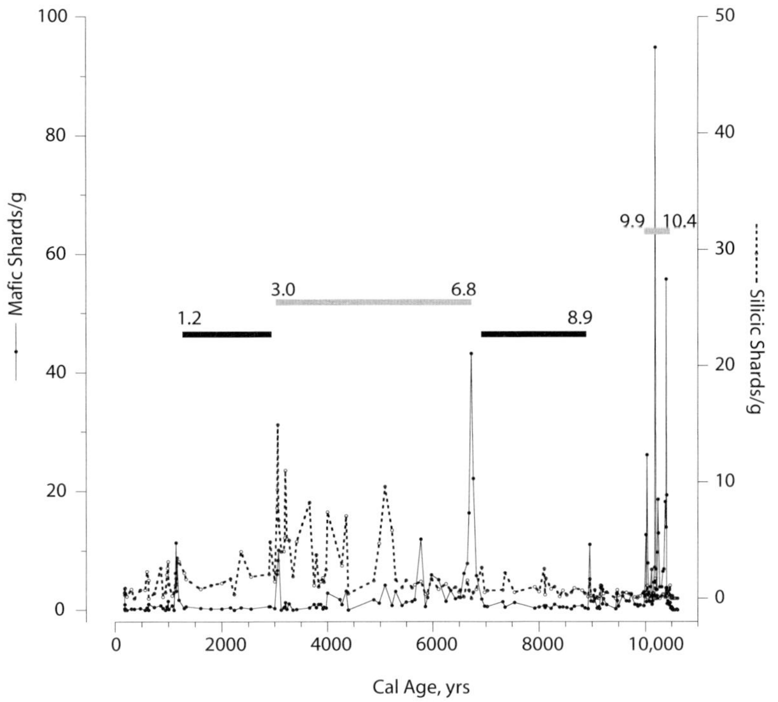

Fig. 16. Mafic (solid line) and silicic (dashed line) shards per gram in MD99-2322 against age in MD99-2322 (note different scales). Black and grey horizontal lines with ages represent low and high intervals of westward atmospheric dispersal of Icelandic eruption plumes to the SE Greenland shelf.

geochemistry (Larsen *et al.* 1999; Meara 2012). These are the only Holocene Hekla tephras with known westward dispersal (Larsen & Eiríksson 2008*a*, *b*). Of these, Hekla B is the best match in geochemistry and age for the Hekla tephra found in MD99-2322. The Settlement layer is the final west-dispersed peak found in MD99-2322; it is also found in the Greenland ice cores (Grönvold *et al.* 1995; Zielinski *et al.* 1997). Given the rarity of westward dispersal of Icelandic tephra, there are a surprising number of volcanic eruptions recorded on the SE Greenland shelf. These events must be the product of favourable easterly or south-easterly winds and/or high explosivity index and high altitude reached by the eruption column. Because the Settlement layer is an historical tephra, it provides an excellent opportunity for future work to document the marine reservoir age in the late Holocene on the SE Greenland shelf.

Conclusions

The Holocene tephrostratigraphic record of westward tephra dispersal from Iceland was investigated using sediment cores from the SE Greenland shelf. Volcanic events from the Bárðarbunga–Veiðivötn, Katla, Hekla, Torfajökull and Grímsvötn volcanic systems were recovered from the cores. Direct comparison of the tephra stratigraphic and compositional data from MD99-2322 (SE Greenland shelf) to MD99-2269 (north Iceland shelf) showed only one volcanic event in common. We conclude that, at least during the Holocene, most eruptions were directed discretely westward (2322) or northward (2269) so that local rather than regional isochrones were formed. The discrete dispersal of many tephra horizons and the similar geochemistry of closely spaced events from the same volcanic system underscore the need for very careful stratigraphy and chronology in all records before an isochronous event can be confirmed. PSV records from high depositional rate areas are useful stratigraphic tools for this purpose.

At least three basaltic volcanic eruptions from Grímsvötn were identified on SE Greenland, with ages of 10.39, 10.19 and 10.03 cal kyr BP. Two additional basaltic peaks at 10.26 and 10.01 are probably Grímsvötn events but were not confirmed geochemically. The earliest of these peaks probably matches the so-called Saksunarvatn ash identified in the Greenland ice cores and on the north Iceland shelf in core MD99-2269. Two additional Grímsvötn peaks in MD99-2269 do not match stratigraphically with those from SE Greenland, suggesting that these two peaks represent separate northward-directed eruptions from Grímsvötn. The 9.9–10.4 timeframe in MD99-2322 and -2269 contains as many as seven basaltic Grímsvötn eruptions.

A new local marker horizon correlative with Katla 6750 on Iceland, named Katla-EG 6.73 in the Kangerlussuaq Trough, was defined in the two SE Greenland cores. This peak closely matches a peak in MD99-2269 in terms of composition and timing, suggesting that, even if these represent different eruptions, they are so close in time that they are functionally a regional isochron that will be useful for correlation among fairly widespread records in the Denmark Strait region.

Within a group of six rhyolitic tephra peaks in the interval between 5.1 and 3 cal kyr BP, three peaks had complex compositions with components that could be identified to specific eruptions: SILK A1, SILK MN and Hekla B. Only Hekla B is confirmed by both geochemistry and stratigraphy as a marker horizon on the SE Greenland shelf. The potential correlation of 2322/358.5 and 2322/555.5 to SILK MN and SILK A1, respectively, will require additional analyses of these tephra layers where they are found in soil profiles on Iceland to determine whether they have an initial rhyolitic phase and in other cores from the SE Greenland shelf.

The *c.* 3600 yr BP Alaskan Aniakchak Eruption formed a tephra abundance peak in JM96-1215. This peak probably has a correlative in MD99-2322 at 414.5 cm (3660 cal yr BP). Geochemical confirmation is needed to confirm its presence in MD99-2322.

Settlement layer (AD 871 ± 2 yr) was documented on the SE Greenland shelf. This will provide an important tie point with other cores in the region as well as with the Greenland ice cores. Future work on radiocarbon dating from this level in MD99-2322 will allow analysis of the marine reservoir correction for sites on the SE Greenland shelf.

Two undergraduate students, S. Johnston and M. Radcliffe helped to pick tephra shards from MD99-2322. Support for this study was provided by The Icelandic Research Fund, Grant of Excellence 070272013, and United States National Science Foundation grant ARCSS 0714074, both entitled: 'Volcanism in the Arctic SysTem (VAST): Geochronology and Climate Impacts'.

References

Abbott, P. M. & Davies, S. M. 2012. Volcanism and the Greenland Ice-cores: the Tephra Record. *Earth-Science Reviews*, **115**, 173–191.

Andrews, J. T., Geirsdóttir, A. *et al.* 2002. Distribution, sediment magnetism, and geochemistry of the Saksunarvatn (10.189/cal ka) tephra in marine, lake, and terrestrial sediments, NW Iceland. *Journal of Quaternary Science*, **17**, 731–745.

Begét, J. E., Mason, O. & Anderson, P. 1992. Age, extent and climatic significance of the *c.* 3400 BP Aniakchak tephra, western Alaska, USA. *The Holocene*, **2**, 51–56.

Birks, H. H., Gulliksen, S., Haflidason, H., Mangerud, J. & Possnert, G. 1996. New radiocarbon dates for the Vedde Ash and Saksunarvatn Ash from western Norway. *Quaternary Research*, **45**, 119–127.

Björck, S., Ingólfsson, O., Haflidason, H., Hallsdóttir, M. & Anderson, N. J. 1992. Lake Torfadalsvatn: a high resolution record of the North Atlantic ash zone 1 and the late glacial-interglacial environmental changes in Iceland. *Boreas*, **21**, 15–22.

Cage, A. G., Davies, S. M., Wastegård, S. & Austin, W. E. N. 2011. Identification of the Icelandic Landnam tephra (AD 871 ± 2) in Scottish fjordic sediment. *Quaternary International*, **246**, 168–176.

Carey, R., Houghton, B. F. & Thordarson, T. 2010. Tephra dispersal and eruption dynamics of wet and dry phases of the 1875 eruption of Askja Volcano, Iceland. *Bulletin of Volcanology*, **72**, 259–278.

Davies, S. M., Wastegård, S. *et al.* 2010. Tracing volcanic events in the NGRIP ice-core and synchronising North Atlantic marine records during the last glacial period. *Earth and Plantetary Science Letters*, **294**, 69–79.

Davis, S. M., Wastegård, S., Rasmussen, T. L., Svensson, A., Johnsen, S. J., Steffensen, J. P. & Andersen, K. K. 2008. Identification of the Fugloyarbanki tephra in the NGRIP ice core: a key tie-point for marine and ice-core sequences during the last glacial period. *Journal of Quaternary Science*, **23**, 409–414.

Dreher, S. T., Eichelberger, J. C. & Larsen, J. F. 2005. The petrology and geochemistry of the Aniakchak caldera-forming ignimbrite, Aleutian Arc, Alaska. *Journal of Petrology*, **46**, 1747–1768, http://dx.doi.org/10.1093/petrology/egi032

Dugmore, A. J. & Newton, A. J. 1998. Holocene tephra layer in the Faroe Islands. *Fródskaparrit*, **46**, 191–204.

Dugmore, A. J., Larsen, G. & Newton, A. J. 1995. Seven tephra isochrones in Scotland. *The Holocene*, **5**, 257–266.

Eiríksson, J., Larsen, G., Knudsen, K., Heinemeier, J. & Símonarson, L. A. 2004. Marine reservoir age variability and water mass distribution in the Iceland Sea. *Quaternary Science Reviews*, **23**, 2247–2268.

Francis, P. & Oppenheimer, C. 2004. *Volcanoes*. Oxford University Press, Oxford.

Grönvold, K. 1984. *Mývatn fires 1724-1729 chemical composition of the lava*. Nordic Volcanological Institute Report **8401**.

Grönvold, K. & Makipaa, H. 1978. Chemical composition of Krafla lavas 1975–77. Nordic Volcanological Institute Report, **7816**.

Grönvold, K., Larsen, G., Einarsson, P., Thorarinsson, S. & Sæmundsson, K. 1983. The Hekla eruption 1980–1981. *Bulletin of Volcanology*, **46**, 349–363.

Grönvold, K., Óskarsson, K., Johnsen, S. J., Clausen, H. B., Hammer, C. U., Bomd, G. & Bard, E. 1995. Ash layers from Iceland in the Greenland GRIP ice core correlated with oceanic and land sediments. *Earth and Planetary Science Letters*, **135**, 149–55.

GUÐMUNDSDÓTTIR, E. R., LARSEN, G. & EIRÍKSSON, J. 2012. Tephra stratigraphy on the North Icelandic shelf: extending tephrochronology into marine sediments off North Iceland. *Boreas*, **41**, 719–734.

GUÐMUNDSSON, Á., GRÖNVOLD, G. *ET AL*. 1992. The 1991 eruption of Hekla, Iceland. *Bulletin of Volcanology*, **54**, 238–246.

GUÐMUNDSSON, M. T., THORDARSON, T. *ET AL*. 2012. *Ash generation and distribution from the April–May 2010 eruption of Eyjafjallajökull, Iceland.* Scientific Reports, **2**, 572, http://dx.doi.org/10.1038/srep00572

GULLIKSEN, S., BIRKS, H. H., POSSNERT, G. & MANGERUD, J. 1998. A calendar age estimate of the younger Dryas–Holocene boundary at Kråkenes, western Norway. *The Holocene*, **8**, 249–259.

HAFLIÐASON, H., EIRÍKSSON, J. & VAN KREVELD, S. 2000. The tephrochronology of Iceland and the North Atlantic region during the Middle and Late Quaternary: a review. *Journal of Quaternary Science*, **15**, 3–22.

HAMMER, C. U., KURAT, G., HOPPE, P., GRUM, W. & CLAUSEN, H. B. 2003. Thera eruption date 1645 BC confirmed by new ice core data? Paper presented at the *SCIEM 2000-EuroConference*. SCIEM, Vienna.

HANNON, G. E. & BRADSHAW, R. H. W. 2000. Impacts and timing of the first human settlement on vegetation on the Faroe Islands. *Quaternary Research*, **54**, 404–413.

HARTLEY, M. E. 2012. *Postglacial volcanism and magmatism on the Askja volcanic system, North Iceland.* PhD thesis, University of Edinburgh.

HASTINGS, A. D. 1960. *Environment of Southeast Greenland.* Headquarters, Quartermaster Research & Engineering Command, US Army, Quartermaster Research & Engineering Center.

HAYWARD, C. L. 2012. High spatial resolution electron probe microanalysis of tephras and melt inclusions without beam-induced chemical modification. *The Holocene*, **22**, 119–125.

HÖSKULDSSON, Á., ÓSKARSSON, N., PEDERSEN, R., GRÖNVOLD, K., VOGFJÖRD, K. & ÓLAFSDÓTTIR, R. 2007. The millennium eruption of Hekla in February 2000. *Bulletin of Volcanology*, **70**, 169–182.

IRVINE, T. N. & BARAGAR, W. R. A. 1971. A guide to the chemical classification of the common volcanic rocks. *Canadian Journal of Earth Science*, **8**, 523–548.

JAGAN, A. 2010. *Tephra stratigraphy and geochemistry from three Icelandic lake cores: a new method for determining source volcano of tephra layers.* MSc thesis, University of Edinburgh.

JENNINGS, A. E., GRÖNVOLD, K., HILBERMAN, R., SMITH, M. & HALD, M. 2002. High-resolution study of Icelandic tephras in the Kangerlussuaq Trough, southeast Greenland, during the last de-glaciation. *Journal of Quaternary Science*, **17**, 747–757.

JENNINGS, A. E., ANDREWS, J. T. & WILSON, L. 2011. Holocene environmental evolution of the SE Greenland Shelf North and South of the Denmark Strait: Irminger and East Greenland current interactions. *Quaternary Science Reviews*, **30**, 980–998.

JÓHANNESSON, H. & SÆMUNDSSON, K. 1998. *Geological Map of Iceland, 1:500.000. Bedrock Geology.* Icelandic Institute of Natural History and Iceland Geodetic Survey, Reykjavík.

JÓHANNSDÓTTIR, G. E. 2007. *Mid-Holocene to late glacial tephrochronology in west Iceland as revealed in three lacustrine environments.* M.S. thesis, University of Iceland, Reykjavík.

JÓHANNSDÓTTIR, G. E., THORDARSON, T., GEIRSDÓTTIR, Á. & LARSEN, G. 2005. The widespread 10 a Saksunarvatn tephra: a product of three large basaltic phreatoplmian eruptions? *European Geosciences Union, Geophysical Research Abstracts*, **7**, 05991.

JÓNSSON, T. 1990. Hvert liggja gjóskugeirar? *Náttúrufræðingurinn*, **60**, 103.

JØRGENSEN, K. A. 1981. *The Eruption of the Þórsmörk Ignimbrite, South Iceland: The Structure of the Magma Chamber Deduced from the Eruption Products.* Nordic Volcanological Institute, Reykjavík, Professional Papers, **8103**, 82.

KARHUNEN, R. 1988. *Eruption mechanism and rheomorphism during the basaltic fissure eruption in Biskupsfell, Kverkfjöll, north-central Iceland.* Research Report, **8802**, Nordic Volcanological Institute, Reykjavík.

KOÇ, N. & JANSEN, E. 1994. Response of the high-latitude Northern Hemisphere to orbital climate forcing: Evidence from the Nordic Seas. *Geology*, **22**, 523–526.

KRISTJÁNSDÓTTIR, G. B., STONER, J. S., JENNINGS, A. E., ANDREWS, J. T. & GRÖNVOLD, K. 2007. Geochemistry of Holocene cryptotephras from the North Iceland Shelf (MD99-2269): intercalibration with radiocarbon and palaeomagnetic chronostratigraphies. *The Holocene*, **17**, 155–176.

KUNO, H. 1968. Differentiation of basaltic magmas. *In*: HESS, H. H. & POLDERVAART, A. (eds) *Basalts: The Poldervaart Treatise on Rocks of Basaltic Composition.* Interscience, New York, **2**, 623–688.

KYLANDER, M. E., LIND, E. M., WASTEGÅRD, S. & LÖWEMARK, L. 2012. Recommendations for using XRF core scanning as a tool in tephrochronology. *The Holocene*, **22**, 371–375.

LABEYRIE, L., JANSEN, E. & CORTIJO, E. 2003. *Les Rapports de Campagnes a la Mer MD114/IMAGES V.* Institut Polaire Francais Paul-Emile Victor, Brest.

LACASSE, C. 2001. Influence of climate variability on the atmospheric transport of Icelandic tephra in the subpolar North Atlantic. *Global and Planetary Change*, **29**, 31–55.

LACASSE, C., KARLSDÓTTIR, S., LARSEN, G., SOOSALU, H., ROSE, W. I. & ERNST, G. G. J. 2004. Weather radar observations of the Hekla 2000 eruption cloud, Iceland. *Bulletin of Volcanology*, **66**, 457–473.

LARSEN, D. J., MILLER, G. H., GEIRSDÓTTIR, Á. & ÓLAFSDÓTTIR, S. 2012. Non-linear Holocene climate evolution in the North Atlantic: a high-resolution, multi-proxy record of glacier activity and environmental change from Hvítárvatn, central Iceland. *Quaternary Science Reviews*, **39**, 14–25.

LARSEN, G. 1982. Gjóskutímatal Jökuldals og nágrennis. *In*: THÓRARINSDÓTTIR, H., ÓSKARSSON, Ó. H., STEINTHÓRSSON, S. & EINARSSON, T. (eds) *Eldur er í Norðri.* Sögufélag, Reykjavík, 51–65.

LARSEN, G. 1984. Recent volcanic history of the Veidivötn fissure swarm, southern Iceland – an approach to volcanic risk assessment. *Journal of Volcanology and Geothermal Research*, **22**, 33–58.

LARSEN, G. 2002. A brief overview of eruptions from ice-covered and ice-capped volcanic systems in Iceland during the past 11 centuries: frequency, periodicity and implications. *In*: SMELLIE, J. L. & CHAPMAN, M. G. (eds) *Volcano–Ice Interaction on Earth and Mars*. Geological Society, London, Special Publications, 81–90.

LARSEN, G. 2010. Katla: tephrochronology and eruption history. *Developments in Quaternary Sciences*, **13**, 23–49.

LARSEN, G. & EIRÍKSSON, J. 2008*a*. Late Quaternanry terrestrial tephrochronology of Iceland – frequency of explosive eruptions, type and volume of tephra deposits. *Journal of Quaternary Science*, **23**, 109–120.

LARSEN, G. & EIRÍKSSON, J. 2008*b*. Holocene tephra archives and tephrochronology in Iceland – a brief overview. *Jökull*, **58**, 229–250.

LARSEN, G. & THORARINSSON, S. 1977. H-4 and other acid Hekla tephra layers. *Jökull*, **27**, 28–46.

LARSEN, G., DUGMORE, A. J. & NEWTON, A. J. 1999. Geochemistry of historical-age silicic tephras on Iceland. *The Holocene*, **9**, 463–71.

LARSEN, G., NEWTON, A. J. & VILMUNDARDÓTTIR, E. G. 2001. Geochemistry, dispersal, volumes and chronology of Holocene slicic tephras from the Katla volcanic system, Iceland. *Journal of Quaternary Science*, **16**, 119–132.

LIND, E. M. & WASTEGÅRD, S. 2011. Tephra horizons contemporary with short early Holocene climate fluctuations: new results from the Faroe Islands. *Quaternary International*, **246**, 157–167.

MACDONALD, G. A. & KATSURA, T. 1964. Chemical composition of Hawaiian Lavas. *Journal of Petrology*, **5**, 82–133, http://dx.doi.org/10.1093/petrology/5.1.82

MANGERUD, J., FURNES, H. & JÓHANSEN, J. 1986. A 9000-year old ash bed on the Faroe Islands. *Quaternary Research*, **26**, 262–265.

MARTIN, E. & SIGMARSSON, O. 2007. Crustal thermal state and origin of silicic magma in Iceland: the case of Torfajökull, Ljósufjöll and Snæfellsjökull volcanoes. *Contributions to Mineralogy and Petrology*, **153**, 593–605.

MEARA, H. D. 2012. *Geochemical fingerprinting of Icelandic silicic Holocene tephra layers*. PhD thesis, University of Edinburgh.

MOORE, J. G. & CALK, L. C. 1991. Degassing and differentiation in subglacial volcanoes, Iceland. *Journal of Volcanology and Geothermal Research*, **46**, 157–180.

MORTENSEN, A., BIGLER, M., GRÖNVOLD, K., STEFFENSEN, J. & JOHNSEN, S. 2005. Volcanic ash layers from the Last Glacial Termination in the NGRIP ice core. *Journal of Quaternary Science*, **20**, 209–219.

MOUNE, S., SIGMARSSON, O., THORDARSON, T. & GAUTHIER, P.-J. 2007. Recent volatile evolution in the magmatic system of Hekla volcano, Iceland. *Earth and Planetary Science Letters*, **255**, 373–389.

NEAL, C. A., MCGIMSEY, R. G., MILLER, T. P., RIEHLE, J. R. & WAYTHOMAS, C. F. 2001. *Preliminary Volcano-Hazard Assessment for Aniakchak Volcano, Alaska. USGS Alaska Volcano Observatory Anchorage*, Alaska Open-File Report, **00-519**.

ODDSSON, B., GUDMUNDSSON, M. T., LARSEN, G. & KARLSDÓTTIR, S. 2012. Monitoring of the plume from the basaltic phreatomagmatic 2004 Grímsvötn eruption – application of weather radar and comparison with plume models. *Bulletin of Volcanology*, **74**, 1395–1407.

ÓLADÓTTIR, B. A., LARSEN, G., THORDARSON, T. H. & SIGMARSSON, O. 2005. The Katla volcano S-Iceland: Holocene tephra stratigraphy and eruption frequency. *Jökull*, **55**, 53–74.

ÓLADÓTTIR, B. A., SIGMARSSON, O., LARSEN, G. & THORDARSON, T. 2008. Katla volcano, Iceland: magma composition, dynamics and eruption frequency as recorded by Holocene tephra layers. *Bulletin of Volcanology*, **70**, 475–493.

ÓLADÓTTIR, B. A., LARSEN, G. & SIGMARSSON, O. 2011*a*. Holocene volcanic activity at Grímsvötn, Bárdarbunga and Kverkfjöll subglacial centres beneath Vatnajökull, Iceland. *Bulletin of Volcanology*, **73**, 1187–1208.

ÓLADÓTTIR, B. A., SIGMARSSON, O., LARSEN, G. & DEVIDAL, J.-L. 2011*b*. Provenance of basaltic tephra from Vatnajökull subglacial volcanoes, Iceland, as determined by major- and trace-element analyses. *The Holocene*, **21**, 1037–1048.

ÓLAFSDÓTTIR, S. 2010. *Holocene marine and lacustrine paleoclimate and paleomagnetic records from Iceland*. PhD thesis, University of Iceland.

OLSEN, J., GUDMUNDSDÓTTIR, E. R., BJØRCK, S., ODGAARD, B. V. & HEINEMEIER, J. 2010. Revised age estimate of the Mjáuvøtn tephra A on the Faroe Islands based on Bayesian modelling of ^{14}C dates from two lake sequences. *Journal of Quaternary Science*, **25**, 612–616.

PEARCE, N. J. G., WESTGATE, J. A., PREECE, S. J., EASTWOOD, W. J. & PERKINS, W. T. 2004. Identification of Aniakchak (Alaska) tephra in Greenland ice core challenges the 1645 BC date for Minoan eruption of Santorini. *Geochemistry, Geophysics, Geosystems*, **5**, Q03005, http://dx.doi.org/10.1029/2003GC000672

PÉTURSSON, H. G. & LARSEN, G. 1992. An early Holocene basaltic tephra bed in North Iceland, a possible equivalent to the Saksunarvatn ash bed. *20th Nordic Winter Meeting*, Reykjavík.

RASMUSSEN, S. O., ANDERSON, K. K. *ET AL*. 2006. A new Greenland ice core chronology for the last glacial termination. *Journal of Geophysical Research*, **111**, D06102, http://dx.doi.org/10.1029/2005JD006079

RASMUSSEN, S. O., VINTHER, B. M., CLAUSEN, H. B. & ANDERSEN, K. K. 2007. Early Holocene climate oscillations recorded in three Greenland ice cores. *Quaternary Science Reviews*, **26**, 1907–1914.

REIMER, P. J., BAILLIE, M. G. L. *ET AL*. 2009. IntCal09 and Marine09 radiocarbon age calibration curves, 0–50,000 years cal BP. *Radiocarbon*, **51**, 1111–1150.

RIEHLE, J. R., MEYER, C. E. & MIYAOK, R. T. 1999. *Data on holocene tephra (volcanic ash) deposits in the Alaska Peninsula and Lower Cook Inlet region of the Aleutian Volcanic Arc, Alaska*. United States Geological Survey Open-file Report **99-135**, www.avo.alaska.edu

SIGURGEIRSSON, M. A. 1992. *Gjóskumyndanir á Reykjanesi (Tephra deposits on the Reykjanes volcanic system)*. MSc thesis, University of Iceland.

STEINÞÓRSSON, S. 1978. Tephra layers in a drill core from the Vatnajökull Ice Cap. *Jökull*, **27**, 2–27.

STEVENSON, J. A., LOUGHLIN, S. *ET AL*. 2012. Distal deposition of tephra from the Eyjafjallajökull 2010 summit eruption. *Journal of Geophysical Research*, **117**, B00C10, http://dx.doi.org/10.1029/2011JB008904

STONER, J. S., JENNINGS, A. E., KRISTJÁNSDÓTTIR, G. B., DUNHILL, G., ANDREWS, J. T. & HARDARÓTTIR, J. 2007. A paleomagnetic approach toward refining Holocene radiocarbon based chronologies: paleoceanographic records from North Iceland (MD99-2269) and East Greenland (MD99-2322) margins. *Paleoceanography*, **22**, PA1209, http://dx.doi.org/10.1029/2006PA001285

STUIVER, M., REIMER, P. J. & REIMER, R. W. 2010. CALIB 6.0, http://calib.qub (accessed 24 June 2011).

SVENSSON, A., ANDERSEN, K. K. *ET AL*. 2008. A 60,000 year Greenland stratigraphic ice core chronology. *Climate of the Past*, **4**, 47–57.

THORARINSSON, S. 1944. Tefrokronologiska studier på Island. Thjórsárdalur och dess förödelse. *Geografiska Annaler*, **26**, 1–217.

THORARINSSON, S. 1954. The tephra-fall from Hekla on March 29th 1947. *In*: EINARSSON, T., KJARTANSSON, G. & THORARINSSON, S. (eds) *The Eruption of Hekla 1947–1948*. Societas Scientarium Islandica, Reykjavík, **II**, 1–68.

THORARINSSON, S. 1967. The eruptions of Hekla in historical times. *In*: EINARSSON, T., KJARTANSSON, G. & ÞÓRARINSSON, S. (eds) *The Eruption of Hekla 1947–1948*. Societas Scientarium Islandica, Vísindafélag Íslendinga, **I**, 1–183.

THORARINSSON, S. 1970. *Hekla*. (In Icelandic.) Almenna Bókafélagid, Reykjavík.

THORARINSSON, S. 1981. The application of tephrochronolgy in Iceland. *In*: SELF, S. & SPARKS, R. S. J. (eds) *Tephra Studies: Proceedings of the NATO Advanced Study Institute 'Tephra Studies as a Tool in Quaternary Research'*. Reidel, Dordrecht, **75**, 109–134.

THORDARSON, T. & HÖSKULDSSON, Á. 2008. Postglacial volcanism in Iceland. *Jökull*, **58**, 197–228.

THORDARSON, T., SELF, S., ÓSKARSSON, N. & HULSEBOSCH, T. 1996. Sulfur, chlorine and fluorine degassing and atmospheric loading by the 1783–1784 AD Laki (Skaftár Fires) eruption in Iceland. *Bulletin of Volcanology*, **58**, 205–225.

THORDARSON, T., MILLER, D. J. & LARSEN, G. 1998. New data on the age and origin of the Leidólfsfell cone group in south Iceland. *Jökull*, **46**, 3–15.

THORDARSON, T., MILLER, D. J., LARSEN, G., SELF, S. & SIGURDSSON, H. 2001. New estimates of sulfur degassing and atmospheric mass-loading by the 934 AD Eldgjá eruption, Iceland. *Journal of Volcanology and Geothermal Research*, **108**, 33–54.

THORDARSON, T., SELF, S., MILLER, D. J., LARSEN, G. & VILMUNDARDÓTTIR, E. G. 2003. Sulphur release from flood lava eruptions in the Veidivötn, Grímsvötn and Katla volcanic systems, Iceland. *In*: OPPENHEIMER, C., PYLE, D. M. & BARCLAY, J. (eds) *Volcanic Degassing*. Geological Society, London, Special Publications, **213**, 103–121.

TOMLINSON, E. L., THORDARSON, T., MÜLLER, W., THIRLWALL, M. & MENZIES, M. A. 2010. Microanalysis of tephra by LA-ICP-MS — strategies, advantages and limitations assessed using the Thorsmörk ignimbrite (Southern Iceland). *Chemical Geology*, **279**, 73–89.

TOMLINSON, E. L., THORDARSON, T. *ET AL*. 2012. Petrogenesis of the Sólheimar Ignimbrite (Katla, Iceland): implications for tephrostratigraphy. *Geochimica et Cosmochimica Acta*, **86**, 318–337.

TRØNNES, R. 1990. Basaltic melt evolution of the Hengill Volcanic System, SW Iceland, and evidence for clinopyroxene assimilation in primitive tholeiitic magmas. *Journal of Geophysical Research*, **95**, 15 893–15 910.

TURNEY, C. S. M., LOWE, J. J. *ET AL*. 2004. Tephrochronology of Last termination sequences in Europe: a protocol for improved analytical precision and robust correlation procedures (a joint SCOTAVE-INTIMATE proposal). *Journal of Quaternary Science*, **19**, 111–120.

WASTEGÅRD, S. 2005. Late Quaternary tephrochronology of Sweden: a review. *Quaternary International*, **130**, 49–62.

WASTEGÅRD, S., BJÖRCK, S., GRAUERT, M. & HANNON, G. E. 2001. The Mjàuvøtn tephra and other Holocene tephra horizons from the Faroe Islands: a link between the Icelandic source region, the Nordic Seas, and the European continent. *The Holocene*, **11**, 101–109.

WASTEGÅRD, S., HALL, V. A., HANNON, G. E., VAN DEN BOGAARD, C., PILCHER, J. R., SIGURGEIRSSON, Á. & HERMANNS-AUDARDÓTTIR, M. 2003. Rhyolitic tephra horizons in northwest Europe and Iceland from the AD 700s–800s: a potential alternative for dating first human impact. *The Holocene*, **13**, 277–283.

WOLFE, C. J., BJARNASON, I. T., VAN DECAR, J. C. & SOLOMON, S. C. 1997. Seismic structure of the Iceland mantle plume. *Nature*, **385**, 245–247.

ZIELINSKI, G. A., GERMANI, M. S., LARSEN, G., BAILLE, M. G. L., WHITLOW, S., TWICKLER, M. S. & TAYLOR, K. C. 1997. Volcanic aerosol records, tephrochronology of the Summit, Greenland, ice cores *Journal of Geophysical Research*, **102**, 26 625–26 640.

Quantifying bioturbation of a simulated ash fall event

JOE A. TODD[1,2]*, WILLIAM E. N. AUSTIN[1,3] & PETER M. ABBOTT[1,4]

[1]*School of Geography and Geosciences, University of St Andrews, St Andrews, Fife, Scotland KY16 9AL, UK*

[2]*Present address: Department of Geography, Scott Polar Research Institute, University of Cambridge, Lensfield Road, Cambridge CB2 1ER, UK*

[3]*Present address: Scottish Marine Institute, Scottish Association for Marine Science, Oban PA37 1QA, UK*

[4]*Present address: Department of Geography, College of Science, Swansea University, Singleton Park, Swansea SA2 8PP, UK*

**Corresponding author (e-mail: jat71@cam.ac.uk)*

Abstract: Tephrochronology allows the establishment of 'isochrons' between marine, lacustrine, terrestrial and ice cores, typically based on the geochemical fingerprint of the tephra. The development of cryptotephrochronology has revealed a vast inventory of isochrons which hold the potential to improve stratigraphic correlation and identify systemic leads and lags in periods of rapid climate change. Unfortunately, bioturbation acts to blur these isochrons, reducing the temporal resolution in marine and lacustrine records. In order to better resolve these event horizons, we require a better understanding of bioturbative processes, and the depth and time over which they operate. To this end, an ash fall event was simulated on the intertidal zone of the Eden Estuary, Fife, Scotland and sediment cores were collected over 10 days. A novel approach to tephra quantification was developed, using the imaging software ImageJ. Our results showed limited bioturbation (mixed depth = 18 mm), most likely owing to the fine grain size, low-energy environment and the resulting faunal composition of the sediments. These results imply a strong ecological control on bioturbation, and suggest that inferences may be made about palaeoenvironments from the observed bioturbation profiles.

Supplementary material: The ImageJ macro used in this study, as well as raw tephra concentration data and details of the method validation are available at http://www.geolsoc.org.uk/SUP18725.

In most marine or lacustrine benthic habitats, the surficial sediment layer is very likely to be mixed and reworked by its epifauna and infauna, creatures living on and in the sediment, respectively (e.g. Berger & Heath 1968), as well as by any fish or mammals which come into contact with it (Biles *et al.* 2002). The collective mixing effect of this biological activity is known as bioturbation. In terms of the processes involved, bioturbation occurs largely through burrowing, feeding, defecation and locomotion.

Bioturbation processes are of interest to researchers in many fields. For marine ecologists, bioturbation is an essential set of processes that enhance nutrient release from the sediment (Solan *et al.* 2004*a*) as well as increasing the surface area of the sediment–water interface (Biles *et al.* 2002). Within the context of evolutionary theory, bioturbation is a classic example of ecosystem engineering, the alteration of a habitat by its inhabitants (Meysman *et al.* 2006). However, to palaeoceanographers, bioturbation presents a problem because the processes of bioturbation act to mix younger material into the older sediments below, and vice versa. Bioturbation therefore acts to obfuscate the geological sedimentary record. This makes it difficult to identify boundaries between strata which may have been relatively clear at the time of deposition.

Previous bioturbation investigations (Solan *et al.* 2004*b*) have focussed on bioturbation at the scale of individual events using, for example, time lapse photography. However, carrying out such an investigation is expensive; furthermore, while individual bioturbation events are of interest to ecologists, geochronology tends to be concerned with the net effect of repeat events. Unlike ecologists, geochronologists are not primarily interested in the individual bioturbation events but, rather, the integrated result of this process through time.

From: Austin, W. E. N., Abbott, P. M., Davies, S. M., Pearce, N. J. G. & Wastegård, S. (eds) 2014. *Marine Tephrochronology*. Geological Society, London, Special Publications, **398**, 195–207.
First published online February 5, 2014, http://dx.doi.org/10.1144/SP398.9

Guinasso & Schink (1975) reviewed bioturbation from an oceanographic perspective, and described it in terms of three principle parameters: an eddy diffusion rate parameter, D; a maximum mixing depth, L; and sediment deposition rate, v. They also defined a dimensionless bioturbation index:

$$\text{bioturbation index} = \frac{D}{Lv}$$

whose value describes the homogeneity of the surface mixed layer, with low values describing limited mixing and vice versa. An important fact highlighted by this bioturbation index is that the extent of mixing is affected not only by the processes of bioturbation, but also by the rate of sediment burial.

Tephrochronology

Tephrochronology uses tephra layers from past volcanic eruptions as isochronous marker horizons, layers of the same age in different geological records. On geological time-scales, tephra deposition by a volcanic eruption is effectively synchronous and can occur over a wide geographical area, making it especially useful in studying the (a)synchronicity of rapid climate transitions, between disparate sequences and where traditional dating methods lack sufficient resolution (Mangerud *et al.* 1984; Haflidason *et al.* 2000; Austin *et al.* 2004; Austin & Abbott 2010). For example, Austin *et al.* (2004) demonstrated how the correlation of the rhyolitic component (II-Rhy-I) of North Atlantic Ash Zone II between the GISP2 ice-core and a Northeast Atlantic marine core (MD95-2006) suggested the regional synchroneity of the abrupt cooling at the end of Greenland Interstadial-15.

One of the great values of tephrochronology lies in the recognition of globally synchronous ash fall events. Unfortunately, an inability to precisely define the position of a tephra isochron in a marine sequence owing to various source, transport, depositional and post-depositional processes, including bioturbation, may sometimes arise. This makes bioturbation particularly problematic for tephrochronologists, especially in the analysis of cryptotephras, whose tephra deposits are invisible to the naked eye. While a non-bioturbated cryptotephra may be detected in a sediment core as a distinct peak in shard concentration, it is often the case that shards from the same population are found through a considerable depth of the core. In these cases, there is some confusion in the literature as to where to define the actual depth of the marker horizon (i.e. isochrons; Lowe 2011; Davies *et al.* 2012).

The situation is further complicated by the fact that bioturbation is a highly variable process; the rate and maximum extent of bioturbation depends not only on the collection of species which make up the benthos and their abundance, but also on the thickness of the ash layer and the subsequent rate of deposition. The work of Carter *et al.* (1995) suggested that thicker (>1 cm) ash layers tend to smother the benthos, minimizing bioturbation. It also suggested that, if rapid deposition occurs after the ash fall event, the tephra layer will be isolated before the benthos can be re-established, also minimizing bioturbation.

By simulating an ash fall event on a marine estuary and then studying the bioturbation of tephra over a period of 10 days, this paper aims to produce a better understanding of the factors which affect the rate and extent of bioturbation, within the wider context of marine tephrochronology.

Methods

Site description

In order to explore the questions outlined above, an ash fall event was simulated in the intertidal zone of the Eden Estuary, Fife (56.3535°N, 2.8390°W, Fig. 1) in July 2010. The decision to study bioturbation in an estuarine environment was made for several reasons. From a practical perspective, intertidal zones are easily accessible and are often well documented; in this case, a commissioned report by Scottish Natural Heritage (Bates *et al.* 2004) outlines the spatial extent of different biotopes on the Eden Estuary, which are shown in Figure 1. This is particularly useful when studying ecosystem processes as it eliminates the need for a full site survey. Furthermore, unlike high-energy coastal systems where bioturbation is limited and disrupted by wave and tide activity (Cadée 2001), estuaries are relatively low-energy environments, analogous to the deep sea-bed.

Our study site is a sheltered estuarine mudflat, with minimal wave activity, making it ideal for investigating bioturbation. The site is exposed for approximately 7 h of every tidal cycle (Austin 2003).

As illustrated in Figure 1, the study site is located near the shoreline, in a biotope classified as 'LMU. HedMac' (Connor *et al.* 1997*a*, *b*). This classification refers to a well-sheltered intertidal zone comprising sandy mud characterized by the presence of Nereis (*Hediste diversicolor*) and Baltic clam (*Macoma baltica*), as well as other oligochaetes, polychaetes and bivalves (Bates *et al.* 2004). A comprehensive database of all biotopes identified in Figure 1 can be found online at www.marlin.ac.uk. Previous efforts to quantify sediment bioturbation

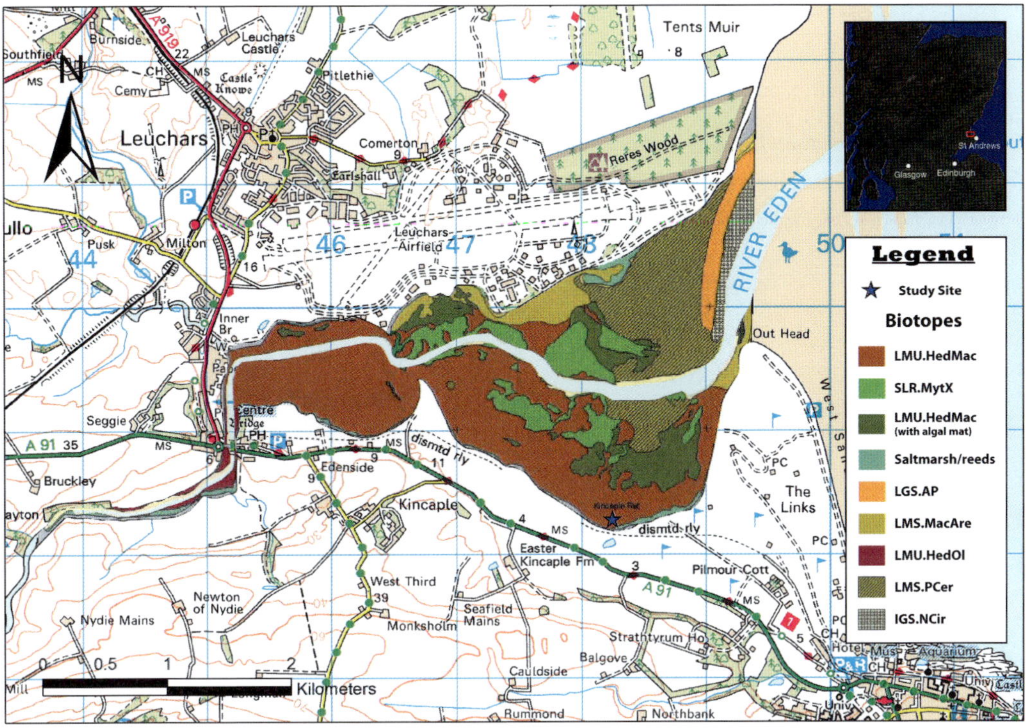

Fig. 1. Eden Estuary Biotopes. Adapted from Bates *et al.* (2004). OS Content: © Crown Copyright/database right 2011. An Ordnance Survey/EDINA supplied service.

have suggested that the annelids (polychaetes and oligochaetes) play a significant role in estuarine bioturbation (Biles *et al.* 2002). Therefore, the dominance of Nereis and the presence of other annelids at this site suggested that bioturbation would indeed occur.

Characterization of tephra

Ash was collected from Seljavellir, South Iceland (63.6°N 19.7°W) in June 2010 following the main eruptions of Eyjafjallajökull earlier in the year. Laser grain size analysis was carried out on a tephra sample using a Beckman Coulter LS 230 Laser Diffraction Particle Size Analyzer at the Facility for Earth and Environmental Analysis at the University of St Andrews. Riffle box splitting was used to ensure a representative sub-sample was obtained. The results (Fig. 2) show that this proximal tephra sample is fairly coarse ($\bar{x} = 299.4$ μm, median = 228.3 μm). This is coarser material than would be expected to occur in distal ash fall deposits (e.g. Abbott *et al.* 2011) as the size of airfall material decreases exponentially with distance from eruptive source (Sparks *et al.* 1981). The tephra was found to be a mix of pumacious and glassy shards containing vesicles. There was no specific or diagnostic morphology; a mixture of spherical and elongate, platy and blocky shards was observed.

Simulating an ash fall event

Ash was spread in 20 × 20 cm square quadrats directly onto the exposed intertidal estuarine sediments. We visited the study site at low tide on 22 July and seeded 18 quadrats with 60 g (37.8 cm^3) of ash each, equivalent to an ash layer of 0.95 mm thickness. So that results were comparable, quadrats were placed in a grid with 1 m spacing between each. This ensured that all quadrats were placed within an ecologically homogenous area and that all experienced the same tide and wave effects. Closer spacing was considered, but it was felt that this might lead to inter-quadrat ash contamination. Additionally, the possibility of a raised edge on the quadrat was considered, to prevent outwashing of tephra by waves and tide; however, it was felt that this would significantly increase the ecosystem disturbance. Furthermore, in such a low-energy environment, tephra loss is likely to be minimal. This point is addressed further in the Discussion.

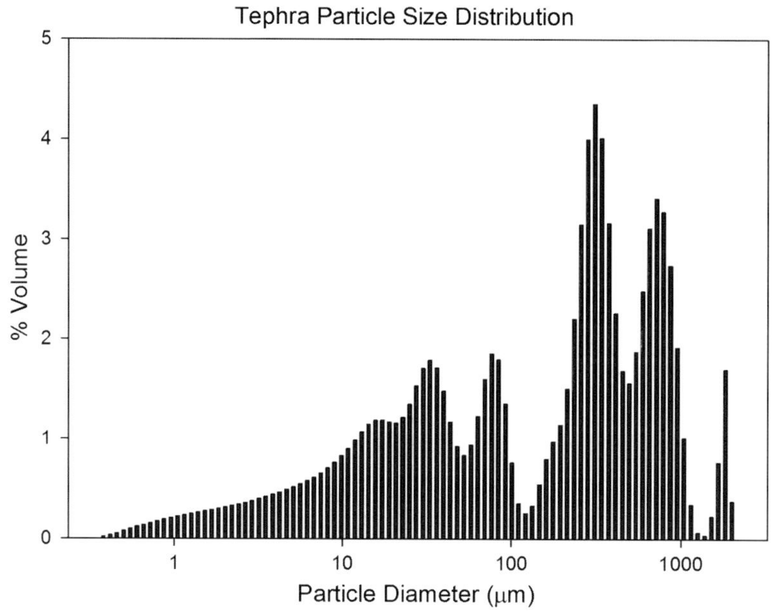

Fig. 2. Particle size distribution for tephra collected from Eyjafjallajökull.

Figure 3 shows a typical quadrat with signs of burrowing clearly visible in the bottom half of the image.

Field sampling and laboratory sub-sampling

Coring was carried out after 0, 2, 5 and 10 days, using a sample tube of 26 mm diameter with the base removed. This relatively simple setup allowed cores of 60 mm depth to be taken, from the middle of each quadrat. On day 0, six quadrats were cored and one control core was taken from an adjacent location on the mudflat, unseeded with tephra. The main purpose of the day 0 cores was to test the coring method; any ash found below the surface section of the day 0 cores will represent disturbance through corer insertion into the sediment. Three of these 'day 0' cores were frozen and three were maintained at room temperature, in order to determine the best method of storage before sub-sampling. However, it quickly became apparent that, if cores were not frozen, bioturbation would continue after collection, invalidating the resultant experimental data. Therefore, the three unfrozen cores were discarded. On days 2, 5 and 10, four replicate cores were taken and frozen immediately after collection. Cores were coded in a 'day-sample number' format (i.e. Core 0-2 refers to the second core taken on day 0). Frozen cores were sliced into 3 mm sections using a three-way clamp, and the sections were prepared for analysis as outlined in Figure 4. The samples were prepared for microscopy by wet sieving (63 μm) to remove the fine fractions; this facilitates the drying, disaggregation and counting of the samples. The aim of this process was to produce samples which could be accurately and rapidly counted by the process outlined below.

Quantifying bioturbation

As discussed above, we treat bioturbation as a diffusive process. Thus, we sought to analyse the presence of tephra such that it could be expressed as a change in concentration through depth and time. Therefore, we determined the number of tephra grains at specific depths in the cores relative to the overall number of mineral grains at these depths and expressed these as percentage counts. Typically, this counting is carried out manually using a picking tray and a tally counter. Repeated counts are carried out until a statistically reliable value is attained for tephra shard concentration. This method is extremely time-consuming; we estimate that to process all 300 sub-sampled sections, each requiring the counting of 15 squares, would take up to 400 h (i.e. 50 standard working days). In order to reduce the analysis time, the possibility of a faster, semi-automated method was investigated.

Costa & Yang (2009) presented a solution to a similar problem involving the counting of pollen grains. Rather than counting manually, they used

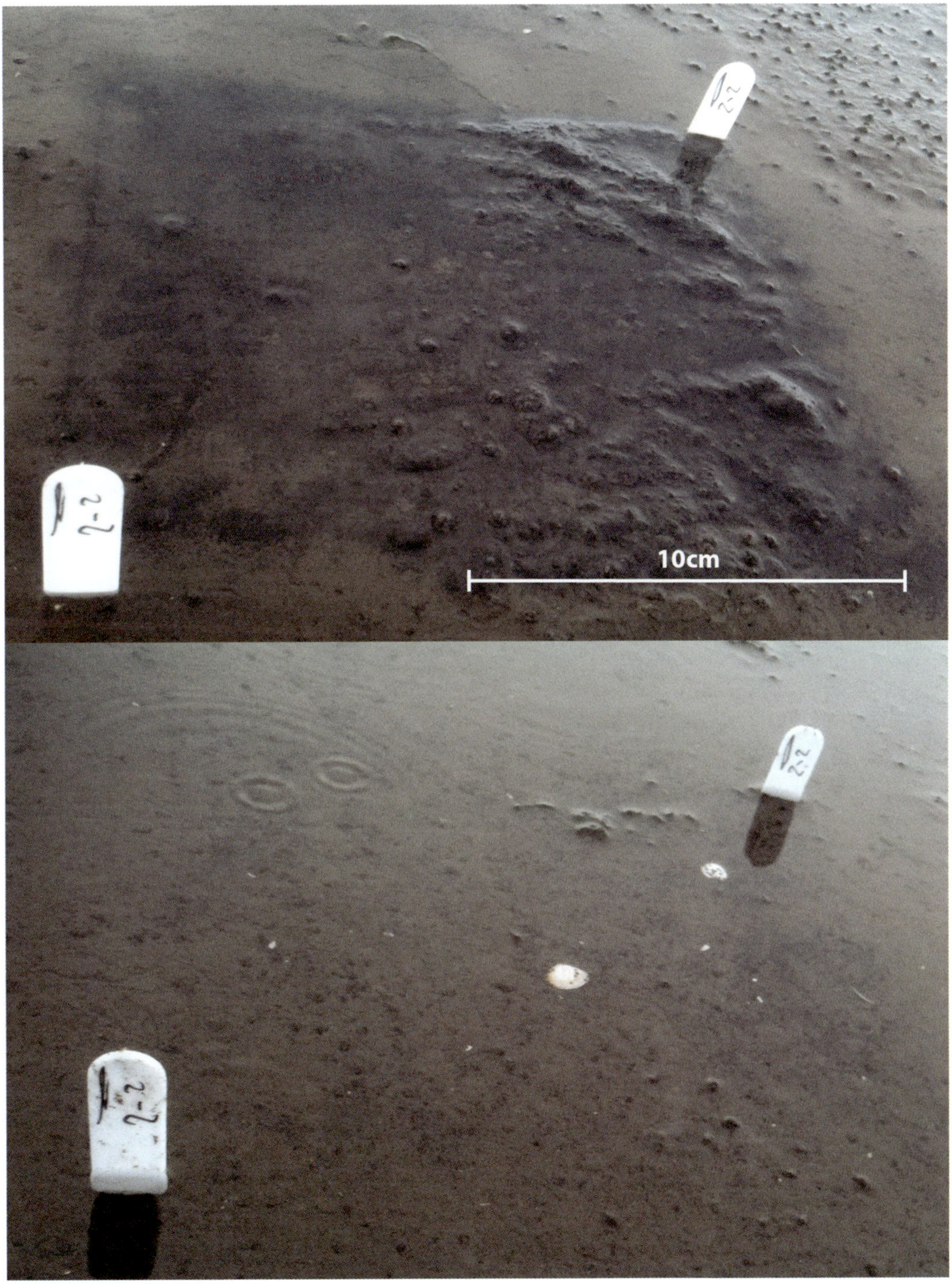

Fig. 3. Microcosm on Eden Estuary (20 × 20 cm) at (a) day 0 and (b) day 2. By day 2, tephra is barely visible at the surface.

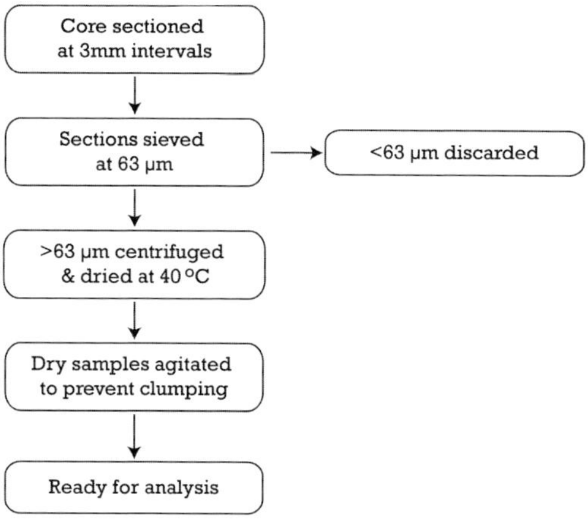

Fig. 4. Laboratory procedure for preparation of sections.

ImageJ, an open-source Java-based image analysis program to count thousands of grains at a time. Furthermore, they developed a macro for the program which would automatically open a specified series of microscope images, count all the pollen grains in each image and record the results in Excel format. They estimated a 5- to 60-fold saving in time using this method, depending on the number of grains in each image.

In order to determine the applicability of ImageJ to the problem of determining tephra shard concentration, a series of trials was carried out. The optimized process resulting from these trials is described below and illustrated in Figure 5, and the parameters and algorithms of choice are outlined in the macro file available in the Supplementary Material.

Using image analysis to quantify tephra concentrations

The first stage of automated counting of particles was image capture. Material from each processed section was placed on a piece of white paper to enhance contrast, then placed under a Zeiss Stemi 2000 optical stereomicroscope. Using an attached Deltapix Infinity X-21 CMOS camera, 10 unique images were taken of each sub-sampled section, each with a field of view of 10.67 × 8.53 mm (1280 × 1024 pixels). Although ImageJ can handle clustered particles, it does not deal well with large amalgamations of grains, and so care was taken to ensure that grain density was sufficiently low (<500 grains per image) that minimal clustering occurred. Care was also taken to ensure that the microscope's light source produced a relatively uniform intensity across the field of view, as the performance of ImageJ is enhanced when analysing particles on a constant background. It should be noted, however, that the program is capable of 'subtracting' the image background.

As illustrated in Figure 5, particle analysis in ImageJ is a multistage process. First, the image is sharpened to better delineate the outline of individual grains. Then the image background is 'subtracted'. ImageJ carries out background subtraction by identifying the predominant trend in colour, saturation and hue across the image, based on the assumption of a light background. Following background subtraction, the image is converted from RGB to 8-bit greyscale. The example in Figure 5a shows an image that has undergone the sharpening, background subtraction and 8-bit conversion, which are prerequisites for 'autothresholding' (Fig. 5b).

Autothresholding describes the use of an algorithm that automatically distinguishes particles from background without user input. It converts the greyscale image in Figure 5a into a binary image (Fig. 5b) where black represents 'particle' and white represents 'background'. This feature is one of the two automatic functions that make ImageJ so useful for particle analysis, the second being the watershed function (Fig. 5c). The watershed function takes the dark patches on a light background produced by autothresholding and determines, based on their geometry, which patches are likely to be multiple particles touching each other. Finally, the program counts, measures and labels the particles (Fig. 5d) based on a preset size threshold (area >100 pixels/6940 μm^2). This threshold corresponds, for a perfectly round particle, to a diameter of 93 µm; this is sufficiently above the 63 µm sieve size to ensure that small particles present owing to incomplete sieving are not counted.

Performing the individual steps shown in Figure 5 on each of the 3000 images in turn would still be very time-consuming; therefore, a macro was written, based on that of Costa & Yang (2009), which first prompts the user to select a directory containing images, and then proceeds to open each image in turn, perform each of the steps in Figure 5 and then save the results in an appropriately named Excel file. In this way, the images are processed at roughly one per second and all 3000 images can be processed in under an hour. Including the few hours required to capture all the images under the microscope, this represents an almost 100-fold saving in laboratory time and introduces an improved uniformity of analysis.

The method outlined above is suitable only for counting the total number of grains; identification of tephra shards in bulk sediment by eye is an acquired skill and could not be readily automated with this image analysis software. As such, the final stage of analysis in ImageJ is manual identification

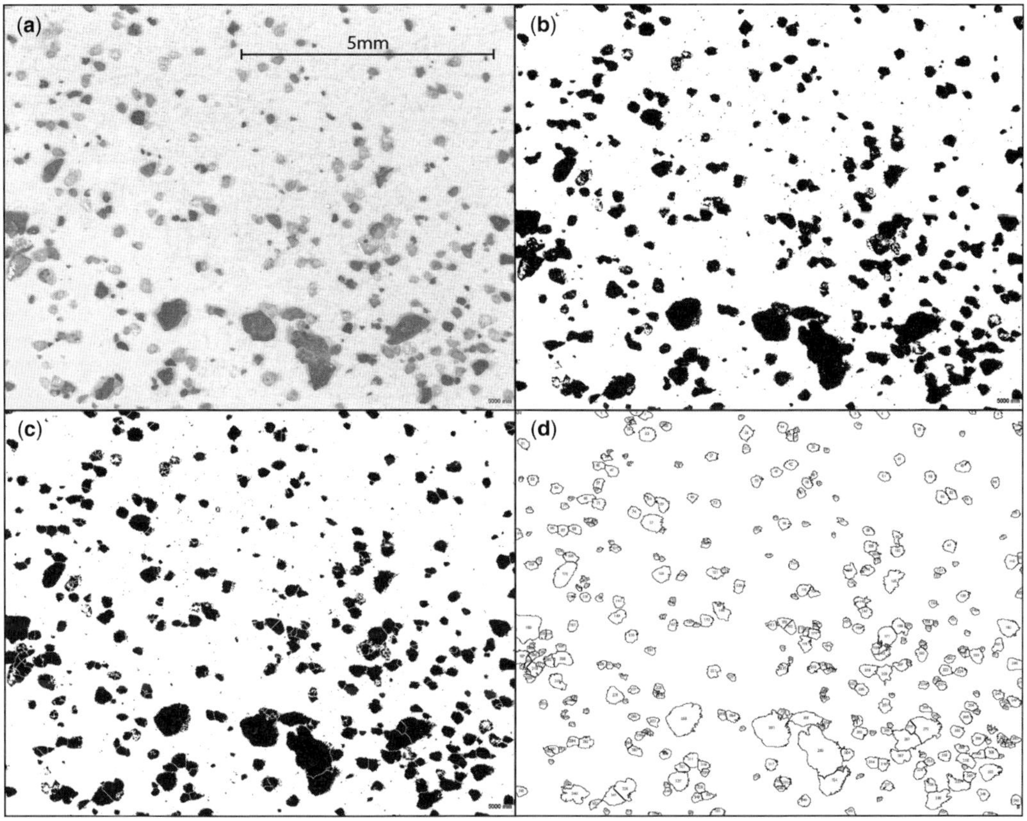

Fig. 5. Particle analysis in ImageJ involves some pre-processing owing to the poor quality of the original photography. First the image is sharpened, converted to 8-bit color and the background subtracted (**a**). Particle detection is carried out using automatic thresholding (**b**). A watershed analysis is performed to split up grains which are touching (**c**), before the particles are counted and labelled (**d**).

of tephra shards in the processed images. ImageJ is equipped with the software equivalent of a tally counter, and so this is simply a process of clicking on all the tephra shards in each image in turn and appending these counts to the Excel files produced by the macro described above.

In addition to automatic grain counting, ImageJ is also able to determine shard size automatically, based on edge detection. Thus, data were collected for shard size distribution for each of the sections, to determine if the rate of bioturbation affects the observed shard size or vice versa. For example, are smaller grains more readily bioturbated, leading to a varying size distribution of tephra through the depth of the core influenced by the mixing process?

Testing the accuracy of the image analysis

The automatic nature of particle analysis in ImageJ has the advantage of being a significantly less labour-intensive alternative to the traditional manual counting method. However, the automatic counting process is not completely accurate; in particular, the watershed function calculates 'probable' breaks, and sometimes cuts large particles into a few smaller particles, as can be seen in the bottom right side of the example in Figure 5d.

Furthermore, accuracy is also affected by the ability of the user to identify tephra shards from a 2D photograph. Sometimes, the distinction between tephra and bulk sediment is not clear and so the traditional counting method often involves inspection at higher magnification or turning particles in order to better distinguish the features thereof. Analysing still 2D images makes this process more difficult and, as such, it is to be expected that the rate of misidentification may be higher.

The above caveats to this new approach to tephra shard concentration analysis require that it be tested and compared with the traditional method to quantify accuracy and ensure validity. To accomplish this, repeat counts were carried out using both the traditional and ImageJ method, to determine the error associated with each. A detailed description of the tests carried out is provided in the Supplementary Material and the results are discussed below.

Accuracy v. efficiency in image analysis of tephra

Our analysis of the ImageJ method of tephra identification, available in the Supplementary Material, suggests that it is less accurate than the traditional method. However, we must consider that the observed error of 23% is relative to the traditional method, which has itself been shown to have an inherent error of around 12% (see Supplementary Material). This figure, therefore, may be an exaggeration of the true error. Undoubtedly, with greater replication and an optimization of the image capture and processing protocols, this error could be considerably reduced. For example, images could be captured at a higher effective resolution, either by photographing at a higher magnification or by using a more specialized microscope camera. In particular, the possibility of taking 'stacks' of images with progressively greater depths of focus has been suggested; this could potentially allow 3D stack analysis to be carried out in ImageJ, which would better represent the morphology of grains and allow easier identification of tephra. This would make the structure of the shards (and their often conchoidal surfaces) more readily identifiable.

While the disadvantage of the ImageJ method is the potential loss of accuracy, the advantage is the speed at which analysis can be performed. The analysis takes roughly two orders of magnitude less time than the traditional method. This vast saving in time means that many more images could be analysed for each section, which would translate into a reduction in the associated error. For example, one could carry out four times as many analyses on each section, which would result in a halving of the margin of error. This would result in a comparable error to the manual counting method, but a 25-fold saving in time. Our results show that errors tend to be large when tephra shard concentrations are low. In these cases, it would be advisable to analyse considerably more images of each section.

Results

Rate and maximum extent of bioturbation

Average tephra concentration profiles for each sampling day were produced by taking the mean tephra concentration from all the replicates at each depth. For each depth, one-way ANOVA was carried out to test for a significant change in tephra shard concentration over time. Where homogeneity of variance was not met, Welch's ANOVA was also carried out. *Post-hoc* analysis was carried out using Tukey's test to determine which sample days have significantly different tephra shard concentrations at each depth. Raw data are available in the Supplementary Material.

Visual inspection of the experimental quadrats on day 2 initially suggested that physical transport (i.e. advection) by wave and tide activity had largely removed the tephra from the quadrat surfaces. Figure 3 shows that, after only 2 days, tephra appears to be almost completely absent. However, the percentage tephra in the surface (i.e. 0–3 mm section) of quadrat 2-2, for example, is 13.41%, which is comparable with the percentage tephra found at the surface of all the day 0 cores (mean = 13.42%).

One would expect that physical mixing by wave and tide action would be associated with at least some degree of horizontal displacement of the tephra. However, the high tephra retention rate seen in the surface layer of successively sampled cores suggests that minimal horizontal displacement has occurred, and that displacement is primarily vertical. This suggests that, rather than having been removed by wave or tide activity, the tephra have largely been bioturbated downwards through the first 3 mm of the sediment. Similarly high tephra shard concentrations were found at the surface of all cores (Fig. 6), suggesting that, despite fears that open quadrats would lead to significant tephra loss, physical transport of tephra was actually a relatively minor process in this environment at the time of the experiment.

Figure 6a shows the tephra concentration profiles for all of the 15 cores investigated; Figure 6b shows tephra concentration profiles averaged for each sampling day. No tephra was found anywhere in the control core. Day 0 cores have a high percentage of tephra in the first 3 mm, which immediately drops to less than 1% at 3–6 mm depth; no tephra was found in day 0 cores below the 6–9 mm section. This suggests that the coring process causes only minimal downward disturbance.

Cores from days 2, 5 and 10 show progressively more tephra at increasing depths, down to a maximum depth of 15–18 mm. This pattern is interpreted as the result of progressive bioturbation and is illustrated in Figure 6. There appears to be little difference, at any depth, between cores from day 5 and day 10. Disregarding the day 0 control cores, the least bioturbation occurs in core 2-4, while the most bioturbation occurs in core 5-1; this implies a degree of spatial heterogeneity in the extent of the bioturbation processes in the intertidal environment.

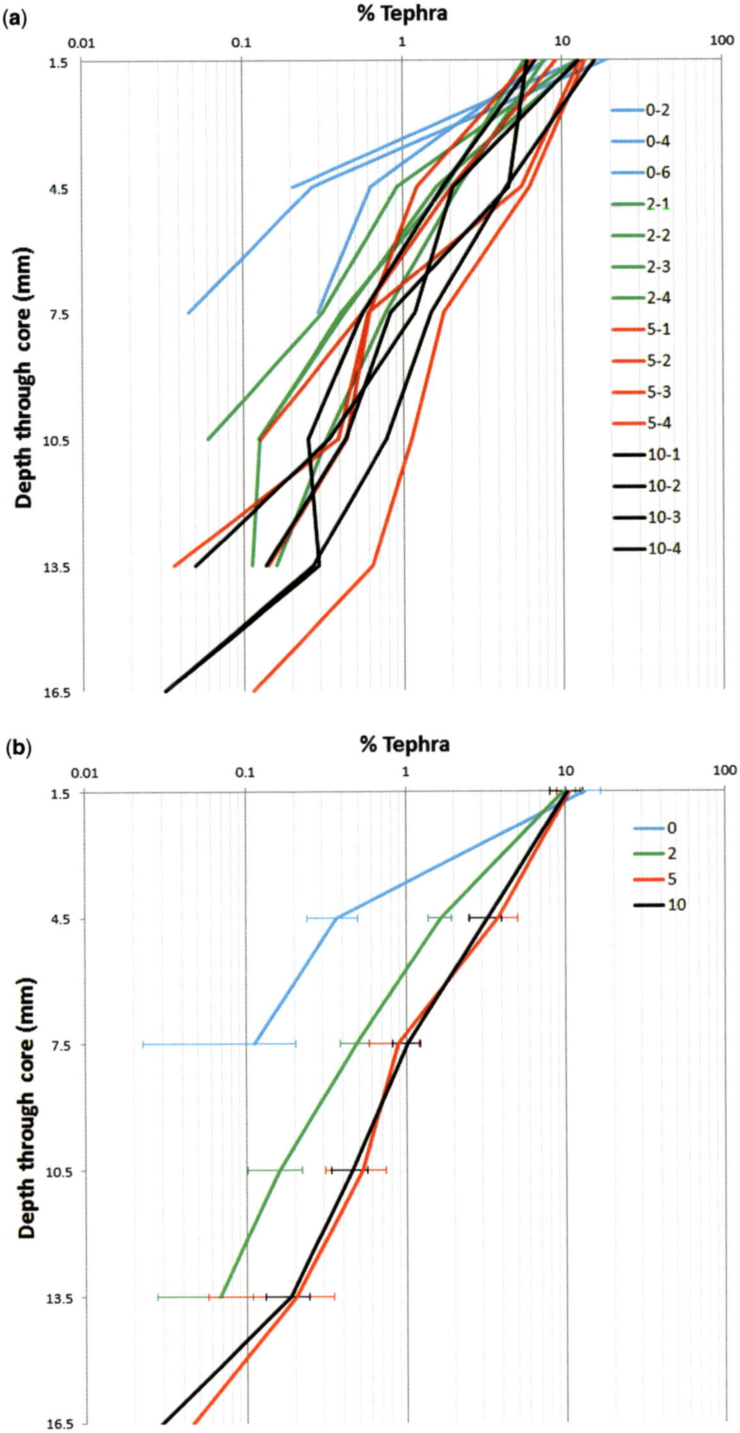

Fig. 6. (**a**) Percentage tephra through the depth of all 15 cores. % Tephra = percentage of identified grains which are tephra. Quoted depths correspond to the top of each section. (**b**) Same as (a), but averaged for each sampling day. Error bars represent standard error of the mean (SEM).

Analysis of variance showed that, at the surface (0–3 mm), there was no significant difference between any of the sampling days. This supports the idea that tephra loss through physical transport away from each quadrat was minimal. At 3–6 mm depth, Welch's one-way ANOVA found a significant difference in means ($F = 9.652$, $p = 0.012$). Tukey's test showed a significant difference (90% confidence interval) between tephra concentration at days 0 and 5 ($p = 0.064$). At 6–9 mm depth, one-way ANOVA found a significant difference in means ($F = 3.587$, $p = 0.050$). Tukey's test showed a significant difference (90% confidence interval) between tephra concentration at days 0 and 10 ($p = 0.056$). Below this depth, no statistically significant differences were found. The tephra concentration data and associated analysis of variance are therefore compatible with bioturbation of tephra from the surface to sub-surface layers at all sites.

Shard size analysis of the tephra grains revealed no statistical difference through the depth of any of the cores, suggesting that the rate of bioturbation is, in this case at least, independent of shard size. The data also suggest that the bioturbation process (downwards mixing) does not preferentially favour the size sorting of tephra.

Discussion

Accurate quantification of bioturbation, in terms of both rate and maximum depth, is essential to the development of the discipline of tephrochronology, as it allows an opportunity to 'work back' from an observed tephra shard concentration profile to identify the point of initial deposition. While previous studies have quantified bioturbation using luminophores (Biles *et al.* 2002; Solan *et al.* 2004*b*) or radioactive tracers (Gerino *et al.* 1998), this study represents the first direct investigation of bioturbation of a simulated ash fall event.

Perhaps the most striking feature of our results is the relatively shallow depth over which bioturbation has occurred. Our work suggests that bioturbation occurs over a maximum 'mixed depth' of 18 mm. The global dataset for bioturbation, however, seems to support the hypothesis of a global mean mixed depth of 9.8 cm (± 4.5 cm, 1σ), largely independent of ecosystem type (Boudreau 1998). Of course, 18 mm lies within 2σ of this global mean and so is not statistically at odds with this observation. However, as described above (under 'Site description'), our study site is dominated by Nereis and other active bioturbators. A previous investigation of bioturbation in an estuarine environment demonstrated the burrowing activity of polychaete worms down to 15 cm (Smith & Schafer 1999), presumably highlighting the highly variable nature of bioturbation and bioturbation depths within estuarine environments.

Bioturbation in low-energy environments

Our site on the Eden Estuary was chosen as an ideal site for studying bioturbation based on the low energy level. Investigating in a low-energy environment minimizes the physical mixing signal which, in the context of a bioturbation study, could be regarded as a significant source of experimental 'noise'. However, the relationship between energy, physical mixing and bioturbation is not straightforward, as illustrated by Figure 7 (Cadée 2001). While it is true that physical mixing does increase linearly with energy, bioturbation is actually greatest in mid-energy environments. If the energy of the environment is too high, vigorous wave and current activity make it difficult for infauna to survive, and so bioturbation is minimal. However, low energy levels also curtail biological activity. Our study site is mostly a very low-energy environment; not only does it lie in a sheltered estuary, but it also lies at the high-intertidal edge. Therefore, despite the presence of several species of active bioturbators, bioturbation depth is apparently severely restricted.

Our study site was chosen for ease of repeat access. However, future investigations into marine bioturbation should choose a sheltered site in the sub-tidal zone. Although this will make data collection significantly more challenging, it is also more likely to represent a clearer analogue to the deep sea, and any potential 'washing out' of tephra material by tidal activity may be reduced.

Bioturbation and sediment grain size

Previous research into bioturbation and nutrient fluxes suggests that grain size exerts a control on the extent of bioturbation. Winston & Anderson (1971) conducted a semi-quantitative investigation of bioturbation in an estuarine environment and showed that more bioturbation occurred at the sandy study sites than at the fine-grained silt sites. Furthermore, in a quantitative study of gas transport in bioturbated sediments, Kristensen & Hansen (1999) showed that the transport of ammonia and carbon dioxide through sediment profiles is dependent on grain size. They found that, in coarser-grained sediments, gas transport fitted an eddy diffusive model. In other words, burrowing and locomotion led to physical movement of porewater, gases and sediment. However, in fine-grained sediments, this eddy diffusive model overestimated gas exchange and so was replaced with a molecular diffusive model. Implicit in the molecular diffusive

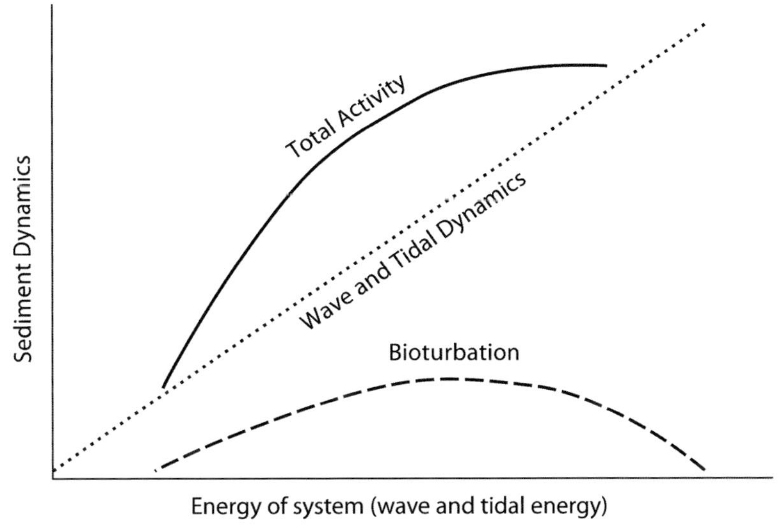

Fig. 7. Relation between energy and both physical mixing and bioturbation. Bioturbation is reduced at both extremes, and most important at mid-energy levels. Modified after Cadée (2001).

model is the *lack* of physical particle mixing by biological activity; the nonporous and cohesive nature of finer-grained sediments therefore tends to restrict vertical mixing.

Qualitatively, the sediment material at the study site clearly contains a high silt content, and the Scottish Natural Heritage ecological survey reports that the entire lower Eden Estuary is mud-flats, while LMU.HedMac is defined as 'sandy mud shores' (Bates *et al.* 2004).

The importance of epifauna

Our results echo those of Solan *et al.* (2004*b*), who carried out bioturbation research in Gullmarsfjord, Sweden. They used time-lapse image analysis to investigate the rate and depth of bioturbation over a 16 h period. They found that, despite the abundance of several active infaunal bioturbators, it was actually the locomotion of a single crab of the species *Hyas araneus* that was responsible for almost all bioturbation. Epifaunal bioturbation of this type takes the form of a few discrete events, as opposed to an ongoing diffusive process, as is usually considered for infaunal bioturbation. No evidence of epifaunal activity was found at any of our experimental quadrats. Both this study and theirs, therefore, suggest that, over the relatively short time-scales considered, those infaunal species typically considered to be active bioturbators are incapable of mixing below a depth of *c.* 20 mm. This would imply that future studies investigating infaunal bioturbation would need to do so over longer time-scales (i.e. months rather than days).

However, Figure 6 appears to imply that bioturbation halted after day 5 of the experiment. This would be at odds with the idea that infaunal bioturbation occurs over longer time periods. In fact, it is unlikely that the similarity of the profiles at days 5 and 10 represents a real cessation of bioturbation; we would expect that, once the maximum mixed depth has been reached, the tephra concentration would homogenize over the mixed-depth. We do not see this, which suggests that, for some reason, bioturbation has 'paused' or significantly slowed between days 5 and 10. This may, in fact, be a coincidental result, and might imply that future investigations would benefit from using a greater number of replicate quadrats or longer field sampling intervals.

The effects of community structure on bioturbation

In this investigation, we purposely chose to investigate a single biotope, LMU.HedMac, which describes sheltered, sandy mud shores dominated by *Hediste diversicolor* and *Macoma balthica*. Investigating a single biotope reduces the biological variable from the investigation, allowing a more detailed focus on how bioturbation progresses over time following a simulated ash fall event. However, the Eden Estuary and similar environments would make ideal study sites for future investigations into the role of ecosystem type on bioturbation. As shown in Figure 1, the estuary contains many different biotopes. Furthermore, the development and successful application of the new, semi-automated method of tephra concentration analysis outlined here makes it feasible to carry out considerably more analyses than would previously have been possible, making a wider ranging study entirely feasible. It would be particularly interesting to consider how grain size varies along the length of the estuary, how this affects community structure and, in turn, the rate and extent of bioturbation. The same, of course, applies to the study of bioturbation in sub-tidal habitats.

Investigating tephra burial

The tephra concentration profiles resulting from the present study represent, in reality, only half of a bioturbation profile; it is equally important to consider how a tephra marker horizon is bioturbated upwards following burial. It would, for example, be incorrect to assume that the original marker horizon occurs at the highest point in the core where tephra is found. Is it therefore correct to assume that the peak tephra concentration represents the original marker horizon? Future research should seek to answer this question. For example, Winston & Anderson (1971) described an investigation where an area of estuarine sediment is excavated, a marker horizon is lain and then re-buried. Their research was not quantitative, but there is no reason why the approach followed in this paper could not, for example, be applied to a buried marker horizon.

Conclusions

Our results demonstrate that bioturbation of tephra begins immediately following deposition, even in a low-energy mid-estuarine environment. The rate of bioturbation does not appear to be dependent on the grain size of the tephra present. It is clear that the processes responsible for bioturbation are significantly depth-limited in the biotope (LMU. HedMac) studied. Future investigations into bioturbation of tephra in low-energy environments should seek to study the process over longer time periods, perhaps over an entire seasonal cycle. Over this time period, we would expect to see tephra penetration to a maximum mixed depth, and this would be confirmed by the progressive homogenization of the tephra concentration profile over said mixed depth.

The extent to which these results are applicable to sub-tidal marine, as opposed to estuarine, bioturbation is uncertain. Certainly, the interrelated factors of sediment grain size distribution and the energy level of the system appear to play a critical role in determining the rate of bioturbation. The sediment at the study site had an oxic layer thickness of the order of 1 mm, underlain by dark anoxic sediments; this is typical of fine sediments with high organic content and a dense and active biofilm. It is likely that this very thin oxic layer impacts bioturbation rates and maximum extent. In this study, the depth of bioturbation may also have been limited by the fact that the study site is above the low-tide line, and is thus exposed for several hours a day; this effect is a subject worthy of further investigation.

Automatic image analysis using ImageJ was previously employed by Costa & Yang (2009) to carry out pollen counts, and it has been adapted here to count all types of particles, as well as assist in measuring and counting tephra particles. Our analysis of the accuracy of this new method suggests that it has a fairly high counting error associated with it. However, this reported error should be taken as a tentative figure, because the traditional method, against which our new method is tested, is far from error-free. Our results also suggest the need for more replication in order to produce statistically robust results. The new particle analysis method demonstrated in this paper has high significance and future applicability.

We measured significantly slower bioturbation rates than previous studies in other marine environments (Boudreau 1998), and we attribute this to the fine grain size and the low energy level. These two factors play a significant role in determining community structure and abundance. This strong ecological control on bioturbation implicitly hints at the possibility of developing a 'bioturbation index', and using it to infer information about palaeoenvironments, which is an exciting possibility for the future development of marine tephrochronology.

References

ABBOTT, P. M., DAVIES, S. M., AUSTIN, W. E. N., PEARCE, N. J. G. & HIBBERT, F. D. 2011. Identification of cryptotephra horizons in a North East Atlantic marine record spanning marine isotope stages 4 and 5a (~60,000–82,000 a b2k). *Quaternary International*, **246**, 177–189.

AUSTIN, H. A. 2003. The biology and ecology of benthic foraminifera inhabiting intertidal mudflats. *PhD thesis*, University of St Andrews.

AUSTIN, W. E. N. & ABBOTT, P. M. 2010. Comment: Were last glacial climate events simultaneous between Greenland and France? A quantitative comparison using non-tuned chronologies. M. Blaauw, B. Wohlfarth, JA Christen, L. Ampel, D. Veres, K. Hughen, F. Preusser and A. Svensson (2009). *Journal of Quaternary Science*, **25**, 1045–1046.

AUSTIN, W. E. N., WILSON, L. J. & HUNT, J. B. 2004. The age and chronostratigraphical significance of North Atlantic Ash Zone II. *Journal of Quaternary Science*, **19**, 137–146.

BATES, C. R., MOORE, C. G., MALTHUS, T., MAIR, J. M. & KARPOUZLI, E. 2004. *Broad scale mapping of habitats in the Firth of Tay and Eden Estuary, Scotland.* Comissioned Report **007**. Scottish Natural Heritage.

BERGER, W. H. & HEATH, G. R. 1968. Vertical mixing in pelagic sediments. *Journal of Marine Research*, **26**, 134–143.

BILES, C., PATERSON, D., FORD, R., SOLAN, M. & RAFFAELLI, D. 2002. Bioturbation, ecosystem functioning and community structure. *Hydrology and Earth System Sciences*, **6**, 999–1005.

BOUDREAU, B. P. 1998. Mean mixed depth of sediments: the wherefore and the why. *Limnology and Oceanography*, **43**, 524–526.

CADÉE, G. 2001. Sediment dynamics by bioturbating organisms. *In*: REISE, K. (ed.) *Ecological Comparisons of Sedimentary Shores*. Springer, Berlin, 128–148.

CARTER, L., NELSON, C. S., NEIL, H. L. & FROGGATT, P. C. 1995. Correlation, dispersal, and preservation of the Kawakawa tephra and other late Quaternary tephra layers in the Southwest Pacific Ocean. *New Zealand Journal of Geology and Geophysics*, **38**, 29–46.

CONNOR, D. W., BRAZIER, D. P., HILL, T. O. & NORTHEN, K. O. 1997*a*. *Marine nature conservation review: marine biotope classification for Britain and Ireland. Volume 1. Littoral biotopes. Version 97.06*. JNCC Report, **229**.

CONNOR, D. W., DALKIN, M. J., HILL, T. O., HOLT, R. H. F. & SANDERSON, W. G. 1997*b*. *Marine nature conservation review: marine biotope classification for Britain and Ireland. Volume 2. Sublittoral biotopes. Version 97.06*. JNCC Report, **230**.

COSTA, C. M. & YANG, S. 2009. Counting pollen grains using readily available, free image processing and analysis software. *Annals of Botany*, **104**, 1005.

DAVIES, S. M., ABBOTT, P. M., PEARCE, N. J. G., WASTEGÅRD, S. & BLOCKLEY, S. P. E. 2012. Integrating the INTIMATE records using tephrochronology: rising to the challenge. *Quaternary Science Reviews*, **36**, 11–27.

GERINO, M., ALLER, R., LEE, C., COCHRAN, J., ALLER, J., GREEN, M. & HIRSCHBERG, D. 1998. Comparison of different tracers and methods used to quantify bioturbation during a spring bloom: 234-thorium, luminophores and chlorophylla. *Estuarine, Coastal and Shelf Science*, **46**, 531–547.

GUINASSO, N. JR. & SCHINK, D. 1975. Quantitative estimates of biological mixing rates in abyssal sediments. *Journal of Geophysical Research*, **80**, 3032–3043.

HAFLIDASON, H., EIRIKSSON, J. & VAN KREVELD, S. 2000. The tephrochronology of Iceland and the North Atlantic region during the Middle and Late Quaternary: a review. *Journal of Quaternary Science*, **15**, 3–22.

KRISTENSEN, K. & HANSEN, K. 1999. Transport of carbon dioxide and ammonium in bioturbated (*Nereis diversicolor*) coastal, marine sediments. *Biogeochemistry*, **45**, 147–168.

LOWE, D. J. 2011. Tephrochronology and its application: a review. *Quaternary Geochronology*, **6**, 107–153.

MANGERUD, J., LIE, S. E., FURNES, H., KRISTIANSEN, I. L. & LØMO, L. 1984. A Younger Dryas ash bed in Western Norway, and its possible correlations with tephra in cores from the Norwegian Sea and the North Atlantic. *Quaternary Research*, **21**, 85–104.

MEYSMAN, F. J. R., MIDDELBURG, J. J. & HEIP, C. H. R. 2006. Bioturbation: a fresh look at Darwin's last idea. *Trends in Ecology and Evolution*, **21**, 688–695.

SMITH, J. & SCHAFER, C. 1999. Sedimentation, bioturbation, and Hg uptake in the sediments of the estuary and Gulf of St. Lawrence. *Limnology and Oceanography*, **44**, 207–219.

SOLAN, M., CARDINALE, B. J., DOWNING, A. L., ENGELHARDT, K. A. M., RUESINK, J. L. & SRIVASTAVA, D. S. 2004*a*. Extinction and ecosystem function in the marine benthos. *Science*, **306**, 1177.

SOLAN, M., WIGHAM, B. D. *ET AL.* 2004*b*. In situ quantification of bioturbation using time-lapse fluorescent sediment profile imaging (f-SPI), luminophore tracers and model simulation. *Marine Ecology Progress Series*, **271**, 1–12.

SPARKS, R., WILSON, L. & SIGURDSSON, H. 1981. The pyroclastic deposits of the 1875 eruption of Askja, Iceland. *Philosophical Transactions of the Royal Society of London. Series A: Mathematical and Physical Sciences*, 241–273.

WINSTON, J. E. & ANDERSON, F. E. 1971. Bioturbation of sediments in a northern temperate estuary. *Marine Geology*, **10**, 39–49.

Index

Page numbers in *italic* denote Figures. Page numbers in **bold** denote Tables.